Collected Algorithms from ACM

Volume I
Algorithms 1-220

A collection of the first 220 ACM Algorithms, including Certifications,
Remarks, and Translations from the Algorithms Department
of Communications of the ACM, 1960-1963.

1980

A Publication of the Association for Computing Machinery, Inc.
1133 Avenue of the Americas
New York, New York 10036

Submittal of an algorithm for publication in the *Collected Algorithms From ACM* implies that unrestricted use of the algorithm within a computer is permissible. General permission to copy the algorithm in fair use, but not for profit, is granted provided ACM's copyright notice is given and reference is made to this publication, its date of issue, and to the fact that copying is by permission of the Association for Computing Machinery.

Price: ACM members $25; others $35. Prices subject to change without notice. For latest prices refer to the current ACM Publications Catalog available free of charge from ACM Order Department, P.O. Box 64145, Baltimore, MD 21264.

ISBN: 0-89791-017-6

The algorithms and other items in this compilation are all excerpted from copyrighted ACM publications unless otherwise noted.

Preface

The Algorithms department of Communications of the ACM (CACM) was established in February 1960, with J. H. Wegstein as editor, for the purpose of publishing algorithms, consisting of procedures and programs, in the Algol language. In 1975 the publication of ACM algorithms material was transferred to ACM Transactions on Mathematical Software (TOMS). A wide variety of algorithms have been published and many of them have been used heavily—either in original form or as translated into other languages. Recognizing the general acceptance of the algorithm material published in CACM and TOMS, the Association for Computing Machinery (ACM) has collected and reprinted the algorithms to make them more readily accessible and more serviceable to a larger group of users.

This collection contains the first 220 algorithms published in the Algorithms department of CACM from 1960 to 1963.

Algorithms 1–220 were originally published as received—without any refereeing whatever. Many of these have since been certified and/or corrected by their authors or by other contributors.

To facilitate the updating and to make this volume convenient to use, an understanding of the page numbering scheme for the algorithms is helpful. The page designation is in a three-part format: the left part is the algorithm number; the middle part is the page number within the algorithm (the first page of each algorithm is P1); and the right part is the number of the revision of that page. All sheets in the original, or first, insertion of an algorithm have "0" for the right part. The first revision of a page will have a page number having the left and middle parts identical with those on the page to be replaced, but the right part will be "R1" instead of "0." The second revision of the same page would read R2, and so on. For example, 123-P2-R1 would mean the first revision of page 2 of Algorithm 123.

Information on submitting algorithms for publication may be found in the introductory section located in the front of the current loose-leaf collection. Included in this material is a cumulative index to all the algorithms published since 1960 as well as the ACM Algorithms Policy, which guides the publication of all algorithms submitted to ACM.

Webb Miller
ACM Algorithms Editor
Department of Mathematics
University of California, Santa Barbara
Santa Barbara, CA 93106

1. QUADI
R. J. Herbold
National Bureau of Standards, Washington 25, D. C.

comment   QuadI is useful when integration of several functions of same limits at same time using same point rule is desired. The interval (a,b) is divided into m equal subintervals for an n-point quadrature integration. p is the number of functions to be integrated. w_k and u_k are normalized weights and abscissas respectively, where $k=1,2,3,\cdots,n$. u_k must be in ascending order. $P(B,j) =: (c)$ is a procedure which must be supplied by the programmer. It evaluates (c) the function (as indicated by j) for B. I_j is the result of integration for function j.;

```
procedure   QuadI (a,b,m,n,p,w_k,u_k,P(B,j) =: (c)) =: (Ij)
            begin
QuadI:            h := (b−a)/m
            for   j := 1(1)p  ;  I_j := 0
                  A := a−h/2
            for   i := 1(1)m
L1          begin A := A+h
            for   k := 1(1)n
L2          begin B := A+(h/2)×u_k
            for   j := 1(1)p
L3:         begin P(B,j) =: (c)
                  I_j := I_j+w_k×c    end L3  ;  end L2
                     end L1
            for   j := 1(1)p
                  I_j := (h/2)×I_j
            return
            integer (j,k,i)
            end   QuadI
```

2. ROOTFINDER

J. Wegstein
National Bureau of Standards, Washington 25, D. C.

comment   This procedure computes a value of g=x satisfying the equation x=f(x). The procedure calling statement gives the function, an initial approximation a≠0 to the root, and a tolerance paramater ϵ for determining the number of significant figures in the solution. This accelerated iteration or secant method is described by the author in *Communications*, June, 1958.;

```
procedure      Root(f( ), a, ε) = : (g)
begin
Root           b := a  ;  c := f(b)  ;  g := c
               if (c=a)  ;  return
               d := a  ;  b := c  ;  e := c
Hob:           c := f(b)
               g := (d×c−b×e)/(c−e−b+d)
               if (abs((g−b)/g)≦ε)  ;  return
               e := c  ;  d := b  ;  b := g  ;  go to Hob
end
```

CERTIFICATION

2. ROOTFINDER, J. Wegstein, *Communications ACM*, February, 1960

Henry C. Thacher, Jr.,* Argonne National Laboratory, Argonne, Illinois

Rootfinder was coded for the Royal-Precision LGP-30 Computer, using an interpretive floating point system with 28 bits of significance. The translation from ALGOL was made by hand. Provision was made to terminate the iteration after ten cycles if convergence had not been secured.

The program was tested against the following functions:

(1) $f(x) = (x + 1)^{1/3}$ (Root = 1.3247180)
(2) $f(x) = \tan x$
(3.α) $f(x) = 2\pi\alpha + \tan^{-1} x$ ($\alpha = 1, 2, 3, 4$)
(4.α) $f(x) = \sinh \alpha x$ ($\alpha = -1.2, -0.5, 0.5, 1.2$)

Selected results were as follows:

$f(x)$	α	ϵ	x_{k-1}	x_k	
1	1.3	$10^{-7}, 10^{-6}$	1.3247233	1.3258637	(1)
1	1.3	10^{-5}		1.3247165	(1)
2	5	10^{-3}	− .4674691	− .36021288	(1, 2)
2	4	10^{-3}	+ .84880381	+ .69496143	(1, 2)
3.1	1	10^{-5}		7.7252531	
3.2	2	10^{-5}		14.066155	
3.3	3	10^{-5}		20.371026	
3.4	4	10^{-5}		26.665767	

(1) No convergence after 10 iterations. Underlined figures are incorrect.
(2) For this function, f'(0) = 1; so convergence is not to be expected at this root. However, the algorithm did not find any other root.

It should be noted that the convergence criterion used fails for a zero root. The provision to terminate after a given number of cycles is therefore essential. Also, double precision is desirable.

* Work supported by the U. S. Atomic Energy Commission.

REMARK ON ALGORITHM 2
ROOTFINDER (J. Wegstein, *Communications ACM*, February, 1960)

Henry C. Thacher, Jr.,* Argonne National Laboratory, Argonne, Illinois

$$\frac{y_k - Y}{y_{k-1} - Y} = \frac{(y_{k-2} - Y)f''}{2(f' - 1) + (y_{k-1} - y_{k-2})f''} + O(y_{k-1} - Y)^3$$

where Y is the desired root, and the derivatives f' and f'' are evaluated there. Convergence is thus second order, provided that $|f''|\,|y_{k-1} - Y| < 2\,|f' - 1|$.

The algorithm is, however, somewhat unstable numerically because of the factor $f(y_{k-1}) - f(y_{k-2}) - y_{k-1} + y_{k-2}$ in the denominator.

Experience has shown that the minimum for ϵ is about one half the precision being used. Provision to indicate when round-off errors are causing random oscillations of g would be a desirable addition.

The criterion used for terminating the iteration renders the algorithm unsuitable for a zero root. A preliminary test for a zero root would be desirable. In addition, the algorithm should include provision for exit after a stated number of iterations. Algorithm 15 appears to offer advantages along these lines.

* Work supported by the U. S. Atomic Energy Commission.
This algorithm has the convergence factor

REMARKS ON ALGORITHMS 2 AND 3 (*Comm. ACM*, February 1960), ALGORITHM 15 (*Comm. ACM*, August 1960) AND ALGORITHMS 25 AND 26 (*Comm. ACM*, November 1960)

J. H. Wilkinson
National Physical Laboratory, Teddington.

Algorithms 2, 15, 25 and 26 were all concerned with the calculation of zeros of arbitrary functions by successive linear or quadratic interpolation. The main limiting factor on the accuracy attainable with such procedures is the condition of the *method* of evaluating the function in the neighbourhood of the zeros. It is this condition which should determine the tolerance which is allowed for the relative error. With a well-conditioned method of evaluation quite a strict convergence criterion will be met, even when the function has multiple roots.

For example, a real quadratic root solver (of a type similar to Algorithm 25) has been used on ACE to find the zeros of triple-diagonal matrices T having $t_{ii} = a_i$, $t_{i+1,i} = b_{i+1}$, $t_{i,i+1} = c_{i+1}$. As an extreme case I took $a_1 = a_2 = \cdots = a_5 = 0$, $a_6 =$

$a_7 = \cdots = a_{10} = 1$, $a_{11} = 2$, $b_i = 1$, $c_i = 0$ so that the function which was being evaluated was $x^5(x - 1)^5(x - 2)$. In spite of the multiplicity of the roots, the answers obtained using floating-point arithmetic with a 46-bit mantissa had errors no greater than 2^{-44}. Results of similar accuracy have been obtained for the same problem using linear interpolation in place of the quadratic. This is because the method of evaluation which was used, the two-term recurrence relation for the leading principal minors, *is a very well-conditioned method of evaluation.* Knowing this, I was able to set a tolerance of 2^{-42} with confidence. If the *same function* had been evaluated from its explicit polynomial expansion, then a tolerance of about 2^{-7} would have been necessary and the multiple roots would have obtained with very low accuracy.

To find the zero roots it is necessary to have an absolute tolerance for $|\,x_{r+1} - x_r\,|$ as well as the relative tolerance condition. It is undesirable that the preliminary detection of a zero root should be necessary. The great power of rootfinders of this type is that, since we are not saddled with the problem of calculating the derivative, we have great freedom of choice in evaluating the function itself. This freedom is encroached upon if we frame the rootfinder so that it finds the zeros of $x = f(x)$ since the true function $x - f(x)$ is arbitrarily separated into two parts. The formal advantage of using this formulation is very slight. Thus, in Certification 2 (June 1960), the calculation of the zeros of $x = \tan x$ was attempted. If the function $(-x + \tan x)$ were used with a general zero finder then, provided the method of evaluation was, for example

$$x = n\pi + y$$
$$\tan x - x = -n\pi + \frac{\dfrac{y^3}{3} - \dfrac{y^5}{30} - \cdots}{\cos y},$$

the multiple zeros at $x = 0$ could be found as accurately as any of the others. With a slight modification of common sine and cosine routines, this could be evaluated as

$$-n\pi + \frac{(\sin y - y) - y(\cos y - 1)}{1 + (\cos y - 1)}$$

and the evaluation is then well-conditioned in the neighbourhood of $x = 0$. As regards the number of iterations needed, the restriction to 10 (Certification 2) is rather unreasonably small. For example, the direct evaluation of $x^{60} - 1$ is well conditioned, but starting with the values $x - 2$ and $x = 1.5$ a considerable number of iterations are needed to find the root $x = 1$. Similarly a very large number are needed for Newton's method, starting with $x = 2$. If the time for evaluating the derivative is about the same as that for evaluating the function (often it is much longer), then linear interpolation is usually faster, and quadratic interpolation much faster, than Newton.

In all of the algorithms, including that for Bairstow, it is useful to have some criterion which limits the permissible change from one value of the independent variable to the next [1]. This condition is met to some extent in Algorithm 25 by the condition S4, that $abs(fprt) < abs(x2 \times 10)$, but here the limitation is placed on the permissible increase in the value of the function from one step to the next. Algorithms 3 and 25 have tolerances on the size of the function and on the size of the remainders r1 and r0 respectively. They are very difficult tolerances to assign since these quantities may take very small values without our wishing to accept the value of x as a root. In Algorithm 3 (Comm. ACM June 1960) it is useful to return to the original polynomial and to iterate with each of the computed factors. This eliminates the loss of accuracy which may occur if the factors are not found in increasing order. This presumably was the case in Certification 3 when the roots of $x^5 + 7x^4 + 5x^3 + 6x^2 + 3x + 2 = 0$ were attempted. On ACE, however, all roots of this polynomial were found very accurately and convergence was very fast using single-precision, but the roots emerged in increasing order. The reference to *slow* convergence is puzzling. On ACE, convergence was fast for all the initial approximations to p and q which were tried. When the initial approximations used were such that the real root $x = -6.35099\,36103$ and the spurious zero were found first, the remaining two quadratic factors were of lower accuracy, though this was, of course, rectified by iteration in the original polynomial. When either of the other two factors was found first, then all factors were fully accurate even without iteration in the original polynomial [1].

REFERENCE

[1] J. H. WILKINSON. The evaluation of the zeros of ill-conditioned polynomials Parts I and II. *Num. Math.* 1 (1959), 150–180.

3. Solution of Polynomial Equation by Bairstow-Hitchcock Method

A. A. Grau

Oak Ridge National Laboratory, Oak Ridge, Tenn.

```
procedure
BAIRSTOW     (n, a[ ], eps0, eps1, eps2, eps3, K) = :
                 (m, x[ ], y[ ], nat[ ], ex[ ]);
comment      The Bairstow-Hitchcock iteration is used to find
```

comment The Bairstow-Hitchcock iteration is used to find successively pairs of roots of a polynomial equation of degree n with coefficients a_i $(i = 0, 1, \cdots, n)$ where a_n is the constant term. On exit from the procedure, m is the number of pairs of roots found, x[i] and y[i] $(i = 1, \cdots, m)$ are a pair of real roots if nat[i]=1, the real and imaginary parts of a complex pair if nat[i]=−1, and ex[i] indicates which of the following conditions was met to exit from the iteration loop in finding this pair:

1. Remainders, r1, r0, become absolutely less than eps1.
2. Corrections, incrp, incrq, become absolutely less than eps2.
3. The ratios, incrp/p, incrq/q, become absolutely less than eps3.
4. The number of iterations becomes K.

In the last case, the pair of roots found is not reliable and no further effort to find additional roots is made. The quantity eps 0 is used as a lower bound for the denominator in the expressions from which incrp and incrq are found.;

```
begin
integer      (i, j, k, n1, n2, m1)   ;
array        (b, c[0 : n+1])   ;
BAIRSTOW     for i := 0(1)n   ;   b_i := a_i
             b_{n+1} := 0   ;   n2 := entire((n+1)/2)
               n1 := 2×n2
             for m1 := 1(1)n2   ;   begin p := 0   ;   q := 0
             for k := 1(1)K   ;   begin
step:        for i := 0(1)n1   ;   c_i := b_i
             for j := n1−2, n1−4   ;   begin
             for i := 0(1)j   ;   begin
             c_{i+1} := c_{i+1} − p × c_i
             c_{i+2} := c_{i+2} − q × c_i end end
             r0 := c_{n1}   ;   r1 := c_{n1−1}
             s0 := c_{n1−2}   ;   s1 := c_{n1−3}
             v0 := −q × s1   ;   v1 := s0 − s1 × p
             det0 := v1 × s0 − v0 × s1
             if (abs(det0)<eps0)   ;   begin
             p := p+1   ;   q := q+1   ;   go to step end
             det1 := s0 × r1 − s1 × r0
             det2 := r0 × v1 − v0 × s1
             incrp := det1/det0   ;   incrq := det2/det0
             p := p + incrp   ;   q := q + incrq
             if (abs (r0) < eps1)   ;   begin
             if (abs (r1) < eps1)   ;   begin
             ex_{m1} := 1   ;   go to next end end
             if (abs (incrp) < eps2)   ;   begin
             if (abs (incrq) < eps2)   ;   begin
             ex_{m1} := 2   ;   go to next end end
             if (abs (incrp/p) < eps3)   ;   begin
             if (abs (incrq/q) < eps3)   ;   begin
             ex_{m1} := 3   ;   go to next end end end
             ex_{m1} := 4
next:        S := p/2   ;   T := S² − q
             if (T ≥ 0)   ;   begin T := sqrt (T)
             nat_{m1} := 1   ;   x_{m1} := S + T
               y_{m1} := S − T end
             if (T < 0)   ;   begin nat_{m1} := −1   ;   x_{m1} := S
               y_{m1} := sqrt(−T) end
             if (ex_{m1} := 4)   ;   go to out
             for j := 0(1) (n1−2)   ;   begin
             b_{j+1} := b_{j+1} − p × b_j
             b_{j+2} := b_{j+2} − q × b_j   ;   end
             n1 := n1 − 2   ;   if  (n1 < 1)
out:         begin m := m1   ;   return end
             if (n1 < 3)   ;   begin
             m1 := m1 + 1   ;   ex_{m1} := 1
             p := b_1/b_0   ;   q := b_2/b_0
             go to next end
end end
```

CERTIFICATION

3. Solution of Polynomial Equation by Bairstow-Hitchcock Method, A. A. Grau, *Communications ACM*, February, 1960.

Henry C. Thacher, Jr.,* Argonne National Laboratory, Argonne, Illinois.

Bairstow was coded for the Royal-Precision LGP-30 computer, using an interpretive floating point system (24.2) with 28 bits of significance. The translation from Algol was made by hand.

The following minor corrections were found necessary.

a. det 2 := r0 × v1 − v0 × s1 should be det 2 := r0 × v1 − v0 × r1

b. S := p/2 should be S := −p/2.

After these were made, the program ran smoothly for the following equations:

$$x^4 - 3x^3 + 20x^2 + 44x + 43 = 0 \qquad x = -.97063897 \pm 1.0058076i$$
$$x = -2.4706390 \pm 4.6405330i$$

$$x^6 - 2x^5 + 2x^4 + x^3 + 6x^2 - 6x + 8 = 0$$
$$x = 0.50000000 \pm 0.86602539i$$
$$x = 1.0000000 \pm 1.0000000i$$
$$x = 1.5000000 \pm 1.3228756i$$

$$x^5 + x^4 - 8x^3 - 16x^2 + 7x + 15 = 0$$
$$x = .000000005,** - 0.99999999$$
$$x = 3.0000000, 0.99999999$$
$$x = -2.0000000 \pm 1.0000000i$$

With the equation $x^5 + 7x^4 + 5x^3 + 6x^2 + 3x + 2 = 0$ convergence was slow, and full accuracy was not obtained. However, the equation with reciprocal roots, $2x^5 + 3x^4 + 6x^3 + 5x^2 + 7x + 1 = 0$, converged rapidly.

---

* Work supported by the U. S. Atomic Energy Commission.

** Spurious zero real roots are introduced for equations of odd order.

CERTIFICATION OF ALGORITHM 3
SOLUTION OF POLYNOMIAL EQUATIONS BY BAIRSTOW HITCHCOCK METHOD (A. A. Grau, *Comm. ACM*, February, 1960)

JAMES S. VANDERGRAFT

Stanford University, Stanford, California

Bairstow was coded for the Burroughs 220 computer using the Burroughs ALGOL. Conversion from ALGOL 60 was made by hand on a statement-for-statement basis. The integer declaration had to be extended to include n, k, n, NAT, EX, and the corrections noted in the certification by Henry C. Thacher, Jr., *Communications ACM*, June, 1960, were incorporated.

By selecting the input parameters carefully, all branches of the routine were tested and the program ran smoothly. The following polynomials equations were solved:

$$x^6 - 14x^4 + 49x^2 - 36 = 0, \quad x = \pm 1.0000000$$
$$x = \pm 1.9999998$$
$$x = \pm 3.0000001$$
$$x^8 - 30x^6 + 273x^4 - 820x^2 + 576 = 0, \quad x = \pm 1.0000000$$
$$x = \pm 2.0000000$$
$$x = \pm 2.9999999$$
$$x = \pm 4.0000001$$

Several minor errors were found in the certification by Mr. Thacher. The constant term in the first polynomial should be 54 instead of 43, the second pair of roots for that polynomial should be $+ 2.470639 \pm 4.6405330\,i$, and the second pair of roots for the second polynomial should be $-1.0 \pm i$.

REMARKS ON ALGORITHMS 2 AND 3 (*Comm. ACM*, February 1960), ALGORITHM 15 (*Comm. ACM*, August 1960) AND ALGORITHMS 25 AND 26 (*Comm. ACM*, November 1960)

J. H. WILKINSON

National Physical Laboratory, Teddington.

Algorithms 2, 15, 25 and 26 were all concerned with the calculation of zeros of arbitrary functions by successive linear or quadratic interpolation. The main limiting factor on the accuracy attainable with such procedures is the condition of the *method* of evaluating the function in the neighbourhood of the zeros. It is this condition which should determine the tolerance which is allowed for the relative error. With a well-conditioned method of evaluation quite a strict convergence criterion will be met, even when the function has multiple roots.

For example, a real quadratic root solver (of a type similar to Algorithm 25) has been used on ACE to find the zeros of triple-diagonal matrices T having $t_{ii} = a_i$, $t_{i+1,i} = b_{i+1}$, $t_{i,i+1} = c_{i+1}$. As an extreme case I took $a_1 = a_2 = \cdots = a_5 = 0$, $a_6 = a_7 = \cdots = a_{10} = 1$, $a_{11} = 2$, $b_i = 1$, $c_i = 0$ so that the function which was being evaluated was $x^5(x - 1)^5(x - 2)$. In spite of the multiplicity of the roots, the answers obtained using floating-point arithmetic with a 46-bit mantissa had errors no greater than 2^{-44}. Results of similar accuracy have been obtained for the same problem using linear interpolation in place of the quadratic. This is because the method of evaluation which was used, the two-term recurrence relation for the leading principal minors, *is a very well-conditioned method of evaluation*. Knowing this, I was able to set a tolerance of 2^{-42} with confidence. If the *same function* had been evaluated from its explicit polynomial expansion, then a tolerance of about 2^{-7} would have been necessary and the multiple roots would have obtained with very low accuracy.

To find the zero roots it is necessary to have an absolute tolerance for $|\,x_{r+1} - x_r\,|$ as well as the relative tolerance condition. It is undesirable that the preliminary detection of a zero root should be necessary. The great power of rootfinders of this type is that, since we are not saddled with the problem of calculating the derivative, we have great freedom of choice in evaluating the function itself. This freedom is encroached upon if we frame the rootfinder so that it finds the zeros of $x = f(x)$ since the true function $x - f(x)$ is arbitrarily separated into two parts. The formal advantage of using this formulation is very slight. Thus, in Certification 2 (June 1960), the calculation of the zeros of $x = \tan x$ was attempted. If the function $(-x + \tan x)$ were used with a general zero finder then, provided the method of evaluation was, for example

$$x = n\pi + y$$
$$\tan x - x = -n\pi + \frac{\dfrac{y^3}{3} - \dfrac{y^5}{30} - \cdots}{\cos y},$$

the multiple zeros at $x = 0$ could be found as accurately as any of the others. With a slight modification of common sine and cosine routines, this could be evaluated as

$$-n\pi + \frac{(\sin y - y) - y(\cos y - 1)}{1 + (\cos y - 1)}$$

and the evaluation is then well-conditioned in the neighbourhood of $x = 0$. As regards the number of iterations needed, the restriction to 10 (Certification 2) is rather unreasonably small. For example, the direct evaluation of $x^{60} - 1$ is well conditioned, but starting with the values $x = 2$ and $x = 1.5$ a considerable number of iterations are needed to find the root $x = 1$. Similarly a very large number are needed for Newton's method, starting with $x = 2$. If the time for evaluating the derivative is about the same as that for evaluating the function (often it is much longer), then linear interpolation is usually faster, and quadratic interpolation much faster, than Newton.

In all of the algorithms, including that for Bairstow, it is useful to have some criterion which limits the permissible change from one value of the independent variable to the next [1]. This condition is met to some extent in Algorithm 25 by the condition S4, that $\mathrm{abs}(fprt) < \mathrm{abs}(x2 \times 10)$, but here the limitation is placed on the permissible increase in the value of the function from one step to the next. Algorithms 3 and 25 have tolerances on the size of the function and on the size of the remainders r1 and r0 respectively. They are very difficult tolerances to assign since these quantities may take very small values without our wishing to accept the value of x as a root. In Algorithm 3 (Comm. ACM June 1960) it is useful to return to the original polynomial and to iterate with each of the computed factors. This eliminates the loss of accuracy which may occur if the factors are not found in increasing order. This presumably was the case in Certification 3 when the roots of $x^5 + 7x^4 + 5x^3 + 6x^2 + 3x + 2 = 0$ were attempted. On ACE, however, all roots of this polynomial were found very accurately and convergence was very fast using single-precision, but the roots emerged in increasing order. The reference to *slow* convergence is puzzling. On ACE, convergence was fast for all the initial approximations to p and q which were tried. When the initial approximations used were such that the real root $x = -6.35099\,36103$ and the spurious zero were found first, the remaining two quadratic factors were of lower accuracy, though this was, of course, rectified by iteration in the original polynomial. When either of the other two factors was found first, then all factors were fully accurate even without iteration in the original polynomial [1].

REFERENCE

[1] J. H. WILKINSON. The evaluation of the zeros of ill-conditioned polynomials Parts I and II. *Num. Math.* 1 (1959), 150–180.

CERTIFICATION OF ALGORITHM 3
SOLUTION OF POLYNOMIAL EQUATION BY BARSTOW-HITCHCOCK (A. A. Grau, *Comm. ACM* Feb. 1960)

JOHN HERNDON
Stanford Research Institute, Menlo Park, California

Bairstow was transliterated into BALGOL and tested on the Burroughs 220. The corrections supplied by Thatcher, *Comm. ACM*, June 1960, were incorporated. Results were correct for equations for which the method is suitable. $x^4 - 16 = 0$ is one of those which gave nonsensical results. Seven-digit results were obtained for 12 test equations, one of which was $x^6 - 2x^5 + 2x^4 + x^3 + 6x^2 - 6x + 8 = 0$.

4. BISECTION ROUTINE

S. Gorn
University of Pennsylvania Computer Center
Philadelphia, Pa.

comment This procedure evaluates a function at the end-points of a real interval, switching to an error exit (fools exit) FLSXT if there is no change of sign. Otherwise it finds a root by iterated bisection and evaluation at the midpoint, halting if either the value of the function is less than the free variable ϵ or two successive approximations of the root differ by less than $\epsilon1$. ϵ should be chosen of the order of error in evaluating the function (otherwise time would be wasted), and $\epsilon1$ of the order of desired accuracy. $\epsilon1$ must not be less than two units in the last place carried by the machine or else indefinite cycling will occur due to round-off on bisection. Although this method is of 0 order, and therefore among the slowest, it is applicable to any continuous function. The fact that no differentiability conditions have to be checked makes it, therefore, an 'old work-horse' among routines for finding real roots which have already been isolated. The free variables y1 and y2 are (presumably) the end-points of an interval within which there is an odd number of roots of the real function F. α is the temporary exit for the evaluation of F.;

```
procedure   Bisec(y1, y2, ε, ε1, F( ), FLSXT) =: (x)
begin
Bisec:        i := 1  ;  j := 1  ;  k := 1  ;  x := y2
α:            f := F(x)  ;  if (abs(f) ≦ ε)  ;  return
              go to γ1
First val:    i := 2  ;  f1 := f  ;  x := y1  ;  go to α
Succ val:     if (sign(f) = sign(f1))  ;  go to δj  ;  go to ηk
Sec val:      j := 2  ;  k := 2
Midpoint:     x := y1/2 + y2/2  ;  go to α
Reg δ:        y2 := x
Precision:    if (abs(y1 − y2) ≧ ε1)  ;  go to Midpoint
              return
Reg η:        y1 := x  ;  go to Precision
              integer (i, j, k)
              switch γ := (First val, Succ val)
              switch δ := (FLSXT, Reg δ)
              switch η := (Sec val, Reg η)
end Bisec
```

CERTIFICATION OF ALGORITHM 4
BISECTION ROUTINE (S. Gorn, *Comm. ACM*, March 1960)

PATTY JANE RADER,* Argonne National Laboratory, Argonne, Illinois

BISEC was coded for the Royal-Precision LGP-30 computer, using an interpretive floating point system (24.2) with 28 bits of significance.

The following minor correction was found necessary.

α: **go to** γ_1 should be **go to** γ_i

* Work supported by the U. S. Atomic Energy Commission.

After this correction was made, the program ran smoothly for F(x) = cos x, using the following parameters:

y1	y2	ϵ	$\epsilon1$	Results
0	1	.001	.001	FLSXT
0	2	.001	.001	1.5703
1.5	2	.001	.001	1.5703
1.55	2	.1	.1	1.5500
1.5	2	.001	.1	1.5625

These combinations test all loops of the program.

---

* Work supported by the U. S. Atomic Energy Commission.

5. BESSEL FUNCTION I, SERIES EXPANSION
Dorothea S. Clarke
General Electric Co., FPLD, Cincinnati 15, Ohio

comment	Compute the Bessel function $I_n(X)$ when n and X are within the bounds of the series expansion. The procedure calling statement gives n, X and an absolute tolerance δ for determining the point at which the terms of the summation become insignificant. Special case: $I_0(0) = 1$;
procedure	$I(n, X, \delta) =: (Is)$
begin	
I:	$s := 0$; $sum := 0$
if	$(n \neq 0)$; go to STRT
if	$(X = 0)$; begin Is := 1 ; return end
	summ := 1 ; go to SURE
STRT:	sfac := 1
if	$(s = 0)$; go to HRE
for	$t := 1\,(1)\,s$
	$sfac := sfac \times t$
HRE:	snfac := sfac
for	$t := s + 1\,(1)\,s + n$
	$snfac := snfac \times t$
	$summ := sum + (X/2)^{n+2 \times s}/(sfac \times snfac)$
SURE: if	$(\delta < abs\,(summ - sum))$
begin	$s := s + 1$; $sum := summ$; go to STRT end
	Is := summ ; return
end	

6. BESSEL FUNCTION I, ASYMPTOTIC EXPANSION
Dorothea S. Clarke
General Electric Co., FPLD, Cincinnati 15, Ohio

```
comment     Compute the Bessel Function I_n(X) when n and X
            are within the bounds of the asymptotic expansion.
            The procedure calling statement gives n, X and an
            absolute tolerance δ for determining the point at
            which the terms of the summation become in-
            significant;
procedure   I(n, X, δ) = : (IA)
begin
I:          r := 1   ;  pe := (4 × n² − 1) / (8 × X)
              sum := − pe
Repeat:     r := r + 1
            pe := pe × ((2 × n)² − (2 × r − 1)²) / (r × 8 × X)
if          (δ < abs (pe))
begin       sum := sum + (−1)ʳ × pe   ;  go to Repeat end
            IA := (1 + sum) × (exp(X) / sqrt (2 × π × X))
                return
end
```

7. EUCLIDIAN ALGORITHM
Robert Claussen
General Electric Co., Cincinnati 15, Ohio

```
comment     Every pair of numbers a, b not both zero have a
            positive greatest common divisor: gcd;
procedure   EUC (a, b) = : (gcd)
begin
EUC:
if          (a = 0)
begin       gcd := b  ;  return end
if          (b = 0)
begin       gcd := a   ;  return end
            r2 := a
            r1 := b
here:       g := r2/r1
comment     Assumption is made that truncation takes place
            in the above statement;
            r := r2 − r1 × g
if          (r = 0)
begin       gcd := r1  ;  return end
begin       r2 := r1
            r1 := r
            go to here end
integer     (g)
end
```

ALGORITHM 8
EULER SUMMATION
3 (May 1960), 318   P. NAUR

procedure euler (fct, sum, eps, tim) ; **value** eps, tim ;
integer tim ; **real procedure** fct ; **real** sum, eps ;
comment euler computes the sum of fct(i) for i from zero up to
infinity by means of a suitably refined euler transformation. The
summation is stopped as soon as tim times in succession the absolute value of the terms of the transformed series are found to be
less than eps. Hence, one should provide a function fct with one
integer argument, an upper bound eps, and an integer tim. The
output is the sum sum. euler is particularly efficient in the case
of a slowly convergent or divergent alternating series   ;
begin integer i, k, n, t ; **array** m[0:15] ; **real** mn, mp, ds ;
i := n := t := 0 ; m[0] := fct(0) ; sum := m[0]/2 ;
nextterm: i := i+1 ; mn := fct(i)   ;
      for k := 0 **step** 1 **until** n **do**
        begin mp := (mn+m[k])/2 ; m[k] := mn ;
         mn := mp **end** means   ;
      if (abs(mn)<abs(m[n]))∧(n<15) **then**
        begin ds := mn/2 ; n := n+1 ; m[n] :=
        mn **end** accept
      else ds := mn   ;
      sum := sum + ds   ;
      if abs(ds)<eps **then** t := t+1 **else** t := 0   ;
      if t<tim **then go to** nextterm
end euler

Errors less than 0.2×10^{-5} were also found for $n = 1, 2, 3, 4, 5,$
6, 7, 8 and 9.

This technique appears to be a useful supplement to direct
telescoping (Algorithms 37 and 38) and to the methods recommended by Clenshaw[1], for slowly convergent power series.

———

[1] Clenshaw, C. W., *Chebyshev Series for Mathematical Functions.*
National Physical Laboratory Math Tables, Vol. 5, London,
H.M.S.O. (1962).

CERTIFICATION OF ALGORITHM 8
EULER SUMMATION [P. Naur et al. *Comm. ACM*
3, May 1960]
HENRY C. THACHER, JR.*
Argonne National Laboratory, Argonne, Ill.
* Work supported by the U. S. Atomic Energy Commission

The body of *euler* was tested on the LGP-30 computer using the
Dartmouth SCALP translator. No errors were detected.

The program gave excellent results when used to derive the coefficients for the expansion of $ln\ (1+x)$ in shifted Chebyshev polynomials from the first ten terms of the power series. For $n = 0, 1,$
2, 3, 4, the coefficient of x^i in the power series was multiplied by
the coefficient of $T_n^*(x)$ in the expression of x^i in terms of the
$T_n^*(x)$. The product, for $i = 1, 2, \cdots, 10$ was used as $fct(i)$ in the
program. Results for $n = 0$ were as follows:

i	$fct(i)$	ds	sum
1	+0.50000000	—	
2	−0.18750000	+0.07812500	+0.3281250
3	+0.10416667	+0.05729166	+0.3854167
4	−0.068359375	−0.005940758	+0.3794759
5	+0.049218750	−0.001928713	+0.3775471
6	−0.037597656	−0.001357019	+0.3761900
7	+0.029924665	+0.0001742393	+0.3763642
8	−0.024547577	+0.0000571311	+0.3764212
9	+0.020607842	+0.0006395427	+0.3764607
10	−0.017619705	−0.0000055069	+0.3764551
True Value[1]			+0.3764528129.....

ALGORITHM 9 RUNGE-KUTTA INTEGRATION
3 (May 1960), 318 P. NAUR

procedure RK(x,y,n,FKT,eps,eta,xE,yE,fi) ; **value** x,y ;
integer n ; **Boolean** fi ; **real** x,eps,eta,xE ; **array**
y,yE ; **procedure** FKT ;
comment : RK integrates the system $y_k' = f_k(x, y_1, y_2, \ldots, y_n)$
$(k=1,2,\ldots,n)$ of differential equations with the method of Runge-
Kutta with automatic search for appropriate length of integration
step. Parameters are: The initial values x and y[k] for x and the unknown functions $y_k(x)$. The order n of the system. The procedure
FKT(x,y,n,z) which represents the system to be integrated, i.e.
the set of functions f_k . The tolerance values eps and eta which
govern the accuracy of the numerical integration. The end of the
integration interval xE. The output parameter yE which represents the solution at x=xE. The Boolean variable fi, which must
always be given the value **true** for an isolated or first entry into
RK. If however the functions y must be available at several mesh-
points $x_0, x_1, \ldots, x_n$, then the procedure must be called repeat-
edly (with x=x_k, xE=x_{k+1}, for k=0, 1, $\ldots$, n−1) and then the
later calls may occur with fi=**false** which saves computing time.
The input parameters of FKT must be x,y,n, the output parameter
z represents the set of derivatives z[k]=f_k(x,y[1], y[2], $\ldots$, y[n])
for x and the actual y's. A procedure comp enters as a non-local
identifier ;
begin
  array z,y1,y2,y3[1:n] ; **real** x1,x2,x3,H ; **Boolean** out ;
  integer k,j ; **own real** s,Hs ;
  procedure RK1ST(x,y,h,xe,ye) ; **real** x,h,xe ; **array**
        y,ye ;
    comment : RK1ST integrates one single RUNGE-KUTTA
        with initial values x,y[k] which yields the output
        parameters xe=x+h and ye[k], the latter being the
        solution at xe. Important: the parameters n, FKT, z
        enter RK1ST as nonlocal entities ;
  begin
    array w[1:n], a[1:5] ; **integer** k,j ;
    a[1] := a[2] := a[5] := h/2 ; a[3] := a[4] := h ;
    xe := x ;
    for k := 1 **step** 1 **until** n **do** ye[k] := w[k] := y[k] ;
    for j := 1 **step** 1 **until** 4 **do**
    begin
      FKT(xe,w,n,z) ;
      xe := x+a[j] ;
      for k := 1 **step** 1 **until** n **do**
      begin
        w[k] := y[k]+a[j]×z[k] ;
        ye[k] := ye[k] + a[j+1]×z[k]/3
      end k
    end j
  end RK1ST ;
Begin of program:
    if fi **then begin** H := xE−x ; s := 0 **end else** H := Hs ;
    out := **false** ;
AA: **if** (x+2.01×H−xE>0)≡(H>0) **then**
    begin Hs := H ; out := **true** ; H := (xE−x)/2
      end if ;
    RK1ST (x,y,2×H,x1,y1) ;
BB: RK1ST (x,y,H,x2,y2) ; RK1ST(x2,y2,H,x3,y3) ;
    for k := 1 **step** 1 **until** n **do**

if comp(y1[k],y3[k],eta)>eps **then go to** CC ;
comment : comp(a,b,c) is a function designator, the value
    of which is the absolute value of the difference of the
    mantissae of a and b, after the exponents of these
    quantities have been made equal to the largest of the ex-
    ponents of the originally given parameters a,b,c ;
x := x3 ; **if** out **then go to** DD ;
for k := 1 **step** 1 **until** n **do** y[k] := y3[k] ;
if s=5 **then begin** s := 0 ; H := 2×H **end if** ;
s := s+1 ; **go to** AA ;
CC: H := 0.5×H ; out := **false** ; x1 := x2 ;
for k := 1 **step** 1 **until** n **do** y1[k] := y2[k] ;
go to BB ;
DD: **for** k := 1 **step** 1 **until** n **do** yE[k] := y3[k]
end RK

[8] This RK-program contains some new ideas which are related
to ideas of S. GILL, A process for the step-by-step integration of
differential equations in an automatic computing machine, *Proc.
Camb. Phil. Soc. Vol. 47* (1951) p. 96; and E. FRÖBERG, On the
solution of ordinary differential equations with digital computing machines, *Fysiograf. Sällsk. Lund, Förhd. 20* Nr. 11 (1950)
p. 136–152. It must be clear, however, that with respect to computing time and round-off errors it may not be optimal, nor has it
actually been tested on a computer.

CERTIFICATION OF ALGORITHM 9 [D2]
RUNGE-KUTTA INTEGRATION [P. Naur et al.,
  Comm. ACM 3 (May 1960), 318]
HENRY C. THACHER, JR. (Recd. 28 July 1964 and 22 Nov.
  1965)
Argonne National Laboratory, Argonne, Ill.

Algorithm 9 was transcribed into the hardware representation
for CDC 3600 ALGOL and run successfully. The following procedure
was used for the global procedure *comp*:
real procedure *comp* (a, b, c); **value** a, b, c; **real** a, b, c;
begin integer AE, BE, CE;
  integer procedure *expon*(x); **real** x;
  comment This function produces the base 10 exponent of x;
  expon := **if** $x = 0$ **then** -999 **else**
    entier $(.4342944819 \times ln(abs(x)) + 1)$;
  comment The number -999 may be replaced by any number
    less than the exponent of the smallest positive number handled
    by the particular machine used, for this algorithm assumes
    that true zero has an exponent smaller than any nonzero
    floating-point number. Users implementing **real procedure**
    comp by machine code should make sure that this condition
    is satisfied by their program;
  $AE := expon(a)$; $BE := expon(b)$; $CE := expon(c)$;
  if $AE < BE$ **then** $AE := BE$; **if** $AE < CE$ **then** $AE := CE$;
  $comp := abs(a - b)/10 \uparrow AE$
end

This has the advantage of machine independence, but is highly inefficient compared to machine code.

The procedure was tested using the two following procedures for *FKT*:

procedure *FKT* (X, Y, N, Z); **real** X; **integer** N; **array** Y, Z;

comment $(dy_1/dx) = z_1 = y_2$, $(dy_2/dx) = z_2 = -y_1$. With $y_1(0) = 0$, $y_2(0) = 1$, the solution is $y_1 = sin\ x$, $y_2 = cos\ x$;

begin $Z[1] := Y[2]$; $Z[2] := -Y[1]$ **end**;

procedure *FKT* (X, Y, N, Z); **real** X; **integer** N; **array** Y, Z;

comment $(dy_1/dx) = 1 + y_1^2$. For $y_1(0) = 0$, $y(x) = tan\ x$; $Z[1] := 1 + Y[1]\uparrow 2$;

The *RK* procedure was used to integrate the differential equations represented by the first *FKT* procedure from $x = 0(0.5)7.0$, with $eps = eta = 10^{-6}$, and with $y_1(0) = 0$, $y_2(0) = 1$. The actual step size h was .0625 for most of the range, but was reduced to .03125 in the neighborhood of $x = k\pi/2$, where one or the other of the solutions is small.

The computed solutions at $x = 7.0$ were: $y_1 = 6.5698602746 \times 10^{-1}$, $y_2 = 7.5390270246 \times 10^{-1}$, with errors -5.71×10^{-7} and 4.48×10^{-7}, respectively.

Results for the second differential equation are summarized in Table I below.

The efficiency of the procedure would be increased slightly on most computers by changing the type of the **own** variable s from **real** to **integer.**

The error is estimated by comparing the results of successive pairs of steps with that of a single double step. This is somewhat more time-consuming than the Kutta-Merson process presented in Algorithm 218 [*Comm. ACM 6* (Dec. 1963) 737–8]. However, the criterion for step-size variation in Algorithm 9 which effectively applies an approximate relative error criterion, *eps*, for $|y| > eta$, and an absolute error criterion $eta \times eps$, for $|y| < eta$, appears superior when the solution fluctuates in magnitude.

TABLE I [ALG. 9]

	η	$x = 0.5$			$x = 1.0$			$x = 1.5$		
		h_{min}	Absolute error	Relative error	h_{min}	Absolute error	Relative error	h_{min}	Absolute error	Relative error
10^{-7}	10^{-3}	.03125	-1×10^{-9}	-2×10^{-9}	.03125	9×10^{-8}	6×10^{-8}	.00390625	-1×10^{-6}	-8×10^{-8}
10^{-5}	10^{-3}	.125	-5×10^{-7}	-9×10^{-7}	.0625	8×10^{-7}	5×10^{-7}	.0078125	2×10^{-4}	1×10^{-5}
10^{-3}	10^{-3}	.25	-1×10^{-5}	-2×10^{-5}	.25	-2×10^{-4}	-1×10^{-4}	.03125	-3×10^{-2}	-2×10^{-3}

10. EVALUATION OF THE CHEBYSHEV POLYNOMIAL $T_n(X)$
BY RECURSION
G. M. Galler
National Bureau of Standards, Washington 25, D. C.

```
comment        This procedure computes the Chebyshev
               polynomial T_n(X) = cos (n × cos⁻¹(X)) for
               any given real argument, X, and any order, n,
               by the recursion formula below;
real procedure Ch(n, X)  ;
real           X  ;  integer n  ;
begin real     a, b, c  ;  integer i  ;
               a := 1  ;  b := X  ;
               if n = 0 then c := a else if n = 1 then
               c := b else for i := 2 step 1 until n do
               begin  c := 2 × X × b − a  ;
                      a := b  ;  b := c
               end
               Ch := c
               end
```

CERTIFICATION OF ALGORITHM 10
CHEBYSCHEV POLYNOMIAL $T_n(x)$ (Galler, *Comm. ACM*, June, 1960)

JOHN HERNDON
Stanford Research Institute, Menlo Park, California

When transliterated into BALGOL and tested on the Burroughs 220, Ch(n, x) gave better than 7-digit accuracy for n = 0, 1, 4, 8 and x = .01, .2, .7. It gave answers when x > 1 which corresponded to the value of the series with x substituted.

11. Evaluation of the Hermite Polynomial $H_n(X)$
by Recursion

G. M. Galler
National Bureau of Standards, Washington 25, D. C.

```
comment          This procedure computes the Hermite poly-
                 nomial
                 H_n(X) = (−1)^n × e^X² × (d^n/dX^n(e^−X²)) for any
                 given real argument, X, and any order, n, by
                 the recursion formula below;
real procedure   He(n, X)  ;
integer          n  ;  real X  ;
begin real       a, b, c  ;  integer i  ;
                 a := 1  ;  b := 2X
                 if n = 0 then c := a else if n = 1 then
                 c := b else for i := 1 step 1 until n−1 do
                 begin  c := 2 × X × b − 2 × i × a  ;
                        a := b  ;  b := c
                 end
                 He := c
                 end
```

12. Evaluation of the Laguerre Polynomial $L_n(X)$
    by Recursion
    G. M. Galler
    National Bureau of Standards, Washington 25 D. C.

comment   This procedure computes the Laguerre polynomial
$$L_n(X) = e^X \times (d^n/dX^n(X^n \times e^{-X}))$$ for any given real argument, X, and any order, n, by the recursion formula below;

```
real procedure  La(n, X) ;
integer         n ; real X ;
begin real      a, b, c ; integer i ;
                a := 1 ; b := 1 − X ;
                if n = 0 then c := a else if n = 1 then
                c := b else for i = 1 step 1 until n−1 do
        begin   c := (1 + 2 × i − X) × b − (i ↑ 2) × a  ;
                a := b ; b := c
        end
        La := c
        end
```

13. EVALUATION OF THE LEGENDRE POLYNOMIAL $P_n(X)$
BY RECURSION
G. M. Galler
National Bureau of Standards, Washington 25 D. C.

comment	This procedure computes the Legendre polynomial $P_n(X) = (1/(2^n \times n!)) \times d^n/dX^n(X^2 - 1)^n$ for any given real argument, X, and any order, n, by the recursion formula below;
real procedure	Le(n, X) ;
integer	n ; **real** X ;
begin real	a, b, c ; **integer** i ;

```
a := 1 ; b := X ;
if n = 0 then c := a else if n = 1 then
c := b else for i := 1 step 1 until n−1 do
begin   c := b × X + (i/(i + 1)) × (X × b − a)   ;
          a := b ; b := c
        end
        Le := c
        end
```

CERTIFICATION OF ALGORITHM 13
LEGENDRE POLYNOMIAL $P_n(x)$ (Galler, *Comm. ACM*, June 1960)
JOHN HERNDON
Stanford Research Institute, Menlo Park, California

When transliterated into BALGOL and tested on the Burroughs 220, Le(n, x) gave 7-digit accuracy for $n = 0, 1, 4, 9$ and $X = .01, .2, .7, 1.9, 5.0$.

ALGORITHM 14
COMPLEX EXPONENTIAL INTEGRAL
A. Beam
National Bureau of Standards, Washington 25, D. C.

procedure EKZ(x,y,k,ϵ,u,v,n) ; **real** x,y,k,ϵ,u,v :
    integer n ;

comment EKZ computes $w(z,k) = u + iv = z^k e^z \int_z^\infty e^{-t} dt/t^k$

from the continued fraction representation found in H. S. Wall, *Continued Fractions*, Chap. 18 (D. Van Nostrand, New York, 1948). Input parameters are x, y, k, and ϵ where z=x+iy. Successive convergents are computed as follows: For n = 2, 3, 4, $\cdots$, $D_n = z/(z + M \times D_{n-1})$, $R_n = (D_n - 1)R_{n-1}$, $C_n = C_{n-1} + R_n$, where M is $k + (n-2)/2$ or $(n-1)/2$ according to whether n is even or odd, and $D_1 = R_1 = C_1 = 1$. Computation is stopped when C_n and C_{n-1} agree to the significance specified by ϵ. The corresponding index n is available after use of the procedure. This method is valid in the entire complex plane except for the origin and the negative real axis. Convergence is too slow to be practical for $|z| < .05$. Also for some range within the half-strip $|y| < 2$, $x < 0$ (this range depends on k). The method is valid for complex k, but only real k is considered in this procedure;

begin **real** t1, t2, t3, M, K, c, a, d, b, g, h, ϵ1 ;
    integer m ;
    comment R = a + ib, D = c + id, C = g + ih ;
    ϵ1 := ϵ↑2 ;
    u := c := a := 1 ; v := d := b := 0 ;
    n := 1 ; K := k - 1 ;
BACK: g := u ; h := v ; n := n + 1 ;
    m := n ÷ 2 ,
    if 2 × m = n **then** M := m + K **else** M := m ;
    t1 := x + M × c ; t2 := y + M × d ;
    t3 := t1↑2 + t2↑2 ;
    c := (x × t1 + y × t2)/t3 ;
    d := (y × t1 - x × t2)/t3 ;
    t1 := c - 1 ; t2 := a ;
    a := a × t1 - d × b ; b := d × t2 + t1 × b ;
    u := g + a ; v := h + b ;
    if (a↑2 + b↑2)/(u↑2 + v↑2) > ϵ1 **then go to**
    BACK ;
end EKZ

and 37), the real and imaginary parts of $E_k(z)$ were computed from u and v. Results are shown in the following table. In all cases, the values agreed with tabulated values within the tolerance specified.

x	y	k	ϵ	n
1×10^{-8}	1.0	1	10^{-1}	7
1×10^{-8}	1.0	1	10^{-2}	14
1×10^{-8}	1.0	1	10^{-3}	24
1×10^{-8}	1.0	1	10^{-4}	37
1×10^{-8}	1.0	1	10^{-5}	52
1×10^{-8}	1.0	1	10^{-6}	70
1×10^{-8}	1.0	1	10^{-7}	90
1×10^{-8}	1.0	1	10^{-8}	114
1×10^{-8}	2.0	1	10^{-6}	37
1×10^{-8}	3.0	1	10^{-6}	26
1×10^{-8}	4.0	1	10^{-6}	21
1.0	1×10^{-8}	1	10^{-6}	40
1.0	1.0	1	10^{-6}	34
1.0	2.0	1	10^{-6}	26
1.0	3.0	1	10^{-6}	21
2.0	1×10^{-8}	1	10^{-6}	23
2.0	1.0	1	10^{-6}	22
2.0	2.0	1	10^{-6}	20
2.0	3.0	1	10^{-6}	17
3.0	1×10^{-8}	1	10^{-6}	17
3.0	1.0	1	10^{-6}	17
3.0	2.0	1	10^{-6}	16
3.0	3.0	1	10^{-6}	15
4.0	0.0	0	10^{-6}	20
4.0	0.0	1	10^{-6}	15
4.0	0.0	2	10^{-6}	16
4.0	0.0	3(1)14	10^{-6}	17
4.0	0.0	15, 16	10^{-6}	16

It thus appears that the algorithm gives satisfactory accuracy, but that in certain ranges of the variables, the time required may be excessive for extensive use.

CERTIFICATION OF ALGORITHM 14
COMPLEX EXPONENTIAL INTEGRAL (A. Beam,
  Comm. ACM, July, 1960)
P. J. RADER AND HENRY C. THACHER, JR.*
Argonne National Laboratory, Argonne, Illinois

EKZ was programmed by hand for the Royal-Precision LGP-30 computer, using a 28-bit mantissa floating-point interpretive system (24.2 modified). To facilitate comparison with existing tables (National Bureau of Standards Applied Mathematics Series 51

ALGORITHM 15
ROOTFINDER II (Modification of Algorithm 2. ROOT-FINDER, J. Wegstein, *Communications ACM*, February, 1960)

HENRY C. THACHER, JR.,* Argonne National Laboratory, Argonne, Illinois

procedure	ROOT II (f, a, eps, n, g, c, m); **integer** n, m; **real procedure** f; **real** a, eps, g, c;

comment ROOT II computes a value of $g = y$ satisfying the equation $y = f(y)$. The iteration will converge to Y providing that at some time in the iteration a g is reached such that $\mathrm{abs}(g - Y) \times \mathrm{abs}(d(df/dy)/dy) < 2 \times \mathrm{abs}((df/dy) - 1)$, where the derivatives are evaluated at Y. Input includes (1) f, a procedure for computing f(y), (2) a, an initial approximation to the root, (3) eps, a tolerance for the relative error in g, and (4) n, a maximum number of iterations to be performed. Output includes: (1) g, the required root, (2) $c = f(g) - g$, (3) m, a parameter indicating the success of the procedure. If the tolerance was not met, $m < 0$. $|m - 1|$ gives the number of times that the correction to g exceeded the preceding one, an indication of instability. ;

begin integer j; **real** b, d, h;

m := 1; **if** f(0) = 0 **then begin** g := 0;

go to return **end**

else g := f(a); b := d := c := a — g;

if c = 0 **then go to** return **else**

for j := 1 **step** 1 **until** n **do begin** c := f(g) — g;

if (abs(c/g) < eps **then go to** return **else** h := b/c;

if h < 0 $\vee$ h > 2 **then** m := m + 1 **else**

d := d/(h — 1); b := c; g := g + d **end** iteration

comment if the system is known to be stable, the **if** clause of the last statement can be omitted;

m := — m return **end**

* Work supported by the U. S. Atomic Energy Commission.

CERTIFICATION OF ALGORITHM 15
ROOTFINDER II (Revision by Henry C. Thacher, Jr., *Communications ACM*, August, 1960)

HENRY C. THACHER, JR.,* Argonne National Laboratory, Argonne, Illinois

The revision of ROOTFINDER suggested in the preceding remark was programmed by hand for the Royal Precision LGP-30 computer, using a 28-bit mantissa floating point interpretive system (24.2).

The program was tested for the following equations:

(1.k) $f(y) = \arctan y + k\pi$ (k = 0, 1, 2, 3, 4, 6, 8)

(2) $f(y) = (y^3 + 1)^{1/3}$ } These both have the root 1.3247180428
(3) $f(y) = y^3 - 1$ }

(4.k) $f(y) = \sinh \alpha_k y$ } $(\alpha_1 = -1.2, \alpha_2 = -0.5, \alpha_3 = 0.5, \alpha_4 = 1.2)$
(5.k) $f(y) = \cosh \alpha_k y$ }

Typical results of these tests were as follows.

f(y)	ϵ	a	g	$[f(g_{-1}) - g_{-1}] \times 10^7$	Remarks
1.0	10^{-8}	1.0000	0.0000000	0.00	
1.1	10^{-8}	3.1415	4.4934094	0.15	
1.2	10^{-8}	6.2832	7.7252518	0.60	
1.3	10^{-8}	9.4248	10.904122	0.00	
1.4	10^{-8}	12.5664	14.066194	0.00	
1.6	10^{-8}	18.8496	20.371303	0.60	
1.8	10^{-8}	25.1327	26.666054	0.60	
1.2	10^{-8}	1.3	1.3247179	0.00	
1.2	10^{-8}	0.5	1.3247179	0.00	
3	10^{-9}	9.0	4.4804900	197.74×10^7	Diverged 2 times, not converged after 20 iter.
3	10^{-9}	5.0	1.3482797	$.51 \times 10^7$	
3	10^{-9}	3.0	1.3247180	0.0	Converged in less than 20 iter.
3	10^{-9}	2.0			Diverged 2 times. Term. with h = 1.
3	10^{-9}	1.1	1.3247180	1384.24	Diverged 9 times. Converged after 20 iter.
3	10^{-9}	1.0			Terminated when g became 0.
3	10^{-9}	0.8	1.3247180	0.00	Diverged 4 times. Conv. in less than 20.
3	10^{-9}	0.6	1.6161598	4.39×10^7	Diverged 2 times. No conv. after 20.
4.k	10^{-9}	1.0	0.00000000	0.00000000	For all k.
5.1 5.4	10^{-8}	1.0	0.09179585	0.793×10^7	Diverged 7 times. No conv. after 20 iter.
5.2, 5.3	10^{-8}	1.0	1.11787755	0.037	

Function (3) is of particular interest, since it does not converge for most algorithms. With the Wegstein iteration, convergence was obtained, or would have been obtained with a few more iterations for a wide range of initial guesses.

* Work supported by the U. S. Atomic Energy Commission.

REMARK ON ALGORITHM 15
ROOTFINDER II (Henry C. Thacher, Jr., *Comm. ACM*, August 1960)

GEORGE E. FORSYTHE AND JOHN G. HERRIOT, Stanford University, Stanford, California

As pointed out by Lieberstein (*Comm. ACM*, January 1959, p. 5), this algorithm is precisely the Newton method of chords or the scant method applied to $g(x) = f(x) - x = 0$. Thus convergence is not of second order but rather (for simple roots) of order $\frac{1}{2}(\sqrt{5} - 1) = 1.618$, as shown by Jeeves (*Comm. ACM*, August 1958, pp. 9–10). In the first portion of the algorithm b, c, d, should be set equal to g— a instead of a— g in order to be consistent with the iteration portion. Doing this will usually cut down the number of iterations. Not only is a preliminary test for a zero root desirable but the possibility that g may be zero at any stage of the iteration should be considered in writing the return criterion. The possibility that h = 1 should also be checked and appropriate action taken. Algorithm 26 takes care of these matters and also corrects some minor errors in Algorithm 15. This method is certainly not the best rootfinder that could be written.

REMARKS ON ALGORITHMS 2 AND 3 (*Comm. ACM*, February 1960), ALGORITHM 15 (*Comm. ACM*, August 1960) AND ALGORITHMS 25 AND 26 (*Comm. ACM*, November 1960)

J. H. WILKINSON
National Physical Laboratory, Teddington.

Algorithms 2, 15, 25 and 26 were all concerned with the calculation of zeros of arbitrary functions by successive linear or quadratic interpolation. The main limiting factor on the accuracy attainable with such procedures is the condition of the *method* of evaluating the function in the neighbourhood of the zeros. It is this condition which should determine the tolerance which is allowed for the relative error. With a well-conditioned method of evaluation quite a strict convergence criterion will be met, even when the function has multiple roots.

For example, a real quadratic root solver (of a type similar to Algorithm 25) has been used on ACE to find the zeros of triple-diagonal matrices T having $t_{ii} = a_i$, $t_{i+1,i} = b_{i+1}$, $t_{i,i+1} = c_{i+1}$. As an extreme case I took $a_1 = a_2 = \cdots = a_5 = 0$, $a_6 = a_7 = \cdots = a_{10} = 1$, $a_{11} = 2$, $b_i = 1$, $c_i = 0$ so that the function which was being evaluated was $x^5(x - 1)^5(x - 2)$. In spite of the multiplicity of the roots, the answers obtained using floating-point arithmetic with a 46-bit mantissa had errors no greater than 2^{-44}. Results of similar accuracy have been obtained for the same problem using linear interpolation in place of the quadratic. This is because the method of evaluation which was used, the two-term recurrence relation for the leading principal minors, *is a very well-conditioned method of evaluation*. Knowing this, I was able to set a tolerance of 2^{-42} with confidence. If the *same function* had been evaluated from its explicit polynomial expansion, then a tolerance of about 2^{-7} would have been necessary and the multiple roots would have obtained with very low accuracy.

To find the zero roots it is necessary to have an absolute tolerance for $|x_{r+1} - x_r|$ as well as the relative tolerance condition. It is undesirable that the preliminary detection of a zero root should be necessary. The great power of rootfinders of this type is that, since we are not saddled with the problem of calculating the derivative, we have great freedom of choice in evaluating the function itself. This freedom is encroached upon if we frame the rootfinder so that it finds the zeros of $x = f(x)$ since the true function $x - f(x)$ is arbitrarily separated into two parts. The formal advantage of using this formulation is very slight. Thus, in Certification 2 (June 1960), the calculation of the zeros of $x = \tan x$ was attempted. If the function $(-x + \tan x)$ were used with a general zero finder then, provided the method of evaluation was, for example

$$x = n\pi + y$$

$$\tan x - x = -n\pi + \frac{\dfrac{y^3}{3} - \dfrac{y^5}{30} - \cdots}{\cos y},$$

the multiple zeros at $x = 0$ could be found as accurately as any of the others. With a slight modification of common sine and cosine routines, this could be evaluated as

$$-n\pi + \frac{(\sin y - y) - y(\cos y - 1)}{1 + (\cos y - 1)}$$

and the evaluation is then well-conditioned in the neighbourhood of $x = 0$. As regards the number of iterations needed, the restriction to 10 (Certification 2) is rather unreasonably small. For example, the direct evaluation of $x^{60} - 1$ is well conditioned, but starting with the values $x = 2$ and $x = 1.5$ a considerable number of iterations are needed to find the root $x = 1$. Similarly a very large number are needed for Newton's method, starting with $x = 2$. If the time for evaluating the derivative is about the same as that for evaluating the function (often it is much longer), then linear interpolation is usually faster, and quadratic interpolation much faster, than Newton.

In all of the algorithms, including that for Bairstow, it is useful to have some criterion which limits the permissible change from one value of the independent variable to the next [1]. This condition is met to some extent in Algorithm 25 by the condition S4, that abs(fprt) < abs(x2 × 10), but here the limitation is placed on the permissible increase in the value of the function from one step to the next. Algorithms 3 and 25 have tolerances on the size of the function and on the size of the remainders r1 and r0 respectively. They are very difficult tolerances to assign since these quantities may take very small values without our wishing to accept the value of x as a root. In Algorithm 3 (Comm. ACM June 1960) it is useful to return to the original polynomial and to iterate with each of the computed factors. This eliminates the loss of accuracy which may occur if the factors are not found in increasing order. This presumably was the case in Certification 3 when the roots of $x^5 + 7x^4 + 5x^3 + 6x^2 + 3x + 2 = 0$ were attempted. On ACE, however, all roots of this polynomial were found very accurately and convergence was very fast using single-precision, but the roots emerged in increasing order. The reference to *slow* convergence is puzzling. On ACE, convergence was fast for all the initial approximations to p and q which were tried. When the initial approximations used were such that the real root $x = -6.35099\ 36103$ and the spurious zero were found first, the remaining two quadratic factors were of lower accuracy, though this was, of course, rectified by iteration in the original polynomial. When either of the other two factors was found first, then all factors were fully accurate even without iteration in the original polynomial [1].

REFERENCE

[1] J. H. WILKINSON. The evaluation of the zeros of ill-conditioned polynomials Parts I and II. *Num. Math.* 1 (1959), 150–180.

ALGORITHM 16
CROUT WITH PIVOTING
GEORGE E. FORSYTHE
Stanford University, Stanford, California

```
real procedure INNERPRODUCT(u,v) index : (k) start : (s)
                  finish : (f);
    value s, f;  integer k, s, f;  real u, v;
comment        INNERPRODUCT forms the sum of u(k) ×
               v(k) for k = s, s+1, . . . , f. If s > f, the value
               of INNERPRODUCT is zero. The substitution
               of a very accurate inner product procedure
               would make CROUT more accurate;
begin
    real h;
    h := 0;  for k := s step 1 until f do h := h + u × v;
    INNERPRODUCT := h
end  INNERPRODUCT;

procedure      CROUT (A, b, n, y, pivot, INNERPRODUCT);
    value n;  array A, b, y, pivot;  integer n, pivot;
    real procedure INNERPRODUCT;
comment        This is Crout's method with row interchanges, as
               formulated in reference [1], for solving Ay = b
               and transforming the augmented matrix [A b]
               into its triangular decomposition LU with all
               L[k, k] = 1. If A is singular we exit to 'singular,'
               a non-local label. pivot[k] becomes the current
               row index of the pivot element in the k-th
               column. Thus enough information is preserved
               for the procedure SOLVE to process a new
               right-hand side without repeating CROUT.
               The accuracy obtainable from CROUT would
               be much increased by calling CROUT with a
               more accurate inner product procedure than
               INNERPRODUCT;
begin
    integer k, i, j, imax, p;  real TEMP, quot;
    for k := 1 step 1 until n do
1:  begin
        TEMP := 0;
        for i := k step 1 until n do
2:      begin
            A[i, k] := A[i, k] − INNERPRODUCT(A[i,p], A[p, k],
                p, 1, k−1);
            if abs(A[i, k]) > TEMP then
3:          begin
                TEMP := abs(A[i, k]);  imax := i
                end 3
        end 2;
        pivot[k] := imax;
        comment  We have found that A[imax, k] is the largest
            pivot in column k. Now we interchange rows k and imax;
        if imax ≠ k then
4:      begin for j := 1 step 1 until n do
5:          begin
            TEMP := A[k,j];  A[k, j] := A[imax,j];
                A[imax,j] := TEMP
            end 5;
        TEMP := b[k]; b[k] := b[imax]; b[imax] := TEMP
```

```
    end 4;
    comment    The row interchange is done. We proceed to the
               elimination;
        if A[k, k] = 0 then go to singular;
        for i := k+1 step 1 until n do
        begin quot := 1.0/A[k, k];  A[i, k] := quot × A[i, k]
            end;
        for j := k+1 step 1 until n do
            A[k, j] := A[k,j] − INNERPRODUCT(A[k, p],
                A[p,j], p, 1, k−1);
            b[k] := b[k] − INNERPRODUCT(A[k,p], b[p], p,
                1, k−1)
    end 1;
    comment    The triangular decomposition is now finished,
               and we do the back substitution;
    for k := n step −1 until 1 do
        y[k] :=  (b[k] − INNERPRODUCT(A[k,p], y[p], p,
            k+1, n)/A[k, k]
end CROUT;

procedure  SOLVE (B, c, n, z, pivot, INNERPRODUCT);
    value n;  array B, c, z, pivot;  integer n, pivot;
    real procedure INNERPRODUCT;
comment  SOLVE assumes that a matrix A has already been
             transformed into B by CROUT, but that a new
             column c has not been processed. SOLVE solves the
             system Az = c, and the output z of SOLVE is pre-
             cisely the same as the output y of the procedure
             statement CROUT (A, c, n, y, pivot, INNER-
             PRODUCT). However, SOLVE is faster, because
             it does not repeat the triangularization of A;
begin
    integer k;  real TEMP;
    for k := 1 step 1 until n do
    begin
        TEMP := c[pivot[k]];  c[pivot[k]] := c[k];  c[k] :=
        TEMP; c[k] := c[k] − INNERPRODUCT(B[k, p],
        c[p], p 1, k − 1)
    end;
    for k := n step −1 until 1 do
        z[k] :=  (c[k] − INNERPRODUCT(B[k,p], z[p], p,
            k+1, n)/B[k, k]
end SOLVE
```

REFERENCE

[1] J. H. WILKINSON, theory and practice in linear systems, pp.
43–100 of JOHN W. CARR III (editor), *Application of Advanced
Numerical Analysis to Digital Computers*, (Lectures given at
the University of Michigan, Summer 1958, College of Engineering, Engineering Summer Conferences, Ann Arbor,
Michigan [1959]).

REMARK ON ALGORITHM 16
CROUT WITH PIVOTING (G. Forsythe, *Communications* ACM, September, 1960)
GEORGE E. FORSYTHE
Stanford University, Stanford, California

QUERY

Perhaps the most basic procedure for an ALGOL library of matrix programs is an inner product procedure. The procedure Innerproduct given on page 311 of [1] is fairly difficult to comprehend, and probably poses great difficulties for most translating routines. I merely copied its form in writing a modified inner product routine for [2].

My query is: How *should* one write an inner product procedure in ALGOL?

REFERENCES

1. PETER NAUR (editor), J. W. BACKUS, ET AL., Report on the algorithmic language ALGOL 60, *Comm. Assoc. Comp. Mach. 3* (1960), 299–314.
2. GEORGE E. FORSYTHE, CROUT with pivoting in ALGOL 60, *Comm. Assoc. Comp. Mach. 3* (1960), 507–508.

REMARK ON ALGORITHM 16
CROUT WITH PIVOTING (G. E. Forsythe, *Comm. ACM, 3* (Sept. 1960), 507–8.)
HENRY C. THACHER, JR.,* Argonne National Laboratory, Argonne, Illinois

This procedure contains the following errors:
a. In SOLVE, the expression
    $c[k] := c[k]$ − INNERPRODUCT
        $(B[k, p], c[p], p\ 1, k − 1)$
should read:
    $c[k] := c[k]$ − INNERPRODUCT
        $(B[k, p], c[p], p, 1, k − 1)$
b. In CROUT, the specification part should read:
    array A, b, y ; **integer** n ; **integer array** pivot ;
c. In SOLVE, the specification part should read:
    array B, c, z ; **integer** n ; **integer array** pivot ;
The efficiency of the algorithm will be improved by the following changes:
a. In the elimination phase of CROUT, replace
    for i := k + 1 **step** 1 **until** n **do**
    begin quote := 1.0/A[k, k] ; A[i, k] := quot XA[i, k] **end** ;
by
    quot := 1.0/A[k, k] ; **for** i := k + 1 **step** 1 **until** n **do**
        A[i, k] := quot XA[i, k] ;
b. Omit INNERPRODUCT from the formal parameter list in both CROUT and SOLVE, and declare INNERPRODUCT either locally, or globally. This avoids any reference to INNERPRODUCT in the calling sequence produced by a compiler.

It is also to be noted that a minor modification of CROUT allows it to be used to evaluate the determinant of A.

All of these suggestions are included in a later algorithm.

* Work supported by the U. S. Atomic Energy Commission.

ALGORITHM 17
TRDIAG
C. F. Sprague III
General Atomic Division of General Dynamics Corp.,
San Diego, California

procedure	trdiag (a,b,c,d) order : (n) result : (x);
value n;	**array** a, b, c, d, x; **integer** n;
comment	this procedure gives the solution to the tridiagonal system of linear algebraic equations:

$$a_1x_2 + b_1x_1 + d_1 = 0$$
$$a_ix_{i+1} + b_ix_i + c_ix_{i-1} + d_i = 0, \quad i = 2,3, \cdots, n-1$$
$$b_nx_n + c_nx_{n-1} + d_n = 0.$$

This method is often used to obtain solutions to second order difference equations;

```
begin array    gamma [1 : n−1];  integer i;  real y;
               gamma [1] := −a[1]/b[1];
               x[1] := −d[1]/b[1];
for            i := 2 step 1 until n−1 do
begin          y = b[i] + c[i] × gamma [i − 1];
               gamma [i] := −a[i]/y;  x[i] := −(c[i] × x[i−1]
               + d[i])/y end;
               x[n] := −(c[n] × x[n−1] + d[n])/(b[n] + c[n]
               × gamma [n−1]);
for            i := n step −1 until 2 do
               x[i − 1] := x[i] × gamma [i − 1] + x[i − 1]
end trdiag
```

ALGORITHM 18
RATIONAL INTERPOLATION BY CONTINUED
FRACTIONS
R. W. Floyd
Armour Research Foundation, Chicago, Illinois

```
comment   This procedure fits to m given points (xᵢ , yᵢ) a con-
              tinued fraction in the form
          a₁+(x−x₁)/(a₂+(x−x₂)/(a₃+(x−x₃)/···(x−xₘ₋₁)/aₘ))···))
          It also simplifies the continued fraction to a rational function
          (N₀+N₁x+···+N_degx^deg)/(D₀+D₁x+···+D_degx^deg),
          where deg is at most m ÷ 2;
procedure   confr(m,x,y,a,N,D);
real array   x,y,a,N,D;   integer m;
begin real   aa, xx, T;   integer i, j, k;   real array P,Q[0 : m ÷ 2];
    switch sw := sw1, sw2;
    for j := 1 step 1 until m do
    begin aa := y[j];   xx := x[j];
        for i := 1 step 1 until j−1 do
        aa := (xx−x[i])/(aa−a[i]);   a[j] := aa
    end;
    k := 1;   P[0] := 1;   Q[0] := a[1];
    mult : for j := 1 step 1 until m ÷ 2 do P[j] := Q[j] := 0;
    for i := 2 step 1 until m do
    begin for j := i ÷ 2 step −1 until 1 do
        begin T := a[i] × Q[j] − x[i−1] × P[j] + P[j−1];
            P[j] := Q[j]; Q[j] := T
        end;   T := a[i] × Q[0] − x[i−1] × P[0];
        P[0] := Q[0];   Q[0] := T
    end;  go  to  sw[k];
    sw1 : for j := 0 step 1 until m ÷ 2 do N[j] := Q[j];
        k := 2;   P[0] := 0; Q[0] := 1;   go to mult;
    sw2 : for j := 0 step 1 until m ÷ 2 do D[j] := Q[j]
end   procedure
```

CERTIFICATION OF ALGORITHM 18
RATIONAL INTERPOLATION BY CONTINUED
FRACTIONS
[R. W. Floyd, *Comm. ACM.*, Sept. 1960]
Henry C. Thacher, Jr.*
Reactor Engineering Div., Argonne National Lab.,
Argonne, Ill.
* Work supported by the U. S. Atomic Energy Commission

The body of procedure *confr* was tested with the Algol translator system written for the LGP-30 computer by the Dartmouth
College Computer Center. No syntactical errors were found in the
procedure body, except for a missing semicolon after the array
delcaration. The translated algorithm gave satisfactory results
when tested on values of $(4x + 1)/(x + 4)$ at any three of the points
$x = 1, 2, 3, 4$. When all four points were used, a division overflow
occurred in the statement **for** $i := 1$ **step** 1 **until** $j−1$ **do** $aa :=
(xx − x[i])/(aa−a[i])$; which forms the reciprocal differences. An
overflow of this type will occur whenever $y[j]$ is approximated to
high accuracy by one of the continued fractions based only on the
points $x[i], i = 1, 2, \cdots , k$ with k less than j. Unless $i = j−1$, the
difficulty may be overcome by setting aa equal to the largest real
representable in the computer whenever division overflow would

occur. When $i = j−1$, the difficulty is irretrievable, and the data
points must be reordered.

ALGORITHM 19
BINOMIAL COEFFICIENTS
RICHARD R. KENYON
Computing Laboratory, Purdue University, Lafayette, Indiana

comment	This procedure computes binomial coefficients $C_m^n = n!/m!(n - m)!$ by the recursion formula $C_{i+1}^n = (n - 1)C_i^n/(i + 1)$ starting from $C_0^n = 1$;
integer procedure	C(m, n) ;
integer	m, n ;
begin	**integer** i, a, b ;
	a := 1 ;
	if $2 \times m > n$ **then** b := n — m **else** b := m ;
	for i := 0 **step** 1 **until** b **do**
	begin a := (n — i) $\times$ a $\div$ (i + 1) **end**
	C := a
end	Binomial Coefficients

REMARK ON ALGORITHM 19
RINOMIAL COEFFICIENTS (Richard R. Kenyon,
Comm. ACM, Oct. 1960)
BICHARD STECK
Armour Research Foundation, Chicago 16, Ill.

The **for** clause of Algorithm 19 should read:

$$\textbf{for } i := 0 \textbf{ step } 1 \textbf{ until } b-1 \textbf{ do}$$

With this correction the algorithm was certified on the Armour Research Foundation UNIVAC 1105.

The recursion formula stated in the **comment** should read:
$$C_{i+1}^n = (n-i) \ C_i^n/(i+1).$$

CERTIFICATION OF ALGORITHM 19
BINOMIAL COEFFICIENTS [Richard R. Kenyon,
Comm. ACM Oct., 1960]
RICHARD GEORGE*
Particle Accelerator Div., Argonne National Lab., Argonne, Ill.

* Work supported by the U. S. Atomic Energy Commission.

This procedure was tested on the LGP-30, using the compiler ALGOL-30 from Dartmouth College Computation Center. The following changes were found necessary:

(1) Within the **comment**, the line

$$C_{i+1}^n = (n - 1)C_i^n/(i + 1)$$

should be

$$C_{i+1}^n = (n - i)C_i^n/(i + 1)$$

(2) The line defining the iteration loop
$$\textbf{for } i := 0 \textbf{ step } 1 \, \textbf{°until } b \textbf{ do}$$

should be

$$\textbf{for } i := 0 \textbf{ step } 1 \textbf{ until } b-1 \textbf{ do}$$

(3) The sequence

$$\text{end} \qquad C := a \qquad \text{end}$$

should be

$$\text{end;} \qquad C := a \qquad \text{end}$$

ALGORITHM 20
REAL EXPONENTIAL INTEGRAL
S. PEAVY
National Bureau of Standards, Washington 25, D.C.

```
real procedure    Expint (x)  ;  real x  ;
comment           −E_i(−x) = ∫_x^∞ (e^−u/u) du is computed for
                  x > 0 by approximation formulas. For
                  0 < x < 1 the approximation is from E. E.
                  Allen, Note 169, MTAC 56, pg 240 (1954).
                  The second approximation formula is for
                  1 ≤ x < ∞ and is from C. Hastings, Jr.,
                  "Approximations For Digital Computers"
                  (Princeton University Press, Princeton,
                  New Jersey, 1955). The absolute error
                  ε(x) is |ε(x)| < 2 × 10^−7 for 0 < x < 1
                  and |ε(x)| < 2 × 10^−8 for 1 ≤ x < ∞  ;
begin             real y, w, z  ;
                  if x < 1 then
                  z := ((((( .00107857 × x − .00976004) × x
                      + .005519968) × x − .24991055) × x
                      + .99999193) × x − .57721566 − ln(x)
                  else begin
                  y := ((( x + 8.5733287401) × x
                      + 18.059016973) × x + 8.6347608925) × x
                      + .2677737343   ;
                  w := ((( x + 9.5733223454) × x
                      + 25.6329561486) × x
                      + 21.0996530827) × x + 3.9584969228   ;
                  z := exp (−x) /x × (y/w) end
                  Expint := z end
```

REMARK ON ALGORITHM 20
REAL EXPONENTIAL INTEGRAL (S. Peavy, *Comm.
ACM*, October 1960)
S. PEAVY
National Bureau of Standards, Washington, D. C.

A printing error has been called to our attention by J. A.
Beutler of E. I. duPont de Nemours and Co. Lines 15 through 17
of Algorithm 20 should read
```
z := (((((.00107857 × x − .00976004) × x
    + .05519968) × x − .24991055) × x
    + .99999193) × x − .57721566 − ln (x)
```

* Work supported by the U. S. Atomic Energy Commission.

CERTIFICATION OF ALGORITHM 20
REAL EXPONENTIAL INTEGRAL (S. Peavy, *Comm.
ACM*, Oct. 1960)
WILLIAM J. ALEXANDER* and HENRY C. THACHER, JR.*
Argonne National Laboratory, Argonne, Illinois

Expint (x) was programmed for the LGP-30 computer, using
both a 7S floating-point compiler (ACT III) and an 8S floating-
point interpretive code (24.2). Constants given to more than 7S
(or to 8S for the 24.2 program) were rounded to 7S (or 8S).

After changing the constant .005519968 to .05519968, both programs gave acceptable accuracy over the range tested.

The 8S (24.2) program was compared with the 9D values given
for −E_i(−x) in Mathematical Tables Project, *Tables of Sine,
Cosine, and Exponential Integrals, Volume* II (1940) for the
set of values x = 0.1(0.1)1.0(1.0)10.0. The largest discrepancy found
was −16 × 10^−8 for x = 0.1. For x greater than 1, all values tested
were good to 8S.

For computing real values of the exponential integral, this
algorithm is much faster than EKZ (Algorithm 13). For x < 1,
the ratio of speeds was of the order of 20.

* Work supported by the U.S. Atomic Energy Commission.

ALGORITHM 21
BESSEL FUNCTION FOR A SET OF INTEGER
   ORDERS
W. Börsch-Supan
National Bureau of Standards, Washington 25, D. C.

```
procedure  BESSELSETINT (x, n, ε, J)  ;  value x, n, ε  ;
   real x, ε  ;  integer n  ;  real array J  ;
   comment:  This procedure computes the values of the Bessel
      functions J_p(x) for real argument x and the set of all integer
      orders from 0 up to n and stores these values into the array J,
      whose subscript bounds should include the integers from 0 up
      to n.  n must be nonnegative.
         The computation is done by applying the recursion formula
      backward from p = k down to p = 0 as described in MTAC 11
      (1957), 255–257.  k is chosen to yield errors less than 10⁻⁵
      approximately after the first application of the recursion. The
      recursion is repeated with a larger k until the difference be-
      tween the results of the two last recursions doesn't exceed the
      given bound ε > 0. The steps in increasing k are chosen in
      such a way that the errors decrease at least by a factor of
      approximately 10⁻⁵. There is no protection against overflow.  ;
begin real  dist, rec0, rec1, rec2, sum, max, err  ;
   integer k, p  ;  Boolean s  ;  real array Jbar[0:n]  ;
   if x = 0 then
   begin J[0] := 1  ;  for p := 1 step 1 until n do J[p] := 0  ;
      go to Exit
   end  ;
   dist := if abs(x) ≧ 8 then 5 × abs(x) ↑ (1/3) else 10  ;
   k := entier ((if abs(x) ≧ n then abs(x) else n) + dist) + 1  ;
   s := false  ;
Rec:  rec0 := 0  ;  rec1 := 1  ;  sum := 0  ;
   for p := k step −1 until 1 do
   begin   J[if p > n + 1 then n else p − 1] := rec2 :=
            2 × p/x × rec1 − rec0  ;
         if p = 1 then sum := sum + rec2
         else if p ÷ 2 × 2 ≠ p then sum :=
            sum + 2 × rec2  ;
         rec0 := rec1  ;  rec1 := rec2
   end      recursion  ;
Norm:  for p := 0 step 1 until n do J[p] := J[p]/sum  ;
   if s then
   begin   max := 0  ;
            for p := 0 step 1 until n do
            begin   err := abs (J[p] − Jbar[p])  ;
                  if err > max then max := err
            end      maximum error  ;
            if max ≦ ε then go to Exit
   end   then
   else   s := true  ;
   for    p := 0 step 1 until n do Jbar[p] := J[p]  ;
   k := entier (k + dist)  ;
   go to Rec  ;
Exit:  end BESSELSETINT
```

CERTIFICATION OF ALGORITHM 21 [S17]
BESSEL FUNCTION FOR A SET OF INTEGER
   ORDERS
[W. Börsch-Supan, *Comm. ACM 3* (Nov. 1960), 600]
J. Stafford (Recd. 16 Nov. 1964)
Westland Aircraft Ltd., Saunders-Roe Division, East
Cowes, Isle of Wight, Eng.

If this procedure is used with a combination of a moderately
small argument and a moderately large order, the recursive evaluation of *rec2* in the last line of the first column can easily lead to
overflow. This occurred, for instance, in trying to evaluate
$J_{10}(0.01)$.

The following alterations correct this:
(i) Declare a **real** variable z and an **integer** variable m;
(ii) After line *rec* insert:
$z := MAX/4 \times abs\ (x/k)$;
   comment MAX is a large positive number approaching in
   size the largest number which can be represented. The numerical value of $MAX/4$ is written into the procedure;
(iii) At the end of the first column insert:
if $abs(rec2) > z$ **then**
begin
   $rec1 := rec1/z$; $rec2 := rec2/z$; $sum := sum/z$;
   for $m := n$ **step** -1 **until** $p - 1$ **do** $J[m] := J[m]/z$
end;

With these alterations the procedure was run on a National-
Elliott 803, for $x = -1, 0, 0.01, 1, 10$ and $n = 0, 1, 2, 10, 20$. The
results agreed exactly with published seven-place tables.

[See also Algorithm 236, Bessel Functions of the First Kind
(*Comm. ACM 7* (Aug. 1964), 479) which is not restricted to integer values. Although it is a much more complicated program,
Algorithm 236 is slightly faster than Algorithm 21 as corrected, at
least in some cases.—Ed.]

ALGORITHM 22
RICCATI-BESSEL FUNCTIONS OF FIRST AND
    SECOND KIND
H. Oser
National Bureau of Standards, Washington 25, D. C.

```
procedure  RICCATIBESSEL (x, n, eps, S, C)  ;
    value x, n, eps  ;
    real x, eps; integer n; real array S, C  ;
    comment:  RICCATIBESSEL computes $S_k(x) = (\pi x/2)^{\frac{1}{2}} J_{k+\frac{1}{2}}(x)$
        and $C_k(x) = -(\pi x/2)^{\frac{1}{2}} Y_{k+\frac{1}{2}}(x)$ for real $x \neq 0$ and all integer
        values of k from 0 through n with a prescribed (absolute)
        accuracy eps. The computation is done by using the recursion
        relations of the cylinder functions. For $\mathrm{abs}(x) > n$ both $S_k(x)$
        and $C_k(x)$ are computed by using the recursions for ascending
        orders. For $n > \mathrm{abs}(x)$ the functions $S_k(x)$ are obtained by
        using the recursion in descending orders. (See Stegun-
        Abramowitz, MTAC 11, 1957, 255-257). Reaching out two
        different intervals beyond the order n, the two vectors $S_k^1(x)$
        and $S_k^2(x)$ are checked if the maximum component of their
        difference meets the tolerance eps. If this is not the case a
        maximum of 10 iterations is set up to achieve the required
        absolute accuracy. Initial values $S_{kmax}$ and $S_{kmax-1}$ for the
        backward iteration are computed from the corresponding
        values $C_{kmax-1}$ and $C_{kmax}$. No check of accuracy is done in
        case $n < \mathrm{abs}(x)$. Both $C_k(x)$ and $S_k(x)$ are affected in this
        case by errors of the same order of magnitude as the sub-
        routines for sin (x) and cos (x)  ;
begin real  r1, r2, r3, r4, r5, r6, step, acc, max, a, b, d1, d2  ;
    integer i, k, l, imax  ;
    real array W[o:n]  ;
    switch P := initial, improve  ;
    acc: = ₁₀6  ;
    step: = ₁₀3  ;
    imax: = 10  ;
    comment:  These constants may be chosen differently, but
        caution has to be taken because of overflow.  acc sets an
        initial iteration to give roughly a 6-place accuracy.
        Subsequent iterations should improve the result to 3 more
        places each  ;
    i := 1  ;
    if x = 0 then go to exit1  ;
    if n < abs(x) then
case1:  begin  r1 := -sin(x)  ;  r2 := r4 := C[0] := cos(x)  ;
                r5 := S[0] := sin(x)  ;
                for k := 1 step 1 until n do
                begin  C[k] := r3 := (2×k-1) × r2/x - r1  ;
                        S[k] := r6 := (2×k-1) × r5/x - r4  ;
                        r1 := r2  ;  r2 := r3  ;
                        r4 := r5  ;  r5 := r6
                end k  ;  go to finish
        end case1  ;
case2:  l := 1  ;  r1 := -sin(x)  ;  r2 := C[0] := cos(x)  ;
        for k := 1 step 1 until n do
        begin  C[k] := r3 := (2×k-1) × r2/x - r1  ;
                r1 := r2  ;
                r2 := r3
        end  ;
        a := n  ;
```

CERTIFICATION OF ALGORITHM 22 [S17]
RICATTI-BESSEL FUNCTIONS OF FIRST AND
SECOND KIND [H. Oser, *Comm. ACM 3* (Nov.
1960), 600]

Thomas Bray (Recd. 9 Mar. 1970)
Boeing Scientific Research Laboratories, Seattle, WA
98124

KEY WORDS AND PHRASES: Ricatti-Bessel functions, Bessel
functions of fractional order, spherical Bessel functions
CR CATEGORIES: 5.12

The procedure was translated into FORTRAN IV and run on
an IBM 360/44 using double precision arithmetic (15 significant
decimal digits). One error was discovered in the algorithm. The
tenth line following the line with the label *"improve"* reads:

for $k := 1$ **step** 1 **until** n **do**

This line should read:

for $k := 0$ **step** 1 **until** n **do**

The results $S_k(x)/x$ and $-C_k(x)/x$ were computed using this correction and compared with Tables 10.1, 10.2 and 10.5 of [1]. The
results agreed to the number of digits given in the tables for:

x	k
0.1	0(1)8
0.5	0(1)8
1.0	0(1)20
2.0	0(1)8
5.0	0(1)50
7.5	0(1)8
10.0	0(1)50
50.0	0(1)100
100.0	0(1)100

References:

1. Abramowitz, M., and Stegun, I. A. *Handbook of Mathematical
   Functions.* Appl. Math. Ser. 55, Nat. Bur. Standards US Govt.
   Print. Off., Washington, D.C., 1964.

ALGORITHM 23
MATH SORT
WALLACE FEURZEIG
Laboratories for Applied Science, University of Chicago,
Chicago, Ill.

```
procedure  MATHSORT (INVEC, OUTVEC, TOTEVEC,
              n, k, SETFUNC)  ;  value n, k  ;
           array INVEC, OUTVEC  ;
             integer array TOTEVEC  ;
             integer procedure SETFUNC  ;
             integer n, k  ;
begin comment  MATHSORT is a fast sorting algorithm which
```

produces a monotone rearrangement of an arbitrarily ordered
set of n numbers (represented by the vector INVEC) by a
surprising though familiar device. The resultant sorted set is
represented by the vector OUTVEC. The key field, i.e. the
ordered set of bits (or bytes) on which the sort is to be done,
is obtained by some extraction-justification function denoted
SETFUNC. The key field allows the representation of k possible values denoted $0, 1, \ldots, k-1$.

The procedure determines first of all the exact frequency
distribution of the set with respect to the key, i.e. the number
of elements of INVEC with key field value precisely equal to
j for all j between 0 and $k-1$. The cumulative frequency distribution TOTEVEC $[i] \equiv \sum_{j=0}^{i}$ (Number of elements of
INVEC with key value = j) is then computed for $0 \leqq i \leqq k-1$.
This induces the direct assignment (storage mapping function) of each element of INVEC to a unique cell in OUTVEC.
This assignment (like the determination of the frequency
distribution) requires just one inspection of each element of
INVEC. Thus the algorithm requires only 2n "look and do"
operations plus $k-1$ additions (to get the cumulative frequency distribution).

The algorithm can be easily and efficiently extended to
handle alphabetic sorts or multiple key sorts. To sort on
another key the same algorithm is applied to each new key
field with the new INVEC designated as the last induced
ordering (i.e. the current OUTVEC). The algorithm has been
used extensively at LAS on binary as well as decimal machines
both for internal memory sorts and (with trivial modification)
for large tape sorts ;

```
for  i := 1 step 1 until n do
     TOTEVEC[SETFUNC(INVEC[i])] := TOTEVEC
       [SETFUNC(INVEC[i])] + 1  ;
for  i := 1 step 1 until k−1 do
     TOTEVEC[i] := TOTEVEC[i] + TOTEVEC[i−1]  ;
for  i := 1 step 1 until n do
     begin  OUTVEC[TOTEVEC[SETFUNC(INVEC[i])]]
            := INVEC[i]  ;
          TOTEVEC[SETFUNC(INVEC[i])] :=
            TOTEVEC[SETFUNC(INVEC[i])] − 1  ;
          end
end  MATHSORT.
```

CERTIFICATION OF ALGORITHM 23
MATHSORT (Wallace Feurzeig, *Comm. ACM*, Nov.,
1960)
RUSSELL W. RANSHAW
University of Pittsburgh, Pittsburgh, Pa.

The MATHSORT procedure as published was coded for the
IBM 7070 in FORTRAN. Two deficiencies were discovered:
1. The TOTVEC array was not zeroed within the procedure.
This led to some difficulties in repeated use of the procedure.
2. Input vectors already in sort on nonsort fields were unsorted.
That is, given the sequence
$$31, 21, 32, 22, 33,$$
Mathsort would produce, for a sort on the 10's digit:
$$22, 21, 33, 32, 31,$$
which is definitely out of sequence.

The following modified form of the procedure corrects these
difficulties. Note the transformation of symbols.

```
procedure  MATHSORT (I, O, T, n, k, S); value n, k;
           array I, O;  integer array T;  integer procedure S;
           integer n, k;
begin      for i := 0 step 1 until k − 1 do T[i] := 0;
           for i := 1 step 1 until n do T[S(I[i])] := T[S(I[i])] + 1;
           for i := k − 2 step −1 until 0 do T[i] := T[i] +
           T[i + 1];
           for i := 1 step 1 until n do
               begin  O[n + 1 − T[S(I[i])]] := I[i];
                    T[S(I[i])] := T[S(I[i])] − 1;
               end
end MATHSORT.
```

Using the MATHSORT procedure ten times and having the
procedure S supply each digit in order, 1000 random numbers of
10 digits each were sorted into sequence in 31 seconds. The method
of locating the lowest element, interchanging with the first element, and continuing until the entire list has been so examined
yielded a complete sort on the same 1000 random numbers in 227
seconds. Using the Table-Lookup-Lowest command in the 7070
yielded 56 seconds for the same set of random numbers.

ALGORITHM 24
SOLUTION OF TRI-DIAGONAL LINEAR EQUA-
TIONS
B. Leavenworth
American Machine & Foundry Co., Greenwich, Conn.

```
procedure  TRIDAG (n, A, B, C, D)  ;  integer n  ;
array  A, B, C, D  ;
comment: This procedure¹ finds the solution of an n × n system
    of linear equations whose matrix is in tridiagonal form, that
    is, a_ij = 0 for |i − j| ≧ 2. Parameters are: the main diagonal
    B_p, the diagonal just below A_r, the diagonal just above C_r,
    the right-hand side D: (where p = 1, ... , n and r = 1, ... ,
    n − 1) and the matrix order n. The solution vector replaces
    the input vector D and the vector B is also destroyed in the
    process  ;
begin
    real w  ;  integer j  ;
    D[1] := D[1]/B[1]  ;  w := B[1]  ;
    for j := 2 step 1 until n do
        begin B[j − 1] := C[j − 1]/w  ;  w := B[j] − A[j − 1]
        × B[j − 1]  ;
            D[j] := (D[j] − A[j − 1] × D[j − 1])/w end  ;
    for j := 1 step 1 until n − 1 do
        D[n − j] := D[n − j] − B[n − j] × D[n − j + 1]
end TRIDAG
```

¹ D. W. Peaceman and H. H. Rachford, Jr., The Numerical
Solution of Parabolic and Elliptic Differential Equations, Journal
of the Soc. for Ind. and Applied Math. Vol. 3 March 1955.

ALGORITHM 25
REAL ZEROS OF AN ARBITRARY FUNCTION

B. Leavenworth

American Machine and Foundary Co., Greenwich, Conn.

```
procedure  ZEROS(n, C, FUNCTION, m, ep1, ep2, ep3, eta)  ;
integer  n, m  ;  real ep1, ep2, ep3, eta  ;  array C  ;
procedure  FUNCTION  ;
comment:  This procedure finds the real zeros of an arbitrary
    function using Muller's method[1, 2] and is adapted from a
    FORTRAN code by Frank.[3] Each iteration determines a zero
    of the quadratic passing through the last three function
    values. Parameters include the number of roots desired n.
    If C_i is zero, starting values are −1, 1, 0 respectively.  If
    C_i = β then the starting values are .9β, 1.1β, β. The procedure
    FUNCTION(rt, frt) must be supplied to evaluate the func-
    tion value frt, given the argument rt. m is the maximum
    number of iterations permitted.  ep1 is the relative conver-
    gence criterion on successive iterates.  ep2 is the absolute
    convergence criterion on the function value. eta is the spread
    for multiple roots, that is, if |rt − C_i| < ep3 where C_i is a
    previously found root, then rt is replaced by rt + eta  ;
begin integer  L, jk, i, mm  ;  real p, p1, p2, x0, x1, x2, rt,
                    frt, fprt, d, dd, di, h, bi, den, dn, dm, tem  ;
    switch S : S1, S2, S3, S4  ;
    for L := 1 step 1 until m do
    begin jk := 0  ;  if C[L] = 0 then go to initial else
    go to assign  ;
initial:  p := −1  ;  p1 := 1  ;  p2 := 0  ;  go to start  ;
assign:  p := .9 × C[L]  ;  p1 := 1.1 × C[L]  ;  p2 := C[L]  ;
start:  rt := p  ;  go to fn  ;
enter:  go to S[if jk < 4 then jk else 4]  ;
S1:  rt := p1  ;  x0 := fprt  ;  go to fn  ;
S2:  rt := p2  ;  x1 := fprt  ;  go to fn  ;
S3:  x2 := fprt  ;  h := if C[L] = 0 then −1
    else −.1 × C[L]  ;  d := −.5  ;
loop:  dd := 1 + d  ;  bi := x0 × d↑2 − x1 × dd↑2 × x2 ×
        (dd + d)  ;
    den := bi↑2 − 4 × x2 × d × dd × (x0 × d − (x1 × dd) + x2)  ;
    if den ≦ 0 then den := 0 else den := sqrt(den)  ;
    dn := bi + den  ;  dm := bi−den  ;
    if abs(dn) ≦ abs(dm) then den := dm else den := dn  ;
    if den = 0 then den := 1  ;
    di := −2 × x2 × dd/den  ;  h := di × h  ;  rt := rt + h  ;
    go to if abs(h/rt) < ep1 then call else fn  ;
S4:  if abs(fprt) < abs(x2 × 10) then
        begin x0 := x1  ;  x1 := x2  ;  x2 := fprt  ;  d := di  ;
        go to loop end else begin di := di × .5  ;  h := h × .5  ;
        rt := rt− h  ;  go to fn end  ;
fn:  jk := jk + 1  ;  if jk < m then mm := 1 else mm := 0  ;
call:  FUNCTION(rt, frt)  ;  if mm = 1 then go to compute
        else go to root  ;
compute:  fprt := frt  ;
        for i := 2 step 1 until L do
        begin tem := rt − C[i − 1]  ;  if abs(tem) < ep3 then go to
        change else fprt := fprt/tem end  ;
test:  if abs(frt) < ep2 ∧ abs(fprt) < ep2 then go to root
        else go to enter  :
change:  rt := rt + eta  ;  jk := jk − 1  ; go to fn  ;
root:  C[L] := rt end L
end ZEROS
```

[1] D. E. Muller, A Method for Solving Algebraic Equations Using an Automatic Computer, *MTAC 10* (1956).

[2] W. L. Frank, Finding Zeros of Arbitrary Functions, *J. ACM 5* (1958).

[3] W. L. Frank, RWGRT, General Root Finder 704 Fortran Source Language Subroutine Share Distribution # 635. Parameters used by Frank are: ep1 = 10^{-6}, ep2 = 10^{-20}, ep3 = 10^{-20}, eta = 10^{-3}.

REMARK ON ALGORITHM 25
REAL ZEROS OF AN ARBITRARY FUNCTION
(B. Leavenworth, *Comm. ACM*, November 1960)

Robert M. Collinge

Burroughs Corporation, Pasadena, California

On attempting to use this algorithm, I discovered the two following errors:

(1) The line following the SWITCH statement should read:
  for L := 1 **step** 1 **until** n **do**
(2) The line starting with the label loop: should read:
  loop: dd := 1 + d ; bi = x0 × d↑2 − x1 × dd↑2
    + x2 × (dd + d) ;

With these two modifications incorporated the algorithm was translated into the language of the Burroughs Algebraic Compiler and has been used successfully on the Burroughs 220 Computer.

REMARKS ON ALGORITHMS 2 AND 3 (*Comm. ACM*, February 1960), ALGORITHM 15 (*Comm. ACM*, August 1960) AND ALGORITHMS 25 AND 26 (*Comm. ACM*, November 1960)

J. H. Wilkinson

National Physical Laboratory, Teddington.

Algorithms 2, 15, 25 and 26 were all concerned with the calculation of zeros of arbitrary functions by successive linear or quadratic interpolation. The main limiting factor on the accuracy attainable with such procedures is the condition of the *method* of evaluating the function in the neighbourhood of the zeros. It is this condition which should determine the tolerance which is allowed for the relative error. With a well-conditioned method of evaluation quite a strict convergence criterion will be met, even when the function has multiple roots.

For example, a real quadratic root solver (of a type similar to Algorithm 25) has been used on ACE to find the zeros of triple-diagonal matrices T having $t_{ii} = a_i$, $t_{i+1,i} = b_{i+1}$, $t_{i,i+1} = c_{i+1}$. As an extreme case I took $a_1 = a_2 = \cdots = a_5 = 0$, $a_6 = a_7 = \cdots = a_{10} = 1$, $a_{11} = 2$, $b_i = 1$, $c_i = 0$ so that the function which was being evaluated was $x^5(x − 1)^5(x − 2)$. In spite of the multiplicity of the roots, the answers obtained using floating-point arithmetic with a 46-bit mantissa had errors no greater than 2^{-44}. Results of similar accuracy have been obtained for the same problem using linear interpolation in place of the quadratic.

This is because the method of evaluation which was used, the two-term recurrence relation for the leading principal minors, *is a very well-conditioned method of evaluation.* Knowing this, I was able to set a tolerance of 2^{-42} with confidence. If the *same function* had been evaluated from its explicit polynomial expansion, then a tolerance of about 2^{-7} would have been necessary and the multiple roots would have obtained with very low accuracy.

To find the zero roots it is necessary to have an absolute tolerance for $| x_{r+1} - x_r |$ as well as the relative tolerance condition. It is undesirable that the preliminary detection of a zero root should be necessary. The great power of rootfinders of this type is that, since we are not saddled with the problem of calculating the derivative, we have great freedom of choice in evaluating the function itself. This freedom is encroached upon if we frame the rootfinder so that it finds the zeros of $x = f(x)$ since the true function $x - f(x)$ is arbitrarily separated into two parts. The formal advantage of using this formulation is very slight. Thus, in Certification 2 (June 1960), the calculation of the zeros of $x = \tan x$ was attempted. If the function $(-x + \tan x)$ were used with a general zero finder then, provided the method of evaluation was, for example

$$x = n\pi + y$$

$$\tan x - x = -n\pi + \frac{\dfrac{y^3}{3} - \dfrac{y^5}{30} - \cdots}{\cos y},$$

the multiple zeros at $x = 0$ could be found as accurately as any of the others. With a slight modification of common sine and cosine routines, this could be evaluated as

$$-n\pi + \frac{(\sin y - y) - y(\cos y - 1)}{1 + (\cos y - 1)}$$

and the evaluation is then well-conditioned in the neighbourhood of $x = 0$. As regards the number of iterations needed, the restriction to 10 (Certification 2) is rather unreasonably small. For example, the direct evaluation of $x^{60} - 1$ is well conditioned, but starting with the values $x = 2$ and $x = 1.5$ a considerable number of iterations are needed to find the root $x = 1$ Similarly a very large number are needed for Newton's method, starting with $x = 2$. If the time for evaluating the derivative is about the same as that for evaluating the function (often it is much longer), then linear interpolation is usually faster, and quadratic interpolation much faster, than Newton.

In all of the algorithms, including that for Bairstow, it is useful to have some criterion which limits the permissible change from one value of the independent variable to the next [1]. This condition is met to some extent in Algorithm 25 by the condition S4, that $\mathrm{abs}(\mathrm{fprt}) < \mathrm{abs}(x2 \times 10)$, but here the limitation is placed on the permissible increase in the value of the function from one step to the next. Algorithms 3 and 25 have tolerances on the size of the function and on the size of the remainders r1 and r0 respectively. They are very difficult tolerances to assign since these quantities may take very small values without our wishing to accept the value of x as a root. In Algorithm 3 (Comm. ACM June 1960) it is useful to return to the original polynomial and to iterate with each of the computed factors. This eliminates the loss of accuracy which may occur if the factors are not found in increasing order. This presumably was the case in Certification 3 when the roots of $x^5 + 7x^4 + 5x^3 + 6x^2 + 3x + 2 = 0$ were attempted. On ACE, however, all roots of this polynomial were found very accurately and convergence was very fast using single-precision, but the roots emerged in increasing order. The reference to *slow* convergence is puzzling. On ACE, convergence was fast for all the initial approximations to p and q which were tried. When the initial approximations used were such that the real root $x = -6.35099\ 36103$ and the spurious zero were found first,

the remaining two quadratic factors were of lower accuracy, though this was, of course, rectified by iteration in the original polynomial. When either of the other two factors was found first, then all factors were fully accurate even without iteration in the original polynomial [1].

REFERENCE

[1] J. H. WILKINSON. The evaluation of the zeros of ill-conditioned polynomials Parts I and II. *Num. Math.* 1 (1959), 150–180.

ALGORITHM 26
ROOTFINDER III (Modification of Algorithm 15.
Rootfinder II, Henry C. Thacher, Jr., *Comm. ACM*,
August 1960)

JOHN G. HERRIOT, Stanford University, Stanford, California

procedure ROOTIII (f, a, eps, n, g, c, m) ; **integer** n, m ;
 real procedure f ; **real** a, eps, g, c ;
comment ROOTIII computes a value of g = y satisfying the
 equation y = f(y). The iteration will converge to Y providing
 that at some time in the iteration a g is reached such that
 $\text{abs}(g - Y) \times \text{abs}(d(df/dy)/dy) < 2 \times \text{abs}((df/dy) - 1)$,
 where the derivatives are evaluated at Y. Input includes:
 (1) f, a procedure for computing f(y), (2) a, an initial approximation to the root, (3) eps, a tolerance for the relative
 error in g, and (4) n, a maximum number of iterations to be
 performed. Output includes: (1) g, the required root, (2)
 c = f(g) − g, (3) m, a parameter indicating the success of
 the procedure. If the tolerance was not met m < 0. The num-
 ber |m| − 1 gives the number of times that the correction to g
 exceeded the preceding one. If f(y) − y has the same value
 for two successive approximations to g, then h = 1, and we
 exit to "alarm", a nonlocal label. Alarm should provide a
 means of deciding whether g is an acceptable root or not. ;
begin integer j ; **real** b, d, h ;
 m := 1 ; j := 0 ; c := 0 ;
 if f(0) = 0 **then begin** g := 0 ;
 go to return **end** ;
 g := f(a) ; b := d := c := g − a ;
 if c = 0 **then go to** return ;
 for j := 1 **step** 1 **until** n **do**
 begin c := f(g) − g ;
  if abs(c) $\leq$ abs(g) $\times$ eps **then go to** return ;
  h := b/c ;
  if h = 1 **then go to** alarm ;
 if h > 0 $\wedge$ h < 2 **then** m := m + 1 ;
 d := d/(h − 1) ; b := c ; g := g + d
end iteration ;
m := − m ; return : **end**

REMARKS ON ALGORITHMS 2 AND 3 (*Comm.
 ACM*, February 1960), ALGORITHM 15 (*Comm.
 ACM*, August 1960) AND ALGORITHMS 25 AND 26
 (*Comm. ACM*, November 1960)

J. H. WILKINSON
National Physical Laboratory, Teddington.

Algorithms 2, 15, 25 and 26 were all concerned with the calculation of zeros of arbitrary functions by successive linear or
quadratic interpolation. The main limiting factor on the accuracy
attainable with such procedures is the condition of the *method
of evaluating* the function in the neighbourhood of the zeros.
It is this condition which should determine the tolerance which is
allowed for the relative error. With a well-conditioned method of
evaluation quite a strict convergence criterion will be met, even
when the function has multiple roots.

For example, a real quadratic root solver (of a type similar to
Algorithm 25) has been used on ACE to find the zeros of triple-
diagonal matrices T having $t_{ii} = a_i$, $t_{i+1,i} = b_{i+1}$, $t_{i,i+1} = c_{i+1}$. As an extreme case I took $a_1 = a_2 = \cdots = a_5 = 0$, $a_6 = a_7 = \cdots = a_{10} = 1$, $a_{11} = 2$, $b_i = 1$, $c_i = 0$ so that the function which was being evaluated was $x^5(x - 1)^5(x - 2)$. In spite
of the multiplicity of the roots, the answers obtained using floating-point arithmetic with a 46-bit mantissa had errors no greater
than 2^{-44}. Results of similar accuracy have been obtained for the
same problem using linear interpolation in place of the quadratic.
This is because the method of evaluation which was used, the two-
term recurrence relation for the leading principal minors, *is a
very well-conditioned method of evaluation*. Knowing this, I was
able to set a tolerance of 2^{-42} with confidence. If the *same function*
had been evaluated from its explicit polynomial expansion, then
a tolerance of about 2^{-7} would have been necessary and the multiple roots would have obtained with very low accuracy.

To find the zero roots it is necessary to have an absolute toler-
ance for $|x_{r+1} - x_r|$ as well as the relative tolerance condition.
It is undesirable that the preliminary detection of a zero root
should be necessary. The great power of rootfinders of this type
is that, since we are not saddled with the problem of calculating
the derivative, we have great freedom of choice in evaluating the
function itself. This freedom is encroached upon if we frame the
rootfinder so that it finds the zeros of x = f(x) since the true function x − f(x) is arbitrarily separated into two parts. The formal
advantage of using this formulation is very slight. Thus, in Certification 2 (June 1960), the calculation of the zeros of x = tan x
was attempted. If the function (−x + tan x) were used with a
general zero finder then, provided the method of evaluation was,
for example

$$x = n\pi + y$$

$$\tan x - x = -n\pi + \frac{\dfrac{y^3}{3} - \dfrac{y^5}{30} - \cdots}{\cos y},$$

the multiple zeros at x = 0 could be found as accurately as any
of the others. With a slight modification of common sine and cosine routines, this could be evaluated as

$$-n\pi + \frac{(\sin y - y) - y(\cos y - 1)}{1 + (\cos y - 1)}$$

and the evaluation is then well-conditioned in the neighbourhood
of x = 0. As regards the number of iterations needed, the restriction to 10 (Certification 2) is rather unreasonably small.
For example, the direct evaluation of $x^{60} - 1$ is well conditioned,
but starting with the values x = 2 and x = 1.5 a considerable
number of iterations are needed to find the root x = 1. Similarly
a very large number are needed for Newton's method, starting
with x = 2. If the time for evaluating the derivative is about the
same as that for evaluating the function (often it is much longer),
then linear interpolation is usually faster, and quadratic interpolation much faster, than Newton.

In all of the algorithms, including that for Bairstow, it is useful to have some criterion which limits the permissible change
from one value of the independent variable to the next [1]. This
condition is met to some extent in Algorithm 25 by the condition
S4 that abs(fprt) < abs(x2 $\times$ 10), but here the limitation is
placed on the permissible increase in the value of the function
from one step to the next. Algorithms 3 and 25 have tolerances on

the size of the function and on the size of the remainders r1 and r0 respectively. They are very difficult tolerances to assign since these quantities may take very small values without our wishing to accept the value of x as a root. In Algorithm 3 (Comm. ACM June 1960) it is useful to return to the original polynomial and to iterate with each of the computed factors. This eliminates the loss of accuracy which may occur if the factors are not found in increasing order. This presumably was the case in Certification 3 when the roots of $x^5 + 7x^4 + 5x^3 + 6x^2 + 3x + 2 = 0$ were attempted. On ACE, however, all roots of this polynomial were found very accurately and convergence was very fast using single-precision, but the roots emerged in increasing order. The reference to *slow* convergence is puzzling. On ACE, convergence was fast for all the initial approximations to p and q which were tried. When the initial approximations used were such that the real root $x = -6.35099\,36103$ and the spurious zero were found first, the remaining two quadratic factors were of lower accuracy, though this was, of course, rectified by iteration in the original polynomial. When either of the other two factors was found first, then all factors were fully accurate even without iteration in the original polynomial [1].

REFERENCE

[1] J. H. WILKINSON. The evaluation of the zeros of ill-conditioned polynomials Parts I and II. *Num. Math.* 1 (1959), 150–180.

ALGORITHM 27
ASSIGNMENT
ROLAND SILVER
MIT Lincoln Laboratory,* Lexington, Massachusetts

```
procedure Assignment(d, n, x)  ;  value n  ;  integer n  ;
            array d  ;  integer array x  ;
comment:  Assignment determines that permutation x of the
          integers [1:n] for which the sum (i := 1(1)n) of
          the elements d[i, x[i]] of the n × n matrix d is a
          minimum. n ≧ 2. For more complete information
          see: An Algorithm for the Assignment Problem,
          Roland Silver, Comm. ACM, Nov. 1960, p. 605  ;
    begin
    switch Switch := NEXT, L1, NEXT1, MARK  ;
    array  a[1:n, 1:n]  ;
    integer array c[1:n], cb[1:n], lambda[1:n], mu[1:n],
              r[1:n], y[1:n]  ;
    integer cbl, cl, cl0, i, j, k, l, rl, rs, sw  ;
    comment:
INITIALIZE  ;
    for i := 1 step 1 until n do
        begin min := d[i, 1]  :
        for j := 2 step 1 until n do if d[i, j] < min then min
            := d[i, j]  ;
        for j := 1 step 1 until n do a[i, j] := d[i, j] − min
        end i  ;
    for j := 1 step 1 until n do
        begin min := a[1, j]  ;
        for i := 2 step 1 until n do if a[i, j] < min then min
            := a[i, j]  ;
        for i := 1 step 1 until n do a[i, j] := a[i, j] − min
        end j  ;
    for i := 1 step 1 until n do x[i] := y[i] := 0  ;
    for i := 1 step 1 until n do
        begin
        for j := 1 step 1 until n do
            begin
            if a[i, j] ≠ 0 ∨ x[i] ≠ 0 ∨ y[j] ≠ 0 then go to J1  ;
            x[i] := j  ;  y[j] := i
J1:         end j  ;
        end i  ;
START:  comment: Start labeling  ;
        rl := cl := 0  ;  rs := 1  ;
        for i := 1 step 1 until n do
            begin mu[i] := lambda[i] := 0  ;
            if x[i] ≠ 0 then go to I1  ;
            rl := rl + 1  ;  r[rl] := i  ;  mu[i] := −1
I1:         end i  ;
LABEL:  comment: Label and scan  ;
        i := r[rs]  ;  rs := rs + 1  ;
        for j := 1 step 1 until n do
            begin if a[i, j] ≠ 0 or lambda[j] ≠ 0 then go
                to J2  ;
            lambda[j] := i  ;  cl := cl + 1  ;  c[cl] := j  ;
            if y[j] = 0 then go to MARK  ;
            rl := rl + 1  ;  r[rl] := y[j]  ;  mu[y[j]] := i
J2:         end j  ;
        if rs ≦ rl then go to LABEL  ;
        comment:
RENORMALIZE  ;
        sw := 1  ;  cl0 := cl  ;  cbl := 0  ;
        for j := 1 step 1 until n do
            begin if lambda[j] ≠ 0 then go to J3  ;
            cbl := cbl + 1  ;  cb[cbl] := j
J3:         end j  ;
        min := a[r[i], cb[i]]  ;
        for k := 1 step 1 until rl do
            begin
            for l := 1 step 1 until cbl do if a[r[k], cb[l]] ≦ min
                then min := a[r[k], cb[l]]
            end k  ;
        for i := 1 step 1 until n do
            begin if mu[i] ≠ 0 then go to I2  ;
            for l := 1 step 1 until cl0 do a[i, c[l]] := a[i, c[l]] + min  ;
            go to I3  ;
I2:         for l := 1 step 1 until cbl do
                begin a[i, cb[l]] := a[i, cb[l]] − min  ;
            go to Switch[sw]  ;
NEXT:   if a[i, cb[l] ≠ 0 ∨ lambda[cb[l]] ≠ 0 then go to L1  ;
        lambda[cb[l]] := i  ;
        if y[cb[l]] = 0 then
                begin j := cb[l]  ;  sw := 2  ;  go to L1 end  ;
        cl := cl + 1  ;  c[cl] := cb[l]  ;
        rl := rl + 1  ;  r[rl] := y[cb[l]]  ;
L1:     end l  ;
I3:         end i  ;
        go to Switch[sw + 2]  ;
NEXT1:  if cl0 = cl then go to LABEL  ;
        for i := cl0 + 1 step 1 until cl do mu[y[c[i]]] := c[i]  ;
        go to LABEL  ;

MARK:   comment: mark new column and permute  ;
        y[j] := i := lambda[j]  ;
        if x[i] = 0 then begin x[i] := j  ;  go to
            START end  ;
        k := j  ;  j := x[i]  ;  x[i] := k  ;
        go to MARK
        end Assignment
```

* Operated with support from the U. S. Army, Navy and Air
Force.

[NOTE: The reader should distinguish between the letter
and the figure 1, both of which appear in the above algorithm.—Ed.]

CERTIFICATION OF ALGORITHM 27
ASSIGNMENT [Roland Silver, Comm. ACM, Nov. 1960]
ALBERT NEWHOUSE
University of Houston, Houston, Texas

The ASSIGNMENT algorithm was translated into MAD and
successfully run on the IBM 709/7094 after the following corrections were made:

All references to array a and d refer to the same array, i.e. change all $a[i, j]$ to $d[i, j]$. Furthermore:

(a) 3rd line after $LABEL$: **comment**: Label and scan;
should read
begin if $d[i, j] \neq 0 \lor$ lambda $[j] \neq 0$ **then go**

(b) first line after $J3$: **end** j;
should read
$min := d[r[1], cb[1]];$

(c) line $I2$:
should read
$I2$: **for** $l := 1$ **step** 1 **until** cbl **do**

Since there is no provision made for this algorithm to end the following additions were made:

(1) in the integer declaration add the variable: $flag$
(2) first line after $START$: **comment**: $\cdots$
add the line
$flag := n;$
(3) first line before $I1$: **end** i;
change to read
$rl := rl + 1;$ $r[rl] := i;$ $mu[i] := -1;$ $flag := flag - 1$
(4) add a line after $I1$: **end** i;
if $flag = n$ **then go to** $FINI$;
(5) change the last line of the algorithm to read:
$FINI$: **end** Assignment

In order to obtain the minimum value of the $\sum_{i=1}^{n} a_{ix_i}$ (in the following called total) the following additions may be made:
Add a real variable $total$ and

(A) new line after $INITIALIZE$;
$total := 0;$
(B) new line after the first **end** i;
$total := total + min;$
(C) new line after the first **end** j;
$total := total + min;$
(D) after the line **end** k; after $J3$: **end** i:
add the line
$total := total + (rl+cbl-n) \times min;$

CERTIFICATION OF ALGORITHM 27
ASSIGNMENT [Roland Silvers, *Comm. ACM 3*, Nov. 1960].
ROBERT D. WITTY
Burroughs Corp., Detroit, Mich.

Assignment was successfully run on the Burroughs B5000 using Burroughs extended ALGOL 60.

Input Array

60	0	0	76	0	0
0	40	18	0	60	24
60	16	2	4	0	40
0	27	18	3	55	75
0	40	62	16	11	53
28	4	10	84	0	16

Solution Vector: $X(6, 4, 3, 1, 5, 2)$

The following changes were made in the algorithm prior to its successful run:

FROM $MIN := a[r[i], cb[i]];$
TO $MIN := a[r[1], cb[1]];$

FROM **if** $X[i] = 0$ **then begin** $X[i] := j;$
go to $START$ **end**;
TO   **if** $X[i] = 0$ **then begin** $X[i] := j;$
for $i := 1$ **step** 1 **until** N **do begin if** $X[i] = 0$ **then go to** $START$;
end; **go to** $EXIT$; **end**;

FROM **end** $ASSIGNMENT$
TO $EXIT$: **end**; $ASSIGNMENT$

ALGORITHM 28

LEAST SQUARES FIT BY ORTHOGONAL POLY-
NOMIALS

John G. Mackinney

General Kinetics Incorporated, Arlington 6, Virginia

```
procedure  LSFIT (f, x1, xm, m, k, alpha, beta, sigma, s, p)  ;
           value x1, xm, m, k  ;  real x1, xm  ;  integer
           m, k  ;
           real array f, alpha, beta, sigma, s, p  ;
comment  LSFIT accepts m values of the function f at equal
         intervals of the abscissa from x1 through xm, and obtains in
         p[0] through p[k] the coefficients of the best polynomial ap-
         proximation of degree k or less (least squares) as programmed
         by George E. Forsythe, Journal SIAM 5, no. 2, June 1957,
         with only minor variations. The output values alpha [1:k],
         beta [0:k], and s [0:k] enable the user to make final adjust-
         ments to the results, according to the statistic sigma [0:k].
         LSFIT uses the procedure POLYX (a, b, c, d, n) to trans-
         form its results from the interval (−2, 2) to the interval (x1,
         xm)  ;
begin integer  i, j  ;  real dummy, x, xone, deltax, delsq,
               omega, lastw, thisw  ;
               real array cthisp, cpoly [0:k], clastp [−1:k],
               lastp, thisp [1:m]  ;
               Boolean swx  ;
comment  Initialization  ;
swx := true  ;  beta [0] := clastp [0] := clastp [−1] :=
   delsq := omega := 0  ;
cthisp [0] := 1  ;  thisw := m  ;
for i := 1 step 1 until m do
begin delsq := delsq + f[i]↑2  ;
      thisp [i] := 1  ;  lastp [i] := 0  ;
      omega := omega + f[i] end  ;
s [0] := cpoly [0] := omega/thisw  ;
delsq := delsq − s [0] × omega  ;
sigma [0] := delsq/(m−1)  ;
comment  Transformation of abscissa  ;  i := m + 2  ;
if 2×i = m then deltax := 4/(m − 1) else deltax :=
   4/m  ;  xone := −2  ;
comment  Main Computation loop  ;
for i := 0 step 1 until k−1 do
begin dummy := 0  ;  x := xone  ;
1:  for j := 1 step 1 until m do
begin dummy := dummy + x × thisp [j] ↑ 2  ;
x := x + deltax end  ;
2:  alpha [i + 1] := dummy/thisw  ;
    lastw := thisw  ;
    thisw := omega := 0  ;
    x := xone  ;
3:  for j := 1 step 1 until m do
begin dummy := beta [i] × lastp [j]  ;
    lastp [j] := thisp [j]  ;
    thisp [j] := (x − alpha [i + 1]) × thisp [j]
       −dummy  ;
    thisw := thisw + thisp [j] ↑ 2  ;
    omega := omega + f [j] × thisp [j]  ;
x := x + deltax end  ;
```

```
4:  beta [i + 1] := thisw / lastw  ;
s[i + 1] := omega / thisw  ;
    delsq := delsq − s[i + 1] × omega  ;
sigma [i + 1] := delsq / (m − i − 1)  ;
if swx then go to 6  ;
5:  cpoly [i + 1] := 0  ;  go to 9  ;
comment  Termination of main loop when higher power will
         not improve fit  ;
6:  if sigma [i + 1] < sigma [i] then go to 7  ;
    swx := false  ;  go to 5  ;
comment  Recursion for polynomial coefficients  ;
7:  for j := 0 step 1 until i do
    begin dummy := clastp [j] × beta [i]  ;
    clastp [j] := cthisp [j]  ;
cthisp [j] := clastp [j − 1] − alpha [i + 1] × cthisp [j] − dummy  ;
cpoly [j] := cpoly [j] + s [i + 1] × cthisp [j] end  ;
8:  cpoly [i + 1] := s [i + 1]  ;
cthisp [i + 1] := 1  ;
9:  clastp [i + 1] := 0 end of main
computation loop, transformation of polynomial follows  ;
    begin real a, b  ;
    a := deltax × (m − 1) / (xm − x1)  ;
    b := xone − a × x1  ;
    POLYX (a, b, cpoly, p, k) end
end of LSFIT
```

REMARK ON ALGORITHM 28

LEAST-SQUARES FIT BY ORTHOGONAL POLY-
NOMIALS (John G. MacKinney, *Comm. ACM 3*
(Nov. 1960))

D. B. MacMillan

Knolls Atomic Power Laboratory, General Electric Co.,
Schenectady, N. Y.

The algorithm obtains the coefficients of the fitted polynomial
of lowest degree such that an increase in the degree would cause an
increase in the statistic sigma (sigma squared in Forsythe's notation). A significant decrease in sigma, as one goes from a fitted
polynomial to one of higher degree, indicates that the increase in
degree causes an improvement in the fit to the function underlying
the data, rather than merely following more closely the random
variations about that function introduced by the physical meas-
urement process.

If one of the orthogonal polynomials, say the one of ith degree,
is missing from the underlying function, and some of the orthogonal polynomials of higher degree are present, then the fitted
polynomial of ith degree will not be a real improvement over that
of $(i − 1)$-th degree, but higher order fitted polynomials will be
a real improvement. For example, in one of our recent routine
problems the coefficient of the second degree orthogonal polynomial was quite small, and the first few values of sigma, starting
with sigma (1), were .255, .264, .062, .046, .048. The algorithm would
have chosen the first degree fitted polynomial as "best", but the
third and fourth degree fitted polynomials were clearly better
than it.

This loophole may be plugged by modifying the algorithm so it computes the coefficients of the polynomial of lowest degree i for which it is true that

$$\text{sigma } (i + 1) \geq \text{sigma } (i)$$

and that

$$\text{sigma } (j) \geq .6 \text{ sigma } (i) \qquad j = i+2, i+3, \cdots, k,$$

(.6 was chosen arbitrarily).

REMARK ON ALGORITHM 28 [E2]
LEAST SQUARES FIT BY ORTHOGONAL
POLYNOMIALS [John G. MacKinney, *Comm. ACM 3*
  (Nov. 1960), 604]
G. J. Makinson (Recd. 30 Sept. 1965, 29 Aug. 1966 and
  7 Nov. 1966)
University of Liverpool, Liverpool 3, England

There are three errors in the published procedure.

Line 32 $i := m + 2;$ should read $i := m \div 2;$

Line 56 $delsq/(m-i-1);$ should read $delsq/(m-i-2);$

Line 69   ; is missing from end of statement $cpoly[i+1] := s[i+1];$

Three improvements can be made to the procedure. In the case of equally spaced points, it is possible to center them about the origin; all alphas are then zero. This is achieved by replacing the statements on lines 32, 33, and 34 by $deltax := 4/(m-1);$ $xone := -2;$ All statements involving alphas can then be revised.

Another improvement can be made by deleting the two statements on line 37 and all of lines 38, 39, and 40. These statements are completely redundant.

The third improvement is to rewrite line 71 to read

$$clastp[i+1] := 0; \quad 9: \textbf{ end } of \ main$$

instead of

$$9: \quad clastp[i+1] := 0 \ \textbf{end } of \ main$$

ALGORITHM 29
POLYNOMIAL TRANSFORMER
JOHN G. MACKINNEY
General Kinetics Inc., Arlington 6, Virginia

```
procedure  POLYX (a, b, c, d, n)  ;  value a, b, n  ;  integer
              n  ;  real a, b  ;
              real array c, d  ;
comment  POLYX computes coefficients d0, d1, ... , dn of the
              transformed polynomial p(t) given c0, c1, ... ,
              cn of p(x) where x = at + b  ;
begin integer i, j, k  ;  real array z, w [0:n]  ;
    w[0] := z[0] := 1  ;  d[0] := c[0]  ;
    for i := 1 step 1 until n do
        begin w[i] := 1  ;  z[i] := b × z[i − 1]  ;
          d[0] := d[0] + c[i] × z[i]
        end of initialization  ;
    for j := 1 step 1 until n do
        begin w[0] := w[0] × a  ;  d[j] := c[j] × w[0]  ;
          k := 1  ;
          for i := j + 1 step 1 until n do
              begin w[k] := a × w[k] + w[k − 1]  ;
                  d[j] := d[j] + c[i] × w[k] × z[k]  ;
                  k := k + 1 end
end
end  of POLYX polynomial transformer
```

ALGORITHM 30
NUMERICAL SOLUTION OF THE POLYNOMIAL
EQUATION

K. W. Ellenberger
Missile Division, North American Aviation, Downey,
California

```
procedure  ROOTPOL (n, a, L, F, u, v, CONV)  ;
           value n, a, L, F  ; integer L, F, n  ;
           array a, u, v, CONV  ;
```

comment The Bairstow and Newton correction formulae are
used for a simultaneous linear and quadratic iterated synthetic
division. The coefficients of a polynomial of degree n are given as
a_i (i = 0, i, ... , n) where a_n is the constant term. The coefficients are scaled by dividing them by their geometric mean.
The Bairstow or Newton iteration method will nearly always
converge to the number of figures carried, F, either to root
values or to their reciprocals. If the simultaneous Newton and
Bairstow iteration fails to converge on root values or their
reciprocals in L iterations, the convergence requirement will be
successively reduced by one decimal figure. This program antici-
pates and protects against loss of significance in the quadratic
synthetic division. (Refer to "On Programming the Numerical
Solution of Polynomial Equations," by K. W. Ellenberger,
Commun. ACM *3* (Dec. 1960), 644–647.) The real and imaginary
part of each root is stated as u[i] and v[i], respectively, together
with the corresponding constant, $CONV_i$, used in the convergence test. This program has been used successfully for over
a year on the Bendix G15-D (Intercard System) and has recently
been coded for the IBM 709 (Fortran System);

```
            begin integer i, j, m  ; array h, b, c, d, e[−2 :n]  ;
            real t, K, ps, qs, pt, qt, s, rev, r  ;
ROOTPOL:    b_-1 := b_-2 := c_-1 := c_-2 := d_-1 := d_-2 := e_-1 :=
            e_-2 := 0  ;
            for j := 0 step 1 until n do h_j := a_j  ; t := 1  ;
            K := 10^F  ;
ZROTEST:    if h_n = 0 then
            begin u_n := 0  ; v_n := 0  ; CONV_n := K  ;
            n := n − 1  ; go to ZROTEST
            end  ;
INIT:       if n = 0 then go to RETURN  ;
            ps := qs := pt := qt := s := 0  ;
            rev := 1  ; K := 10^F  ;
            if n = 1 then
            begin r := − h_1/h_0  ;  go to LINEAR
            end  ;
            for j := 0 step 1 until n do
            begin
            if h_j = 0 then s := s else s := s + log(abs(hj))
            end  ; s := s^10  ;
            for j := 0 step 1 until n do h_j := h_j/s  ;
            if abs (h_1/h_0) < abs (h_n−1/h_n) then
REVERSE:    begin t := −t  ; m := entier ((n+1)/2)  ;
            for j := 0 step 1 until m do
            begin s := h_j  ; h_j := h_n−j  ; j_n−j := s
            end
            end  ;
            if qs ≠ 0 then
            begin p := ps  ; q := qs  ; go to ITERATE
            end  ;
            if h_n−2 = 0 then
            begin q := 1  ; p := −2
            end else
            begin q := h/h_n−2  ; p := (h_n−1 − q × h_n−3)/h_n−2
            end  ;
            if n = 2 then go to QADRTIC  ; r := 0  ;
ITERATE:    for i := 1 step 1 until L do
            begin
BAIRSTOW:   for j := 0 step 1 until n do
            begin b_j := h_j − p × b_j−1 − q × b_j−2  ;
               c_j := b_j−p × c_j−1 − q × c_j−2
            end  ;
            if n_n−1 = 0 then go to BNTEST  ;
            if b_n−1 = 0 then go to BNTEST  ;
            if abs (h_n−1/b_n−1)<K then go NEWTON  ;
               b_n := h_n−q × b_n−2  ;
BNTEST:     if b_n = 0 then go to QADRTIC  ;
            if K < abs (h_n/b_n) then go to QADRTIC  ;
NEWTON:     for j := 0 step 1 until n do
            begin d_j := h_j + r × d_j−1  ; e_j := d_j + r × e_j−1
            end  ;
            if d_n = 0 then go to LINEAR  ;
            if K < abs (h_n/d_n) then go to LINEAR  ;
               c_n−1 := −p × c_n−2−q × c_n−3  ;
               s := c_n−2^2 − c_n−1 × c_n−3  ;
               if s=0 then
            begin p := p − 2  ; q := q × (q + 1)
            end else
            begin p := p + (b_n−1 × c_n−2 − b_n × c_n−3)/s  ;
               q := q + (−b_n−1 × c_n−1 + b_n × c_n−2)/s
            end  ;
            if e_n−1 = 0 then r := r−1 else r := r − d_n/e_n−1
            end  ; ps := pt  ; qs := qt  ; pt := p  ;
            qt := q  ;
            if rev < 0 then K := K/10  ; rev = −rev  ;
               go to REVERSE  ;
LINEAR:     if t < 0 then r := 1/r  ; u_n := r  ; v_n := 0  ;
            CONV_n := K  ; n := n−1  ;
            for j := 0 step 1 until n do h_j := d_j  ;
            if n = 0 then go to RETURN  ;
               go to BAIRSTOW  ;
QADRTIC:    if t < 0 then
            begin p := p/q  ; q := 1/q
            end  ;
            if 0 < (q − (p/2)²) then
            begin u_n := u_n−1 := −p/2  ;
               s := sqrt (q − (p/2)³)  ; v_n := s  ;
               v_n−1 := −s
            end else
            begin s := sqt ((p/2)²) − q)  ;
            if p < 0 then u_n := −p/2 + s
               else u_n := −p/2−s  ; u_n−1 := q/u_n  ;
               v_n := v_n−1 := 0
            end  ; CONV_n := CONV_n−1 := K  ;
            n := n−2  ;
            for j := 0 step 2 until n do h_j := b_j  ;
               go to INIT  ;
RETURN:     end
```

CERTIFICATION OF ALGORITHM 30
NUMERICAL SOLUTION OF THE POLYNOMIAL
EQUATION (K. W. Ellenberger, *Comm. ACM*, Dec.
1960)
WILLIAM J. ALEXANDER
Argonne National Laboratory,* Argonne, Ill.

ROOTPOL was coded by hand for the LGP-30 using the ACT-III Compiler with 24 bits of significance. The following corrections were found necessary.

(a) $b_{-1} := b_{-2} := c_{-1} := c_{-2} := d_{-1} := d_{-2} := e_{-1} := e_{-2} := 0$
should be
$b_{-1} := b_{-2} := c_{-1} := c_{-2} := d_{-1} := e_{-1} := h_{-1} := 0$

(b) m := entier $((n + 1)/2)$ *should be*
m := entier $((n - 1)/2)$

(c) $j_{n-j} := s$ *should be* $h_{n-j} := s$

(d) $q := h/h_{n-2}$ *should be* h_n/h_{n-2}

(e) $cj := b_j - p \times c_j - 1 - q \times c_{j-2}$ *should be*
$c_j := b_j - p \times c_{j-1} - q \times c_{j-2}$

(f) **if** $n_{n-1} = 0$ **then go to** BNTEST   *should be*
if $h_{n-1} = 0$ **then go to** BNTEST

(g) s := sqrt $(q - (p/2)^3)$ *should be*
s := sqrt $(q - (p/2)^2)$

(h) **for** j := 0 **step** 2 **until** n **do** $h_j := b_j$ *should be*
for j := 0 **step** 1 **until** n **do** $h_j := b_j$

(i) **go to** BAIRSTOW   *should be*   **go to** ITERATE

The following correction was found necessary in the given example (Refer to "On Programming the Numerical Solution of Polynomial Equations," by K. W. Ellenberger, *Comm. ACM 3*, Dec., 1960):
$f(x) = (.10098). 10^8 x^4 - (.98913) 10^6 x^2 + (.10000) 10^6 x + (.10000) 10^1 = 0$ *should be*
$f(x) = (.10098) 10^8 x^4 - (.98913) 10^6 x^3 - (.10990) 10^6 x^2 + (.10000) 10^6 x + (.10000) 10^1 = 0$

With these corrections the results obtained agree with those given in the example.

For equations of higher order it was found necessary to avoid repeated scaling of the reduced equation in order to prevent floating point overflow. The range on the exponent in the ACT III system is $-32 \leqq e \leqq 31$.

Further floating point overflow difficulties were experienced when certain coefficients in the reduced equation became small but not zero. The following additions were made to avoid this fault:

(a) **for** j := 0 **step** 1 **until** n **do** $h_j := d_j$ *was replaced by*
for j := 0 **step** 1 **until** n **do begin if** abs $(h_j/d_j) < K$ **then** $h_j := d_j$ **else** $h_j := 0$ **end**

(b) **for** j := 0 **step** 1 **until** n **do** $h_j := b_j$ *was replaced by*
for j := 0 **step** 1 **until** n **do** begin **if** abs $(h_j/b_j) < K$ **then** $h_j := b_j$ **else** $h_j := 0$ **end**

With the above changes the following results were obtained:
$$x^4 - 3 x^3 + 20 x^2 + 44x + 54 = 0$$
$$x = -.9706390 \pm 1.005808i$$
$$x = 2.470639 \pm 4.640533i$$
$$x^6 - 2 x^5 + 2 x^4 + x^3 + 6x^2 - 6x + 8 = 0$$
$$x = -.9999999 \pm .9999999i$$
$$x = 1.500000 \pm 1.322876i$$
$$x = .5000002 \pm .8660251i$$
$$x^5 + x^4 - 8x^3 - 16x^2 + 7x + 15 = 0$$
$$x = 3.000001$$
$$x = -2.000000 \pm 1.000001i$$
$$x = -.9999997$$
$$x = .9999998$$

* Work supported by the U. S. Atomic Energy Commission

CERTIFICATION OF ALGORITHM 30
NUMERICAL SOLUTION OF THE POLYNOMIAL
EQUATION [K. W. Ellenberger, *Comm. ACM 3*
(Dec. 1960), as corrected in the previous Certification
by William J. Alexander, *Comm. ACM 4* (May 1961)]
KALMAN J. COHEN
Graduate School of Industrial Administration, Carnegie
Institute of Technology, Pittsburgh, Pa.

The ROOTPOL procedure originally published by Ellenberger as corrected and modified by Alexander was coded for the Bendix G20 in 20-GATE. Some serious errors were found in the third and fourth lines above the statement labelled "REVERSE" in Ellenberger's Algorithm which were not mentioned in Alexander's Certification. First, the function "log" is not a standard function in ALGOL 60; it is clear from the context, however, that Ellenberger intends this to be the logarithm function to the base 10. Second, Ellenberger's Algorithm failed to divide the accumulated sum of the logarithms by n+1 before taking the antilogarithm.

The correct, and slightly simplified, manner in which the third and fourth lines above the statement labelled "REVERSE" should read is:
if $h_j \neq 0$ **then** s := ln(abs(h_j))
end; s := s/(n+1); s := exp(s);

With these corrections, the numerical results obtained essentially agree with those reported by Alexander.

CERTIFICATION OF ALGORITHM 30 [C2]
NUMERICAL SOLUTION OF THE POLYNOMIAL
EQUATION [K. W. ELLENBERGER, *Comm. ACM
3* (Dec. 1960), 643]
JOHN J. KOHFELD (Recd. 31 Aug. 1964, 18 Nov. 1964 and
10 Nov. 1966)
Computing Center, United Technology Center, Sunnyvale, Calif. 94088

The *ROOTPOL* procedure was found to use the identifiers p, q, without declaring them. They should be declared **real**.

The first ALGOL statement in Cohen's Certification [*Comm. ACM 5* (Jan. 1962), 50] which reads:

if $h_j \neq 0$ **then** s := *ln* (abs(h_j))

should read:

if $h_j \neq 0$ **then** s := *ln* (abs(h_j)) + s.

The next line could be simplified to read:

end; s := *exp*(s/(n+1));

The above corrections, as well as Algorithm 30 itself, are in publication language ALGOL. In order to translate the algorithm to reference language ALGOL, which is now used in CACM, 10^F would need to be replaced by 10 $\uparrow$ F, and h_j would need to be replaced by h [j].

With these corrections and those contained in Alexander's Certification [*Comm. ACM 4* (May 1961), 238], Ellenberger's Algorithm was adapted to B-5000 ALGOL and successfully executed on the Burroughs B-5000 computer at United Technology Center. The results from the four examples used by Alexander are given below.

Example 1

$$(1.0098)10^7x^4 - (9.8913)10^5x^3 - (1.0990)10^5x^2 + 10^5x + 1 = 0.$$
The roots are:
$$x = -0.201080185406$$
$$x = 0.149521622653 \pm 0.163989609283i$$
$$x = (-9.99989011230)10^{-6}.$$

Example 2

$$x^4 - 3x^3 + 20x^2 + 44x + 54 = 0$$
$$x = 2.47063897001 \pm 4.64053316164i$$
$$x = -0.970638970010 \pm 1.00580758903i$$

Example 3

$$x^6 - 2x^5 + 2x^4 + x^3 + 6x^2 - 6x + 8 = 0$$
$$x = -0.999999999990 \pm 1.000000000000i$$
$$x = 1.500000000000 \pm 1.32287565553i$$
$$x = 0.500000000000 \pm 0.866025403780i$$

Example 4

$$x^5 + x^4 - 8x^3 - 16x^2 + 7x + 15 = 0$$
$$x = 3.00000000000$$
$$x = -2.00000000000 \pm 1.00000000003i$$
$$x = -0.999999999990$$
$$x = 1.000000000000$$

These results agree substantially with those given in Alexander's
Certification.

ALGORITHM 31
GAMMA FUNCTION
Robert M. Collinge
Burroughs Corporation, Pasadena, California

real procedure Gamma (x); **real** x;
comment For x in the range $2 \leq x \leq 3$ an approximating polynomial is used. In this range the maximum absolute error $\epsilon(x)$
is $| \epsilon(x) | < 0.25 \times 10^{-7}$. For $x > 3$ we write $\Gamma(x) = (x-1)(x-2)$
$...(x-n)\Gamma(x-n)$ where $2 \leq (x-n) \leq 3$, and for $x < 2$ we write
$$\Gamma(x) = \frac{\Gamma(x+n)}{x(x+1)...(x+n-1)} \text{ where } 2 \leq (x-n) \leq 3. \text{ For } x = 0$$
or a negative integer $\Gamma(x)$ is set equal to a large value 10^{50}.
begin
  real h, y;
  h := 1.0; y := x;
A1: **if** y = 0 **then** h := 10^{50}
    else if y = 2.0 **then go to** A2
    else if y < 2.0 **then begin**
    h := h/y; y := y + 1.0; **go to** A1 **end**
    else if y $\geq$ 3.0 **then begin**
    y := y − 1.0; h := h $\times$ y; **go to** A1 **end**
    else begin y := y − 2.0;
    h := $(((((((.0016063118 \times$ y $+ .0051589951) \times$ y
      $+ .0044511400) \times$ y $+ .0721101567) \times$ y
      $+ .0821117404) \times$ y $+ .4117741955) \times$ y
      $+ .4227874605) \times$ y $+ .9999999758) \times$ h **end**;
A2: Gamma := h **end** Gamma.

CERTIFICATION OF ALGORITHM 31
GAMMA FUNCTION [R. M. COLLINGE, *Comm.
  ACM*, Feb. 61]
Peter G. Behrenz
Mathematikmaskinnämnden, Stockholm, Sweden

  GAMMA was successfully run on FACIT EDB using Facit-
Algol 1, which is a realization of Algol 60 for FACIT EDB.
No changes in the program were necessary. The relative error
was as stated in the comment of GAMMA about 10^{-8}.

CERTIFICATION OF ALGORITHM 31
GAMMA FUNCTION [R. M. Collinge, *Comm. ACM*,
  Feb. 61]
Peter G. Behrenz
Mathematikmaskinnämnden, Stockholm, Sweden

  GAMMA was successfully run on FACIT EDB using Facit-
Algol 1, which is a realization of Algol 60 for FACIT EDB.
No changes in the program were necessary. The relative error
was as stated in the comment of GAMMA about 10^{-8}

ALGORITHM 32
MULTINT
R. Don Freeman Jr.
Michigan State University, East Lansing, Michigan

real procedure MULTINT (n, Low, Upp, Funev, s, P, u, w);
        value n;
        real procedure Low, Upp, Funev; **array** s, u,
        w; **integer** n;
comment MULTINT will perform a single, double, triple,...,
T-order integration depending on whether n=1, 2,..., T. The
result is:

$$\text{MULTINT} = \int_{\text{Low}(1)}^{\text{Upp}(1)} \text{Funev}(1, x_1)\, dx_1 \int_{\text{Low}(2, x_1)}^{\text{Upp}(2, x_1)} \text{Funev}(2, x_1, x_2)\, dx_2 \ldots$$

$$\int_{\text{Low}(n, x_1, \ldots, x_{n-1})}^{\text{Upp}(n, x_1, \ldots, x_{n-1})} \text{Funev}(x_1, \ldots, x_n)\, dx_n$$

The variable of integration is x[j]. j=1 refers to the outermost
integral, j=n, the innermost integral. The code divides each
interval equally into s[j] subintervals and performs a P-point
Gaussian integration on each subinterval with weight functions w[k[j]] and abscissas u[k[j]]. P is the size of the arrays of
weight functions and abscissas and must be provided by the
main code along with these arrays.

    Since the values x[1], x[2],..., x[n], are stored in an array, as
are a, b, c, d, r, it is necessary to substitute an integer for the
upper bound T of these arrays before the program is executed.
This means, for example, if 3 is substituted for T, then the
procedure will not do a 4th order integral unless it is retrans-
lated with T ≥ 4.

    The values of the lower and upper bounds and functions must
of course be specified at the time of use. If each of these con-
stituted a separate procedure, it would require writing and
translating 3n different procedures. This is eliminated by group-
ing them into Low, Upp, and Funev which compute the lower
and upper bounds and value of the functions respectively in
each of the jth integrals. Since these are each essentially a collection of "subprocedures," the first statement of each should
be a switch directing the code to the "subprocedure" which is
used in the jth integral. Note that, for example, Low(3,x) is
formally a function of x[1], x[2],..., x[T]; this is done merely
because it is more convenient to make Low(j,x) formally a function of the whole array x for all j. Actually of course Low(3, x)
would be a function of x[1] and x[2] only;

```
begin    real array a, b, c, d, r, x[1:T];
         integer array k, h[1:T];  real f;  integer j, m;
         for j :=1 step 1 until T do
                x[j] := 0.0;
         m := 1;
         r[n+1] := d[n+1] := 1.0;
setup:   for j := m step 1 until n do
         begin
             a[j] := Low(j,x);
             b[j] := Upp(j,x);
             d[j] := (b[j]− a[j])/s[j];
             c[j] := a[j] + 0.5 × d[j];
             x[j] := c[j] + 0.5 × d[j] × u[1];
             r[j] := 0.0;
             h[j] := k[j] := 1;  end;
         j := n;
sum:     f := Funev(j,x);
```

```
         r[j] := r[j] + r[j+1] × d[j+1] × f × w[k[j]];
         if (k[j] < P) then go to labk;
         if (h[j] < s[j]) then go to labh;
         j := j−1;
         if (j = 0) then go to exit;
         go to sum;
labh:    h[j] := h[j] + 1;
         c[j] := a[j] + (h[j] − 0.5) × d[j];
         k[j] := 1;
         go to initalx;
labk:    k[j] := k[j] + 1;
initalx: x[j] := c[j] + 0.5 × d[j] × u[k[j]];
         if (j=n) then go to sum;
         m := j+1;
         go to setup;
exit:  MULTINT := r[1] × d[1] × 0.5 ↑ n;  end
```


CERTIFICATION OF ALGORITHM 32
MULTINT [R. Don Freeman, *Comm. ACM*, Feb. 1961]
Henry C. Thacher, Jr.*
Reactor Engineering Div., Argonne National Laboratory,
Argonne, Ill.
* Work supported by the U. S. Atomic Energy Commission.

The procedure was transcribed into the ACT-III language for
the LGP-30 computer, and was tested on the integrals:

$$(1)\quad \int_0^1 \int_0^1 \int_0^1 \int_0^1 k[\cos u - 7u \sin u$$

$$- 6u^2 \cos u + u^3 \sin u]\, dw\, dx\, dy\, dz = \sin k$$

where $u = kwxyz$, and

$$(2)\quad \int_0^1 \int_0^{\sqrt{1-x^2}} \int_0^{\sqrt{1-x^2-y^2}} \frac{dz\, dy\, dx}{x^2 + y^2 + (z - k)^2}$$

$$= \pi \left(2 + \frac{1}{2} \left(\frac{1}{k} - k \right) \log \left| \frac{1 + k}{1 - k} \right| \right).$$

The Algol procedures for the second integral are:

```
real procedure Low (j,x);
Low := 0;
real procedure Upp(j,x);   comment z ≡ x[3],  y ≡ x[2],  x ≡
  x(1);
begin
integer i;  real temp;
temp := 1.0;
for i := j−1 step − 1 until 1 do
temp := temp − x[j] × x[j];
Upp := sqrt(temp)
end;
real procedure Funev(j,x);
comment  The real parameter k is global;
Funev := if j < 3 then 1.0 else 1/(x[1]×x[1]+x[2]×x[2]+(x[3]−k)
  ↑ 2);
```

The first integral was tested only with s[j] = 1, and with various
Gaussian formulas for integrals over the interval (−1,+1). Results were as follows:

k	$\pi/2$	π	$3\pi/2$	2π
true	1.0000000	0.0000000	-1.0000000	0.0000000
p = 2	0.993704	-0.0333603	$+0.020166$	6.881490
p = 3	1.000032	0.0000848	-1.061651	-0.597419
p = 4	0.999999	0.0000001	-0.998407	$+0.0027035$
p = 5	1.000000	-0.0000002	-1.000028	-0.0007857

For the second integral, two values of $s = s[1] = s[2] = s[3]$ were used, and two values of p. Results were as follows:

k	1/2		2	
true	11.46027376		1.10609687	
s	1	2	1	2
p = 2	5.454460	11.838651	1.0368770	1.1184305
p = 3	9.361666	12.408984	1.1343551	1.1094278

The effect of the pole at $(0,0,k)$ is obvious.

For the algorithm to run in any compiler, the semicolon following $x[T]$; in the fourth line above the end of the comment must be deleted. The array bounds on the arrays r and d must be increased to $[1 : T+1]$.

For a system which permits variable array bounds, the introduction of the integer T appears superfluous. For such a system, T may be replaced by n throughout with a probable gain in efficiency. For most translators, the presence of undefined elements in an array will not cause difficulties, provided these elements do not appear in an expression before they are assigned a value.

The statement "**for** $j := 1$ **step** 1 **until** T **do** $x[j] := 0.0$;" is thus superfluous. The semicolon before the **end** which precedes the label "*sum*" also appears unnecessary.

In spite of these minor corrections, the algorithm appears to be extremely convenient for multiple quadratures over arbitrary regions using the Cartesian product of any explicit one-dimensional formula (and not merely a Gaussian formula) for integrating over the range $[-1,1]$. If endpoints are used in the formula, it will, of course, repeat the calculation for each section of the range.

REMARKS ON ALGORITHM 32 [D1]
MULTINT [R. Don Freeman, Jr., *Comm. ACM 4* (Feb. 1961), 106]
AND
CERTIFICATION OF ALGORITHM 32 [Henry C. Thacher, Jr., *Comm. ACM 6* (Feb. 1963), 69]
K. S. Kölbig
Data Handling Division, European Organization for Nuclear Research (CERN), 1211 Geneva 23, Switzerland

KEY WORDS AND PHRASES: numerical integration, multidimensional integration, Gaussian integration
CR CATEGORIES: 5.16

The real procedure *MULTINT* was corrected according to the certification. It was then compiled on a CDC 3800 computer and tested on the second integral given in the certification. It became apparent that

(i) Equation (2) of the certification should read

$$\int_{-1}^{1} \int_{-\sqrt{1-x^2}}^{\sqrt{1-x^2}} \int_{-\sqrt{1-x^2-y^2}}^{\sqrt{1-x^2-y^2}} \frac{dz\,dy\,dx}{x^2 + y^2 + (z-k)^2} = \pi \left(2 + \left(\frac{1}{k} - k\right) \log\left|\frac{1+k}{1-k}\right| \right) \tag{2}$$

It should be noted that the right-hand side of equation (2) as printed in the certification does not correspond either to the original limits or to those given above.
(ii) the statement

$$Low := 0;$$

in the real procedure *Low* should be replaced by

$$Low := -Upp(j, x);$$

(iii) the second line of the **for** statement in the real procedure *Upp* should read

$$temp := temp - x[i] \times x[i];$$

After making these corrections, it is possible to obtain results corresponding to a permuted version of the table given in the certification, which should be replaced by the following:

k	$\frac{1}{2}$		2	
true	11.46027375		1.10609686	
s	1	2	1	2
P = 2	5.454466	9.361670	1.0368787	1.1184317
P = 3	11.838664	12.408983	1.1343568	1.1094294

In addition, since several compilers require specifications, it would be desirable
(i) to change the last specification in the heading of *MULTINT* to read

$$\textbf{integer } n, P;$$

(ii) to insert the specifications

$$\textbf{integer } j; \quad \textbf{array } x;$$

in the heading of the real procedures *Low*, *Upp*, and *Funev*.
Some of these additions were necessary in order to ensure correct results with the compiler used for the tests.

ALGORITHM 33
FACTORIAL
M. F. Lipp
RCA Digital Computation and Simulation Group,
Moorestown, New Jersey

```
real procedure Factorial (n) ;
        value n ;  integer n ;
comment  This procedure makes use of the implicitly defined
    recursive property of Algol to compute n!;
begin  Factorial := if n = 0 then 1. else n× Factorial (n−1)
        end
```

ALGORITHM 34
GAMMA FUNCTION
M. F. LIPP
RCA Digital Computation and Simulation Group,
Moorestown, New Jersey

real procedure Gamma (x) ; **real** x ;
comment This procedure generalizes the recursive factorial
routine, finding $\Gamma(1+x)$ for reasonable values of x. Accuracy
vanishes for large $x(|x| > 10)$ and for negative x with small
fractional parts. For x being a negative integer the impossible
value zero is given;
begin test: **if** x < 0 **then go to** minus **else if** x < 1 **then**
   begin integer i ; **real** y ; **array** a [1:8] ;
   a [1] := −.57719165 ;
   a [2] := .98820589 ; a [3] := −.89705694 ;
   a [4] := .91820686 ;
   a [5] := −.75670408 ; a [6] := .48219939 ;
   a [7] := −.19352782 ;
   a [8] := .03586834 ; y := a [1];
   for i := 2 **step** 1 **until** 8 **do** y := y × x + a [i] ;
   Gamma := y **end** hastings
   else Gamma := x × Gamma (x−1) ; **go to** endgam;
   minus: **if** x = −1 **then** Gamma := 0 **else**
   Gamma := Gamma (x+1) / x ;
   endgam : **end** gam

REMARK ON ALGORITHM 34
GAMMA FUNCTION [M. F. Lipp, *Comm. ACM 4*
 (Feb. 1961)]
MARGARET L. JOHNSON AND WARD SANGREN
Computer Applications, Inc., San Diego, Calif.

The coefficients used in the calculation of the Hasting's polynomial are used in reverse order. The algorithm should have
a[1]=−.19352782; a[2]=.48219939; a[3]=−.75670408;
a[4]=.91820686; a[5]=−.89705694; a[6]=.98820589;
a[7]=−.57719165; a[8]=1.0;
y=.03586834;
for i := 1 **step** 1 **until** 8 **do** y := y×x+a[i];.
Further, since Gamma $(x)=\Gamma(1+x)$, the divisor x in the
statement labeled minus should be $x+1$.

REMARKS ON:
ALGORITHM 34 [S14]
GAMMA FUNCTION
 [M. F. Lipp, *Comm. ACM 4* (Feb. 1961), 106]
ALGORITHM 54 [S14]
GAMMA FUNCTION FOR RANGE 1 TO 2
 [John R. Herndon, *Comm. ACM 4* (Apr. 1961), 180]
ALGORITHM 80 [S14]
RECIPROCAL GAMMA FUNCTION OF REAL
ARGUMENT
 [William Holsten, *Comm. ACM 5* (Mar. 1962), 166]
ALGORITHM 221 [S14]
GAMMA FUNCTION
 [Walter Gautschi, *Comm. ACM 7* (Mar. 1964), 143]
ALGORITHM 291 [S14]
LOGARITHM OF GAMMA FUNCTION
 [M. C. Pike and I. D. Hill, *Comm. ACM 9* (Sept. 1966),
 684]
M. C. PIKE AND I. D. HILL (Recd. 12 Jan. 1966)
Medical Research Council's Statistical Research Unit,
University College Hospital Medical School,
London, England

Algorithms 34 and 54 both use the same Hastings approximation, accurate to about 7 decimal places. Of these two, Algorithm
54 is to be preferred on grounds of speed.

Algorithm 80 has the following errors:
(1) *RGAM* should be in the parameter list of *RGR*.
(2) The lines
 if $x = 0$ **then begin** *RGR* := 0; **go to** *EXIT* **end**
and
 if $x = 1$ **then begin** *RGR* := 1; **go to** *EXIT* **end**
should each be followed either by a semicolon or preferably by an
else.
(3) The lines
 if $x = 1$ **then begin** *RGR* := 1/y; **go to** *EXIT* **end**
and
 if $x < -1$ **then begin** $y := y × x$; **go to** *CC* **end**
should each be followed by a semicolon.
(4) The lines
 BB: **if** $x = -1$ **then begin** *RGR* := 0; **go to** *EXIT* **end**
and
 if $x > -1$ **then begin** *RGR* := *RGAM(x)*; **go to** *EXIT* **end**
should be separated either by **else** or by a semicolon and this
second line needs terminating with a semicolon.
(5) The declarations of **integer** i and **real array** $B[0:13]$ in *RGAM*
are in the wrong place; they should come immediately after
 begin real z;

With these modifications (and the replacement of the array B
in *RGAM* by the obvious nested multiplication) Algorithm 80 ran
successfully on the ICT Atlas computer with the ICT Atlas
ALGOL compiler and gave answers correct to 10 significant digits.

Algorithms 80, 221 and 291 all work to an accuracy of about 10
decimal places and to evaluate the gamma function it is therefore
on grounds of speed that a choice should be made between them.
Algorithms 80 and 221 take virtually the same amount of computing time, being twice as fast as 291 at $x = 1$, but this advantage
decreases steadily with increasing x so that at $x = 7$ the speeds are
about equal and then from this point on 291 is faster—taking only
about a third of the time at $x = 25$ and about a tenth of the time
at $x = 78$. These timings include taking the exponential of *loggamma*.

For many applications a ratio of gamma functions is required
(e.g. binomial coefficients, incomplete beta function ratio) and the
use of algorithm 291 allows such a ratio to be calculated for much
larger arguments without overflow difficulties.

ALGORITHM 35
SIEVE
T. C. Wood
RCA Digital Computation and Simulation Group, Moorestown, New Jersey

procedure Sieve (Nmax) Primes: (p) ;
       integer Nmax; **integer array** p ;
comment Sieve uses the Sieve of Eratosthenes to find all prime numbers not greater than a stated integer Nmax and stores them in array p. This array should be of dimension 1 by entier $(2 \times \text{Nmax}/ \ell n \text{ (Nmax)})$;
begin integer n, i, j ;
       p[1] := 1 ; p[2] := 2 ; p[3] := j := 3 ;
       for n := 3 **step** 2 **until** Nmax **do**
begin   i := 3 ;
       L1: **go to if** p[i] $\leq$ sqrt (n) **then** a1 **else** a2 ;
       a1: **go to if** n/p[i] = n $\div$ p[i] **then** b1 **else** b2 ;
       b2: i := i + 1 ; **go to** L1 ;
       a2 : p[j] := n ; j := j + 1 ;
       b1: **end end**

CERTIFICATION OF ALGORITHM 35
SIEVE (T. C. Wood, *Comm. ACM*, March 1961)
P. J. Brown
University of North Carolina, Chapel Hill, N. C.

SIEVE was transliterated into GAT for the Univac 1105 and successfully run for a number of cases.

The statement:
    go to if n/p[i] = n $\div$ p[i] **then** b1 **else** b2;
was changed to the statement:
    go to if n/p[i] − n $\div$ p[i] < .5/Nmax **then** b1 **else** b2;
Roundoff error might lead to the former giving undesired results.

CERTIFICATION OF ALGORITHM 35
SIEVE [T. C. Wood, *Comm. ACM*. Mar. 1961]
J. S. Hillmore
Elliott Bros. (London) Ltd., Borehamwood, Herts., England

The statement:
    go to if $n/p[i] = n \div p[i]$ **then** $b1$ **else** $b2$;
was changed to the statement:
    go to if $(n \div p[i]) \times p[i] = n$ **then** $b1$ **else** $b2$;
This avoids any inaccuracy that might result from introducing real arithmetic into the evaluation of the relation.

The modified algorithm was successfully run using the Elliott Algol translator on the National-Elliott 803.

REMARKS ON:
ALGORITHM 35 [A1]
SIEVE [T. C. Wood, *Comm. ACM 4* (Mar. 1961), 151]
ALGORITHM 310 [A1]
PRIME NUMBER GENERATOR 1 [B. A. Chartres, *Comm. ACM 10* (Sept. 1967), 569]
ALGORITHM 311 [A1]
PRIME NUMBER GENERATOR 2 [B. A. Chartres, *Comm. ACM 10* (Sept. 1967), 570]

B. A. Chartres (Recd. 13 Apr. 1967)
Computer Science Center, University of Virginia, Charlottesville, Virginia

The three procedures $Sieve(m,p)$, $sieve1(m,p)$, and $sieve2(m,p)$, which all perform the same operation of putting the primes less than or equal to m into the array p, were tested and compared for speed on the Burroughs B5500 at the University of Virginia. The modification of *Sieve* suggested by J. S. Hillmore [*Comm. ACM 5* (Aug. 1962), 438] was used. It was also found that *Sieve* could be speeded up by a factor of 1.95 by avoiding the repeated evaluation of $sqrt(n)$. The modification required consisted of declaring an integer variable s, inserting the statement $s := sqrt(n)$ immediately after $i := 3$, and replacing $p[i] \leq sqrt(n)$ by $p[i] \leq s$.

The running times for the computation of the first 10,000 primes were:

Sieve (Algorithm 35)	845 sec
Sieve (modified)	434 sec
sieve1	220 sec
sieve2	91 sec

The time required to compute the first k primes was found to be, for each algorithm, remarkably accurately represented by a power law throughout the range $500 \leq k \leq 50,000$. The running time of *Sieve* varied as $k^{1.40}$, that of *sieve1* as $k^{1.53}$, and that of *sieve2* as $k^{1.35}$. Thus the speed advantage of *sieve2* over the other algorithms increases with increasing k. However, it should be noted that *sieve2* took approximately 33 minutes to find the first 100,000 primes, and, if the power law can be trusted for extrapolation past this point (there is no reason known why it should be), it would take about 12 hours to find the first million primes.

ALGORITHM 36
TCHEBYCHEFF
A. J. GIANNI
RCA Digital Computation and Simulation Group, Moorestown, New Jersey

```
procedure  tchebycheff (t, x, m, ℓ)  ;
real array  t, x  ;  integer ℓ, m  ;
comment    given a set of m+1 values of x stored in a one-
           dimensional array whose subscripts run from 0
           thru m at least, construct a table of tₙ(x), n =
           0, 1, ⋯,ℓ and store it in the two-dimensional
           array t, where you find tₙ(x[m]) as t[n, m]  ;
begin integer i, k, n   ;
           for k := 0 step 1 until m do begin t[0, k] := 1  ;
           t[1, k] := x[k] end   ;
           for   n := 2 step 1 until ℓ do for i = 0   step  1
           until m do
           t[n,  i] := 2 × x[i] × t[n − 1, i] − t[n − 2,  i]
end   tcheby
```

ALGORITHM 37
TELESCOPE 1
K. A. Brons
RCA Advanced Programming Group, Pennsauken, N. J.

```
procedure  Telescope 1 (N, L, eps, limit, c)  ;  value limit, L  ;
           integer N  ;  real L, eps, limit  ;  array c  ;
comment:   Telescope 1 takes an Nth degree polynomial approxi-
```

comment: Telescope 1 takes an Nth degree polynomial approximation $\sum_{k=0}^{N} c_k x^k$ to a function which was valid to within $eps \geq 0$ over an interval $(0, L)$ and reduces it, if possible, to a polynomial of lower degree, valid to within limit > 0. The initial coefficients c_k are replaced by the final coefficients, and the deleted coefficients are replaced by zero. The initial eps is replaced by the final bound on the error. N is replaced by the degree of the reduced polynomial. N and eps must be variables.

This procedure computes the coefficients given in the Techniques Department of the ACM Communications, Vol. 1, No. 9, from the recursion formula

$$a_{k-1} = -a_k \cdot \frac{k \cdot L \cdot (2k - 1)}{2(N + k - 1) \cdot (N - k + 1)} \; ;$$

```
begin integer k  ; array d[0:N]  ;
start:  if N < 1 then go to exit  ;  d[N] := -c[N]  ;
        for k := N step - 1 until 1 do
        d[k - 1] := -d[k] × L × k × (k - 0.5)/
          ((N + k - 1) × (N - k + 1))  ;
        if eps + abs (d[0]) < limit then
        begin eps := eps + abs (d[0])  ;
        for k := N step - 1 until 0 do c[k] := c[k] + d[k] ;
        N := N - 1  ; go to start end  ;
exit:   end
```

CERTIFICATION OF ALGORITHM 37
TELESCOPE 1 [K. A. Brons, *Comm. ACM*, Mar., 1961]
Henry C. Thacher, Jr.*
Reactor Engineering Div., Argonne National Lab., Argonne, Ill.

* Work supported by the U. S. Atomic Energy Commission.

The body of *Telescope 1* was compiled and tested on the LGP-30 using the Algol 60 translator system developed by the Dartmouth College Computer Center. No syntactical errors were found, and the program ran satisfactorily. The 10th degree polynomial obtained by truncating the exponential series was telescoped using $\lim^! = .1_{10} - 2$ and $L = 1.0$. The result was $N = 3$, $eps = .2103005_{10} - 3$, and coefficients $+.9997892$, $-.9930727$, $+.4636493$, $-.1026781$. The error curve for the telescoped polynomial was computed for $x = 0(.02)1.0$. The error extrema were bounded by eps to within 0.5%. The discrepancy is within the range of input conversion and round-off error.

CERTIFICATION OF ALGORITHM 37
TELESCOPE 1 [K. A. Brons, *Comm. ACM*, Mar. 1961]
James F. Bridges
Michigan State University, East Lansing, Mich.

This procedure was tested on the CDC 160A, using 160A Fortran. The 10th degree polynomial obtained by truncating the series exp $(-x)$ was telescoped using $L = 1$ and lim = 0.001. The result was $N = 3$, eps = $0.21061862_{10} - 3$ and coefficients $+0.99978965$, -0.99307236, $+0.46364955$, -0.10267767. The error curve was computed for $x = 0(0.02)1.0$ and no error exceeded eps, the worst error being 2% of eps less than eps.

This result is in close agreement with that of Henry C. Thatcher. Jr. in his Certification (*Comm. ACM*, Aug. 1962). Mr. Thatcher has pointed out that he inadvertantly referred to the series for exp $(-x)$ as the "exponential series" thereby inferring the positive series exp $(+x)$. There is also a typographical error in his eps. It should be $+0.2103505_{10} - 3$.

ALGORITHM 38
TELESCOPE 2
K. A. BRONS
RCA Advanced Programming, Pennsauken, N. J.

procedure Telescope 2 (N, L, eps, limit, c) ; **value** limit, L ;
    integer N ; **real** L, eps, limit ; **array** c ;
comment Telescope 2 takes an Nth degree polynomial approximation $\sum_{k=0}^{N} c_k x^k$ to a function which was
valid to within eps $\geq$ 0 over an interval $(-L, L)$
and reduces it, if possible, to a polynomial of
lower degree, valid to within limit >0. The initial
coefficients c_k are replaced by the final coefficients,
and deleted coefficients are replaced by zero. The
initial eps is replaced by the final bound on the
error, and N is replaced by the degree of the reduced polynomial. N and eps must be variables.
This procedure computes the coefficients given in
the Techniques Department of the ACM Communications, Vol. 1, No. 9, from the recursion
formula

$$a_{k-2} = -a_k \frac{k \cdot L^2(k-1)}{(N+k-2) \cdot (N-k+2)} \;\; ;$$

```
          begin integer k  ;  real s  ;  array d[0: N]  ;
start:    if N < 2 then go to exit  ;  d[N] := −c[N]  ;
          for k := N step − 2 until 2 do
          d[k − 2]  := −d[k] × L ↑ 2 × k × (k − 1)/
          ((N + k − 2) × (N − k + 2))  ;
          if (N/2) − entier (N/2) = 0 then s := d[0] else
          s := d[1]/N  ;
          if eps + abs(s) < limit then begin
          eps := eps + abs(s)  ;
          for k := N step − 2 until 0 do
          c[k] := c[k] + d[k]  ;
          N := N − 1  ;  go to start end  ;
exit:     end
```

CERTIFICATION OF ALGORITHM 38
TELESCOPE 2 [K. A. Brons, *Comm. ACM*, Mar., 1961]
JAMES F. BRIDGES
Michigan State University, East Lansing, Mich.

This procedure was tested on the CDC 160A using 160A FOR-
TRAN. The 10th degree polynomial obtained by truncating the
series expansion of exp $(+x)$ was telescoped using $L = 1.0$ and
lim = 0.001. The result was $N = 4$, eps = $0.59159949_{10} - 3$ and
coefficients +1.0000447, +0.99730758, +0.49919675, +0.17734729,
+0.043793910. Errors were calculated for $x = -1.0(0.02)1.0$. The
only error to exceed cps was at $x = 1.0$ and was within 0.6% of eps.

ALGORITHM 39
CORRELATION COEFFICIENTS WITH MATRIX
  MULTIPLICATION

PAPKEN SASSOUNI
Burroughs Corporation, Pasadena, California

procedure NORM (x) number of rows: (m) number of columns:
(n) normalized output: (y) standard deviations:
(s) ;

value m, n ; **integer** m, n ; **array** x, y, s ;

comment Given an observation matrix [x] consisting of observations x_{ij} on a population, NORM will calculate

$$y_{ij} = \frac{x_{ij} - \bar{x}_j}{\sqrt{\sum_{i=1}^{m} (x_{ij} - \bar{x}_j)^2}} \quad \begin{array}{l} \text{for } i = 1, \cdots, m \\ j = 1, \cdots, n \end{array}$$

and the standard deviations

$$s_j = \sqrt{\frac{\sum_{i=1}^{m} (x_{ij} - \bar{x}_j)^2}{m}}$$

where $\bar{x}_j$ is the mean of observations on the j-th factor ;

begin **integer** i, j ; **real** r, h, c, b ;
r := sqrt (m) ; **for** j := 1 **step** 1 **until** n **do**

1: **begin** h := 0 ;
for i := 1 **step** 1 **until** m **do**
h := h + x[i, j] ; h := h/m ; b := 0 ;
for i := 1 **step** 1 **until** m **do**

2: **begin** c := x[i, j] − h ; b := b + c↑2 ; y[i, j] := c
end 2 ;
b := sqrt (b) ;
for i := 1 **step** 1 **until** m **do**
y[i, j] := y[i, j]/b ; s[j] := b/r
end 1
end NORM ;

comment The normalization is now completed, and we are
ready to compute the correlation matrix ;

procedure TRANSMULT (y) number of rows: (m) number of
columns: (n) symmetrical square matrix result:
(z) ;

value m, n ; **integer** m, n ; **array** y, z ;

comment This procedure multiplies two matrices, the first
being the transpose of the second. The result is a
symmetrical matrix with respect to the main diagonal, therefore only the lower part of it, including
the main diagonal, is computed. The upper half is
obtained by equating corresponding elements;

begin **integer** i, j, k ; **real** h ;
for j := 1 **step** 1 **until** n **do**
for i := j **step** 1 **until** n **do**

begin h := 0 ;
for k := 1 **step** 1 **until** m **do**
h := h + y[k, i] × y[k, j] ; z[i, j] := h ;
if i ≠ j **then** z[j, i] := h
end i

---

end TRANSMULT. [z] is the square matrix of the
correlation coefficients of the initial observation
matrix [x]

ALGORITHM 40
CRITICAL PATH SCHEDULING
B. LEAVENWORTH
American Machine & Foundry Co., Greenwich, Conn.

```
procedure  CRITICALPATH (n, I, J, DIJ, ES, LS, EF, LF, TF,
              FF)  ;
   integer n  ;  integer array I, J, DIJ, ES, LS, EF, LF, TF,
              FF  ;
   comment:  Given the total number of jobs n of a project, the
      vector pair Iₖ , Jₖ representing the kth job, which is thought
      of as an arrow connecting event Iₖ to event Jₖ(Iₖ < Jₖ ,
      k = 1 ··· , n), and a duration vector (DIJ)ₖ , CRITICAL-
      PATH determines the earliest starting time (ES)ₖ , latest
      starting time (LS)ₖ , earliest completion time (EF)ₖ , latest
      completion time (LF)ₖ , the total float (TF)ₖ , and the free
      float (FF)ₖ .  I₁ must be 1 and the Iₖ , Jₖ must be in ascending
      order.  For example, if the first three jobs are labelled (1, 2),
      (1, 3), (3, 4), then the I, J vectors are (1, 1, 3) and (2, 3, 4)
      respectively. The critical path is given by each arrow whose
      total float is zero. The following non-local labels are used for
      exits: out1 — Iₖ not less than Jₖ  ;  out2 — Iₖ out of se-
      quence  ;  out3 — Iₖ missing;
begin
   integer k, index, max, min  ;  integer array ti, te [1:n]  ;
   index := 1  ;
   for k := 1 step 1 until n do
   begin
      if I[k] ≧ J[k] then go to out1  ;
      if I[k] < index then go to out2  ;
      if I[k] > index ∧ I[k] ≠ index + 1 then go to out3  ;
      if I[k] = index + 1 then index := I[k]  ;
C:  end  ;
   for k := 1 step 1 until n do
      ti[k] := te[k] := 0  ;
   for k := 1 step 1 until n do
   begin
      max := ti[I[k]] + DIJ[k]  ;
      if ti[J[k]] = 0 ∨ ti[J[k]] < max then
      ti[J[k]] := max  ;
A:  end ti  ;
      te[J[n]] := ti[J[n]]  ;
      for k := n step −1 until 1 do
   begin
      min := te[J[k]] − DIJ[k]  ;
      if te[I[k]] = 0 ∨ te[I[k]] > min then
      te[I[k]] := min  ;
B:  end te  ;
   for k := 1 step 1 until n do
    begin
      ES[k] := ti[I[k]]  ;
      LS[k] := te[J[k]] − DIJ[k]  ;
      EF[k] := ti[I[k]] + DIJ[k]  ;
      LF[k] := te[J[k]]  ;
      TF[k] := te[J[k]] − ti[I[k]] − DIJ[k]  ;
      FF[k] := ti[J[k]] − ti[I[k]] − DIJ[k]
    end
end  CRITICALPATH
```

REFERENCES
(1) JAMES E. KELLEY, JR. AND MORGAN R. WALKER, "Critical-
   Path Planning and Scheduling," 1959 Proceedings of the
   Eastern Joint Computer Conference.
(2) M. C. FRISHBERG, "Least Cost Estimating and Scheduling
   — Scheduling Phase Only," IBM 650 Program Library
   File No. 10.3.005.

CERTIFICATION OF ALGORITHM 40
CRITICAL PATH SCHEDULING (B. Leavenworth,
   Comm. ACM, Mar. 1961)
NEAL P. ALEXANDER
Union Carbide Olefins Company, South Charleston,
West Virginia

The Critical Path Scheduling algorithm was coded in FORTRAN
for the IBM 7070. The following changes were made:
   (a) ti [k] := te [k] := 0;
should be
      ti [k] := 0;
      te [k] := 9999;
   (b) if te [I[k]] = 0 ∨ te [I[k]] > min then
should be
      if te [I[k]] > min then
This change permits a value of 0 to be calculated for te [I[k]] and
remain as the minimum value.
   In the statement
         if ti [J[k]] = 0 ∨ ti [J[k]] < max then
the part of the statement "ti [J[k]] = 0" is redundant and can be
omitted.

CERTIFICATION OF ALGORITHM 40
CRITICAL PATH SCHEDULING [B. Leavenworth,
   Comm. ACM (Mar. 1961)]
LARS HELLBERG
Facit Electronics AB, Solna, Sweden.

The Critical Path Scheduling algorithm was transliterated into
FACIT-ALGOL-1 and tested on the FACIT EDB. The modifications
suggested by Alexander [*Comm. ACM* (Sept. 1961)] were included.
Results were correct in all tested schedules.

CERTIFICATION OF ALGORITHM 40
CRITICAL PATH SCHEDULING [B. Leavenworth,
   Comm. ACM 4 (Mar. 1961), 152; *4* (Sep. 1961), 392;
   5 (Oct. 1962), 513]
IRVIN A. HOFFMAN   (Recd 7 Feb. 1964)
Woodward Governor Co., Rockford, Ill.

The Critical Path Scheduling algorithm was coded in FAST for
the NCR315. The modifications suggested by Alexander [*Comm.
ACM 4* (Sept. 1961)] were included. Results were correct in all
tested cases. However, the example of the *I*, *J* vectors given in

the comment is incorrect, as it would cause the exit $out3 - I_k$ missing.

[EDITOR's NOTE. There are also two semicolons which should be removed from the comment of Algorithm 40.—G.E.F.]

ALGORITHM 41
EVALUATION OF DETERMINANT
JOSEF G. SOLOMON
RCA Digital Computation and Simulation Group, Moorestown, New Jersey

```
real procedure   Determinant (A,n);
real array       A;  integer n;
comment  This procedure evaluates a determinant by triangu-
    larization;
begin real       Product, Factor, Temp;  array B[1 : n, 1 : n],
                 C[1 : n, 1 : n];
integer          Count, Sign, i, j, r, y;
                 Sign := 1;  Product := 1;
for              i := 1 step 1 until n do for j := 1 step 1 until
                 n do
begin            B[i,j] := A[i,j];  C[i,j] := A[i,j] end;
for              r := 1 step 1 until n−1 do
begin            Count := r−1;
    zerocheck:   if B[r,r] ≠ 0 then go to resume;
                 if Count < n−1 then Count := Count + 1
                 else go to zero;
for              y := r step 1 until n do
begin            Temp := B[Count+1,y];  B[Count+1,y] :=
                 B[Count,y]; B[Count,y] := Temp end;
                 Sign := − Sign;  go to zerocheck;
    zero:        Determinant := 0;  go to return;
    resume:      for i := r+1 step 1 until n do
begin            Factor := C[i,r] / C[r,r];
                 for j := r+1 step 1 until n do
begin            B[i,j] := B[i,j] − Factor × C [r,j] end end;
for              i := r+1 step 1 until n do
                 for j := r+1 step 1 until n do C[i,j] := B[i,j]
                 end;
for              i := 1 step 1 until n do Product := Product
                 × B[i,i];  Determinant := Sign × Product;
    return:      end
```

ALGORITHM 41, REVISION
EVALUATION OF DETERMINANT [Josef G. Solomon, RCA Digital Computation and Simulation Group, Moorestown, N. J.]
BRUCE H. FREED
Dartmouth College, Hanover, N. H.

```
real procedure determinant (a,n);
real array a;  integer n;  value a,n;
comment  This procedure evaluates a determinant by triangu-
    larization;
begin real product, factor, temp;
array b[1:n,1:n];
integer count, ssign, i, j, r, y;
ssign := product := 1;
for i := 1 step 1 until n do
for j := 1 step 1 until n do
b[i,j] := a[i,j];
for r := 1 step 1 until n−1 do
begin count := r−1;
zerocheck: if b[r,r] ≠ 0 then go to resume;
if count < n−1 then count := count +1 else go to zero;
for y := r step 1 until n do
begin temp := b[count+1,y];
    b[count+1,y] := b[count,y];
    b[count,y] := temp end;
ssign := −ssign;
go to zerocheck;
zero: determinant := 0;  go to return;
resume:  for i := r+1 step 1 until n do
begin factor := b[i,r]/b[r,r];
for j := r+1 step 1 until n do
b[i,j] := b[i,j] − factor × b[r,j] end end;
for i := 1 step 1 until n do
product := product × b[i,i];
determinant := ssign × product;
return: end
```

CERTIFICATION OF ALGORITHM 41
EVALUATION OF DETERMINANT [Josef G. Solomon, RCA Digital Computation and Simulation Group, Moorestown, N. J.]
BRUCE H. FREED
Dartmouth College, Hanover, N. H.

When Algorithm 41 was translated into SCALP for running on the LGP-30, the following corrections were found necessary:
1. In the "y" loop after "B[Count,y] := Temp" and before the "end" insert
    "Temp := C[Count+1,y];
    C[Count +1,y] := C[Count,y];
    C[Count,y] := Temp"
2. "Sign" is an ALGOL word when uncapitalized. However, many systems (if not all) do not recognize the difference between small and capital letters. For this reason "Sign" was changed to "ssign" for the LGP-30 run (and in the revision which follows later).

The following addition might be made in the specification as a concession to efficiency: "value A,n;".

The following changes might be made to make the Algorithm less wordy:
1. for "Ssign := 1; Product := 1;"
   put "Ssign := Product := 1;"
2. for "begin B[i,j] := A[i,j]; C[i,j] := A[i,j] end;"
   put "B[i,j] := C[i,j] := A[i,j];"
3. for "begin B[i,j] := B[i,j] − Factor × C[r,j] end end;"
   put "B[i,j] := B[i,j] − Factor × C[r,j] end;"

The above corrections and changes were made and the program was run with the correct results, as follows:

$$A = \begin{pmatrix} 10.96597 & 35.10765 & 96.72356 \\ 2.35765 & -84.11256 & .87932 \\ 18.24689 & 22.13579 & 1.11123 \end{pmatrix}$$

Determinant = $.1527313_{10}06$

Hand calculation on a desk calculator gives the value of the determinant for the above matrix as 152,731.3600.

$$A = \begin{pmatrix} 1.0 & 3.0 & 3.0 & 1.0 \\ 1.0 & 4.0 & 6.0 & 4.0 \\ 1.0 & 5.0 & 10.0 & 10.0 \\ 1.0 & 6.0 & 15.0 & 20.0 \end{pmatrix} \quad \text{Determinant} = .9999999_{10}+00$$

The above matrix, being a finite segment of Pascal's triangle, has determinant equal to 1.000000000.

$$A = \begin{pmatrix} 0.0 & 0.0 & 0.0 \\ 5.0 & 9.0 & 2.0 \\ 7.0 & 5.0 & 4.0 \end{pmatrix} \quad \text{Determinant} = .0000000_{10}+00$$

This is, of course, exactly correct.

Finally, one major change can be made which does away with several instructions and reduces variable storage requirements by n^2. This change is the complete removal of matrix C from the program. It is extraneous.

The revised Algorithm was translated into SCALP and run on the LGP-30 with exactly the same results as above.

The revised Algorithm 41 follows.

REMARK ON REVISION OF ALGORITHM 41
EVALUATION OF DETERMINANT [Josef G. Solomon, *Comm. ACM 4* (Apr. 1961), 176; Bruce H. Freed, *Comm. ACM 6* (Sept. 1963), 520]
LEO J. ROTENBERG (Recd 7 Oct. 63)
Box 2400, 362 Memorial Dr., Cambridge, Mass.

While desk-checking the program an error was found. For example, the algorithm as published would have calculated the value zero as the determinant of the matrix

$$\begin{bmatrix} 0 & 0 & 0 & 1 \\ 0 & 0 & 1 & 0 \\ 0 & 1 & 0 & 0 \\ 1 & 0 & 0 & 0 \end{bmatrix}$$

The error lies in the search for a nonzero element in the rth column of the matrix b.

Editor's Note. Apparently the best general determinant evaluators in this section are imbedded in the linear equation solvers Algorithm 43 [*Comm. ACM 4* (Apr. 1961), 176, 182; and 6 (Aug. 1963), 445] and Algorithm 135 [*Comm. ACM 5* (Nov. 1962), 553, 557]. They search each column for the largest pivot in absolute value. Algorithm 41 searches only for a nonzero pivot in each column, and will therefore fail for the matrix

$$\begin{bmatrix} 2^{-t} & 1 & 1 \\ 1 & 1 & 2 \\ 1 & 1 & 1 \end{bmatrix}$$

if $t \gg s$, for a machine with s-bit floating point.

It is hoped that soon a good determinant evaluator will be published to take the place of Algorithm 41.—G. E. F.

CERTIFICATION OF:
ALGORITHM 41 [F3]
EVALUATION OF DETERMINANT
[Josef G. Solomon, *Comm. ACM 4* (Apr. 1961), 171]
ALGORITHM 269 [F3]
DETERMINANT EVALUATION
[Jaroslav Pfann and Josef Straka, *Comm. ACM 8* (Nov. 1965), 668]
A. BERGSON (Recd. 4 Jan. 1966 and 4 Apr. 1966)

Computing Lab., Sunderland Technical College, Sunderland, Co. Durham, England

Algorithms 41 and 269 were coded in 803 ALGOL and run on a National-Elliott 803 (with automatic floating-point unit).

The following changes were made:
(i) **value** n; was added to both Algorithms;
(ii) In Algorithm 269, since procedure *EQUILIBRATE* is only called once, it was not written as a procedure, but actually written into the **procedure** *determinant* body.

The following times were recorded for determinants of order N (excluding input and output), using the same driver program and data.

N	T_1 Algorithm 41 (minutes)	T_2 Algorithm 269
10	0.87	0.78
15	2.77	2.18
20	6.47	4.78
25	12.47	8.99
30	21.37	14.98

From a plot of $\ln(T_1)$ against $\ln(N)$ it was found that

$$T_1 = 0.00104 N^{2.92}.$$

Similarly,

$$T_2 = 0.00153 N^{2.70}.$$

From a plot of T_1 against T_2, it was found that Algorithm 269 was 30.8 percent faster than Algorithm 41, but Algorithm 41 required less storage.

ALGORITHM 42
INVERT
T. C. WOOD
RCA Digital Computation and Simulation Group.
  Moorestown, New Jersey

procedure Invert (A) order: (n) Singular: (s) Inverse: (A1);
        array A, A1; **integer** n,s, **value** n;
comment This procedure inverts the square matrix A of order
  n by applying a series of elementary row operation to the matrix
  to reduce it to the identity matrix. These operations when
  applied to the identity matrix yield the inverse A1. The case
  of a singular matrix is indicated by the value s := 1;
begin   **comment** augment matrix A with identity matrix;
    array a[1:n, 1:2 × n]; **integer** i,j;
    for i := 1 **step** 1 **until** n **do**
    for j := 1 **step** 1 **until** 2 × n **do**
    if j $\leq$ n **then** a[i,j] := A[i,j] **else**
    if j = n+1 **then** a[i, j] := 1.0 **else** a [i,j] := 0.0;
    comment begin inversion;
    for i := 1 **step** 1 **until** n **do**
begin   **integer** k, ℓ, ind; j := ℓ := i; ind := s := 0;
  L1: **if** a[ℓ,j] = 0 **then**
    begin ind := 1; **if** ℓ < n **then begin** ℓ := ℓ + 1;
    go to L1 **end**
        else begin s := 1; **go to** L2 **end**
        end;
    if ind = 1 **then for** k := 1 **step** 1 **until** 2 × n **do**
    begin real temp;
        temp := a[ℓ,k];
        a[ℓ,k] := a [i,k];
        a[i,k] := temp **end** k loop;
    for k := j **step** 1 **until** 2 × n **do**
    a[i,k] := a[i,k] / a[i,j];
    for ℓ := 1 **step** 1 **until** n **do**
    if ℓ $\neq$ i **then for** k := 1 **step** 1 **until** 2 × n **do**
    a[ℓ,k] := a[ℓ,k] − a[i,k] × a[ℓ,j];
    end i loop;
    for i := 1 **step** 1 **until** n **do**
    for j := 1 **step** 1 **until** n **do**
    A1[i,j] := a[i,n+j];
  L2: **end** of procedure

CERTIFICATION OF ALGORITHM 42
INVERT (T. C. Wood, *Comm. ACM*, Apr., 1961)
ANTHONY W. KNAPP AND PAUL SHAMAN
Dartmouth College, Hanover, N. H.

INVERT was hand-coded for the LGP-30 using machine language and the 24.0 floating-point interpretive system, which carries 24 bits of significance for the fractional part of a number and
five bits for the exponent. The following changes were found
necessary:

(a) **if** j = n+1 **then** a[i, j] := 1.0 **else** a[i, j] := 0.0;
    should be
    if j = n+i **then** a[i, j] := 1.0 **else** a[i, j] := 0.0;

(b) **for** k := j **step** 1 **until** 2 × n **do**
    a[i, k] := a[i, k]/a[i, j];
    should be
    for k := 2 × n **step** −1 **until** i **do**
    a[i, k] := a[i, k]/a[i, i];

(c) **if** l $\neq$ i **then for** k := 1 **step** 1 **until** 2 × n **do**
    a[l, k] := a[l, k] − a[i, k] × a[l, j];
    should be
    if l $\neq$ i **then for** k := 2 × n **step** −1 **until** i **do**
    a[l, k] := a[l, k] − a[i, k] × a[l, i];

Given these changes, j becomes superfluous in the second i loop,
and the other references to j may be changed to references to i.
    INVERT obtained the following results:
    The computer inverted a 17-by-17 matrix whose elements were
integers less than ten in absolute value. When the matrix and its
inverse were multiplied together, the largest nondiagonal element
in the product was −.00003. Most nondiagonal elements were less
than .00001 in absolute value.
    INVERT was tested using finite segments of the Hilbert matrix.
The following results were obtained in the 4 × 4 case:

16.005	−120.052	240.125	−140.082
−120.052	1200.584	−2701.407	1680.917
240.126	−2701.411	6483.401	−4202.217
−140.082	1680.920	−4202.219	2801.446

The exact inverse is:

16	−120	240	−140
−120	1200	−2700	1680
240	−2700	6480	−4200
−140	1680	−4200	2800

INVERT was also coded for the LGP-30 in machine language
and the 24.1 extended range interpretive system. This system,
which uses 30 significant bits for the fraction, obtained the following as the inverse of the 4 × 4 Hilbert matrix:

16.000	−120.001	240.001	−140.001
−120.001	1200.006	−2700.015	1680.010
240.001	−2700.016	6480.037	−4200.024
−140.001	1680.010	−4200.024	2800.016

The program coded in the 24.0 interpretive system successfully
inverted a matrix consisting of ones on the minor diagonal and
zeros everywhere else.

REMARKS ON ALGORITHM 42
INVERT [T. C. Wood, *Comm. ACM*, Apr. 1961]
P. NAUR
Regnecentralen, Copenhagen, Denmark

INVERT cannot be recommended since it does not search for
pivot and therefore will give poor accuracy. This is confirmed by
the figures quoted by Knapp and Shaman in their certification
[*Comm. ACM* 4 (Nov. 1961), 498]. The results obtained by them
using 30 significant bits for the fraction may be compared directly
with those obtained using INVERSION II (Algorithm 120) and
gjr with the GIER ALGOL system (see certification below). Inverting the 4 × 4 segment of the Hilbert matrix, the largest error
in any element is found to be:

	Subscripts	Error
INVERT (Knapp and Shaman)	3,3	0.037
INVERSION II ⎫ (see certification of	3,3	0.0306
gjr ⎭ Alg. 120)	4,3	0.00010

In view of this basic shortcoming of Algorithm 42, it is unnecessary to report on other features of it.

CORRECTION TO EARLIER REMARKS ON ALGORITHM 42 INVERT, ALG. 107 GAUSS'S METHOD, ALG. 120 INVERSION II, AND gjr [P. Naur, *Comm. ACM*, Jan. 1963, 38–40.]

P. NAUR

Regnecentralen, Copenhagen, Denmark

George Forsythe, Stanford University, in a private communication has informed me of two major weaknesses in my remarks on the above algorithms:

1) The computed inverses of rounded Hilbert matrices are compared with the exact inverses of unrounded Hilbert matrices, instead of with very accurate inverses of the rounded Hilbert matrices.

2) In criticizing matrix inversion procedures for not searching for pivot, the errors in inverting positive definite matrices cannot be used since pivot searching seems to make little difference with such matrices.

It is therefore clear that although the figures quoted in the earlier certification are correct as they stand, they do not substantiate the claims I have made for them.

To obtain a more valid criterion, without going into the considerable trouble of obtaining the very accurate inverses of the rounded Hilbert matrices, I have multiplied the calculated inverses by the original rounded matrices and compared the results with the unit matrix. The largest deviation was found as follows:

	Maximum deviation from elements of the unit matrix		
Order	*INVERSION II*	*gjr*	*Ratio*
2	$-1.49_{10}-8$	$-1.49_{10}-8$	1.0
3	$-4.77_{10}-7$	$-8.34_{10}-7$	0.57
4	$-9.54_{10}-6$	$-3.43_{10}-5$	0.28
5	$-7.32_{10}-4$	$-4.58_{10}-4$	1.6
6	$-1.61_{10}-2$	$-1.42_{10}-2$	1.1
7	$-5.78_{10}-1$	$-5.47_{10}-1$	1.1
8	$-1.20_{10}-2$	$-1.38_{10}1$	8.7
9	$-4.91_{10}1$	$-2.22_{10}1$	2.2

This criterion supports Forsythe's criticism. In fact, on the basis of this criterion no preference of INVERSION II or *gjr* can be made.

The calculations were made in the GIER ALGOL system, which has floating numbers of 29 significant bits.

ALGORITHM 43
CROUT WITH PIVOTING II
HENRY C. THACHER, JR.*
Argonne National Laboratory, Argonne, Illinois

```
real procedure INNERPRODUCT (u,v) index : (k) start : (s)
    finish : (f);
value s, f;  integer k, s, f;  real u, v;
comment  INNERPRODUCT forms the sum of u(k) × v(k) for
    k = s, s+1, . . . , f. If s > f, the value of INNERPRODUCT is
    zero. The substitution of a very accurate inner product proce-
    dure would make CROUT more accurate;
comment  INNERPRODUCT may be declared in the head of
    any block which includes the block in which CROUT is de-
    clared. It may be used independently for forming the inner
    product of vectors;
begin
        real h;
        h := 0;  for k := s step 1 until f do h := h+u × v;
        INNERPRODUCT := h
end    INNERPRODUCT;
procedure CROUT II (A, b, n, y, pivot, det, repeat)
comment  This procedure is a revision of Algorithm 16, Crout
    With Pivoting by George E. Forsythe, Comm. ACM 3, (1960)
    507-8. In addition to modifications to improve the running of
    the program, and to conform to proper usage, it provides for
    the computation of the determinant, det, of the matrix A. The
    solution is obtained by Crout's method with row interchanges,
    as formulated in reference [1], for solving Ay = b and transform-
    ing the augmented matrix [A b] into its triangular decomposi-
    tion LU with all L(k,k) = 1. If A is singular we exit to 'singular,'
    a nonlocal label.  pivot (k) becomes the current row index of
    the pivot element in the k-th column. Thus enough information
    is preserved for the procedure to process a new right-hand
    side without repeating the triangularization, if the boolean pa-
    rameter repeat is true. The accuracy obtainable from CROUT
    would be much increased by calling CROUT with a more accu-
    rate inner product procedure than INNERPRODUCT.
        The contributions of Michael F. Lipp and George E. Forsythe
    by prepublication review and pointing out several errors are
    gratefully acknowledged;
comment  Nonlocal identifiers appearing in this procedure are:
    (1) The nonlocal label 'singular', to which the procedure exits
    if det A=0, and (2) the real procedure 'INNERPRODUCT'
    given above;
        value n;  array A, b, y;  integer n;  integer array
            pivot;  real det;  Boolean repeat;
begin
        integer k, i, j, imax, p;  real TEMP, quot;
        det := 1;  if repeat then go to 6;
        for k := 1 step 1 until n do
1:      begin
            TEMP := 0;
            for i := k step 1 until n do
2:          begin
            A[i,k] := A[i,k] − INNERPRODUCT (A[i,p], A[p,k],
                p, 1, k−1);
            if abs(A[i,k]) > TEMP then
3:          begin
            TEMP := abs(A[i, k]);  imax := i
```

```
        end 3
        end 2;
        pivot [k] := imax;
comment  We have found that A[imax, k] is the largest pivot in
    column k. Now we interchange rows k and imax;
        if imax ≠ k then
4:      begin det := − det;  for j := 1 step 1 until n do
5:      begin
            TEMP := A[k,j];  A[k,j] := A[imax, j];  A[imax, j]
                := TEMP
        end 5;
        TEMP := b[k];  b[k] := b[imax]; b[imax] := TEMP
        end 4;
comment  The row interchange is done. We proceed
    to the elimination;
        if A[k,k] = 0 then go to singular;
        quot := 1.0/A[k,k];
        for i . = k+1 step 1 until n do
        A[i,k] := quot × A[i,k];
        for j := k+1 step 1 until n do
        A[k,j]  :=  A[k,j]  −  INNERPRODUCT  (A[k,p]
            A[p,j], p, 1, k−1);
        b[k] := b[k] − INNERPRODUCT (A[k,p], b[p]
            p, i, k−1)
        end 1;  go to 7;
comment  The triangular decomposition is now finished,
    and we skip to the back substitution;
6:      begin comment  This section is used when the formal
        parameter repeat is true, indicating that the matrix A
        has previously been decomposed into triangular form by
        CROUT II, with row interchanges specified by pivot,
        and that it is desired to solve the linear system with a
        new vector b, without repeating the triangularization;
        for k := 1 step 1 until n do
        begin
            TEMP := b[pivot[k]];  b[pivot[k]] := b[k];  b[k] :=
                TEMP;  b[k]  :=  b[k]  −  INNERPRODUCT
                (A[k, p], b[p], p, 1, k−1) end;
        end 6;
7:      for k := n step − 1 until 1 do
8:      begin if ¬ repeat then det := A[k,k] × det;
        y[k]  :=  (b[k] − INNERPRODUCT (A[k,p], y[p], p,
            k+1, n)/A[k,k]
        end 8;
end CROUT II;
```

REFERENCE:

(1) J. H. WILKINSON, Theory and practice in linear systems. In
    John W. Carr III (editor), Application of Advanced Nu-
    merical Analysis to Digital Computers, pp. 43–100 (Lectures
    given at the University of Michigan, Summer 1958, College
    of Engineering, Engineering Summer Conferences, Ann
    Arbor, Michigan [1959]).

* Work supported by the U. S. Atomic Energy Commission.

CERTIFICATION OF ALGORITHM 43
CROUT II (Henry C. Thacher, Jr., *Comm. ACM*, 1960)

HENRY C. THACHER, JR.*
Argonne National Laboratory, Argonne, Illinois

CROUT II was coded by hand for the Royal Precision LGP-30 computer, using a 28-bit mantisa floating point interpretive system (24.2 modified).

The program was tested against the linear system:

$$A = \begin{pmatrix} 12.1719 & 27.3941 & 1.9827 & 7.3757 \\ 8.1163 & 23.3385 & 9.8397 & 4.9474 \\ 3.0706 & 13.5434 & 15.5973 & 7.5172 \\ 3.0581 & 3.1510 & 6.9841 & 13.1984 \end{pmatrix} \quad b = \begin{pmatrix} 6.6355 \\ 6.1304 \\ 4.6921 \\ 2.5393 \end{pmatrix}$$

with the following results:

$$A' = \begin{pmatrix} 12.171900 & 27.394100 & 1.9827000 & 7.3756999 \\ 0.25226957 & 6.6327021 & 15.097125 & 5.6565352 \\ 0.25124262 & -0.56260107 & 14.979620 & 14.527683 \\ 0.66680633 & 0.76468695 & -0.20207132 & -1.3606142 \end{pmatrix}$$

$$b' = \begin{pmatrix} 6.6354999 \\ 3.0181653 \\ 2.5702026 \\ -0.082780734 \end{pmatrix} \quad \text{pivot} = \begin{pmatrix} 1 \\ 3 \\ 4 \\ 4 \end{pmatrix} \quad y = \begin{pmatrix} 0.15929120 \\ 0.14691771 \\ 0.11257482 \\ 0.060840712 \end{pmatrix}$$

det $= -1645.4499$. All elements of Ab $-$ y were less than 10^{-7} in magnitude. Identical results were obtained with the same b, and repeat **true**. With the same b and the last row vector of A replaced by (19.1927, 33.4409, 25.1298, 5.2811), i.e. A 4, j = A 1 j, + 2A 2, j − 3A 3, j, the results were:

det $= 0.10924352 \times 10^{-3}$,

y $= (0.29214425 \times 10^{8}, -0.12131172 \times 10^{8}, 0.72411923 \times 10^{7}, -0.51018392 \times 10^{7})$

Failure to recognize this singular matrix is due to roundoff, either in the data input or in the calculation.

---

* Work supported by the U.S. Atomic Energy Commission.

CERTIFICATION OF ALGORITHM 43
CROUT II [Henry Thacher, Jr., *Comm. ACM* (1960), 176]

C. DOMINGO AND F. RODRIGUEZ-GIL
Universidad Central, Caracas, Venezuela

CROUT II was coded in PUC-R2 and tested in the IBM-1620. Two types of INNERPRODUCT subroutines were used. The first one finds the scalar product in fixed-point arithmetic to increase accuracy, using an accumulator of 32 digits. The second one uses ordinary floating-point with eight significative figures.

Using a unit matrix as right-hand side, a 6×6 segment of Hilbert matrix was inverted. The inverse was inverted again.

The maximum difference between this result and the original segment of Hilbert matrix was:

Using fixed-point INNERPRODUC......... 8.2426 $\times 10^{-4}$
(Value of determinant.................... 4.7737088 $\times 10^{-18}$)
Using floating-point INNERPRODUC...... 3.014016 $\times 10^{-2}$
(Value of determinant.................. 4.4950721 $\times 10^{-18}$)

Two typographical errors were observed in the algorithm:

The statement:

$b[k] := g[k] - \text{INNERPRODUCT } (A[k,p], b[p], p,i,k-1)$

should be:

$b[k] := b[k] - \text{INNERPRODUCT } (A[k,p], b[p], 1,k-1)$

The statement:

$y[k] := (b[k] - \text{INNERPRODUCT } (A[k,p], y[p], p,k+1,n)/A[k,k]$

should be:

$y[k] := (b[k] - \text{INNERPRODUCT } (A[k,p], y[p], p,k+1,n))/A[k,k]$

Storage may be saved eliminating the array y and using instead the array b, in which the solution is formed.

A previous certification of this algorithm [*Comm. ACM* 4, 4 (Apr. 1961), 182] was tested again with the same results. Two errors were detected in the certification: The row that must replace the last row of A in order to obtain a singular matrix must be:
19,1927   33.4409   −251298   −5.2811

ALGORITHM 44
BESSEL FUNCTIONS COMPUTED RECURSIVELY
Maria E. Wojcicki
RCA Digital Computation and Simulation Group,
Moorestown, New Jersey

```
procedure  Bessfr(N, FX, LX, Z) Result: (J, Y);
     value  LX, FX, N;
     real    FX, LX, Z;  real array J, Y;  integer N;
comment  Bessel Functions of the first and second kind, $J_P(X)$
   and $Y_P(X)$, integral order P, are computed by recursion for
   values of X,  FX ≤ X ≤ LX, in steps of Z. The functions are
   computed for values of P,  0 ≤ P ≤ N. M[SUB], the initial
   value of P being chosen according to formulae in Erdelyi's
   Asymptotic Expansions. The computed values of $J_P(X)$ and
   $Y_P(X)$ are stored as column vectors for constant argument in
   matrices J, Y of dimension (N+1) by entier ((LX − FX)/Z + 1);
begin real  PI, X, GAMMA, PAR, LAMDA, SUM, SUM1;
     integer  P, SUB, MAXSUB;
             PI := 3.14159265;
             GAMMA := .57721566;
             PAR := 63.0 − 1.5 × ℓn (2 × PI);
             MAXSUB := entier ((LX − FX)/Z);
begin real array  JHAT [0:N, 0:MAXSUB];
     integer array  M[0:MAXSUB];
          SUB := 0;
        for X := FX step Z until LX do
begin if (X > 0) ∧ (X < 10) then M [SUB] := 2 × entier (X) + 9
   else
begin real ALOG;
        ALOG := (PAR − 1.5 × ℓn (X))/X;
        M [SUB]  :=  entier  (X × (exp (ALOG) + exp
        (−ALOG))/2) end;
        if N > M [SUB] then
begin for P := M [SUB] + 1 step 1 until N do
        J [P, SUB] := 0 end;
        JHAT [M [SUB], SUB] := 10 ↑ (−9);
comment  Having set the uppermost $\hat{J}_P(X)$ to a very small
   number we are now going to compute all the $\hat{J}_P(X)$ down to
   P = 0;
        for P := M [SUB] step −1 until 1 do
        JHAT [P−1, SUB] := 2 × P/X × JHAT [P, SUB] −JHAT
        [P+1, SUB];
        SUM := SUM1 := 0;
        for P := 2 step 2 until (M [SUB] ÷ 2) do
        SUM := SUM + JHAT [P, SUB];
        LAMDA := JHAT [0, SUB] + 2 × SUM;
        for P := 0 step 1 until N do
        J [P, SUB] := JHAT [P, SUB] /LAMDA;
comment  $J_P(X)$ have been computed by use of $\hat{J}_P(X)$;
        for P := 2 step 2 until (M [SUB] ÷ 2) do
        SUM1 := SUM1 + (−1) × (−1) ↑ P ÷ J [2 × P, SUB]
        /2/P;
        Y [0, SUB] := 2/PI × (J [0, SUB] × (GAMMA + ℓn(X/2))
        + 4 × SUM1);
        for P := 0 step 1 until (M[SUB]−1) do
        Y [P+1, SUB] := (−2/PI/P + J [P+1, SUB] × Y [P,
        SUB])/J [P, SUB];
        SUB := SUB + 1 end end end
```

ALGORITHM 45
INTEREST
PETER Z. INGERMAN
University of Pennsylvania, Philadelphia, Pa.

procedure monpay (i, B, L, t, k, m, tol, goof)
comment This procedure calculates the periodic payment
necessary to retire a loan when the interest rate on the loan
varies (possibly from period to period) as a function of the as-
yet-unpaid principal.

The formal parameters are: i, array identifier for the vector
of interest rates; $-B$, array identifier for the minimum amounts
at which the corresponding i applies; $-L$, the amount to be
borrowed; $-t$, the number of periods for which the loan is to
be taken out; $-k$, the number of different interest rates (and
upper limit for vectors i and B); $-m$, the desired periodic payment; $-$tol, the allowable deviation of m from some ideal;
and goof, the error exit to use if convergence fails. The only
output parameter is m. For further discussion, see *Comm.
ACM* 3 (Oct. 1960), 542;
begin array h, S [1:k, 1:t], M, X [1:k];
integer array T, a, b [1:k];
integer p, q, r, sa, sb, I, ib, mb, nb;
comment This section sets up the procedure;
for p := 1 **step** 1 **until** k **do**
begin for q := 1 **step** 1 **until** t **do**
    begin $h_{p,q} := i_p{}^q$;
        $S_{p,q} := (h_{p,q} - 1)/(i_p - 1)$ **end**;
    if p = 1 **then** $X_p := 0$ **else** $X_p := B_p \times (i_{p-1} - i_p)$;
    $M_p := L \times (h_{p,t}/S_{p,t})$ **end**;
sa := sb := ib := mb := 0; nb := t;
for p := 1 **step** 1 **until** k **do**
begin $a_p :=$ entier $(B_{p+1}/M_{p+1} + 0.5) - $ sa;
    sa := sa + a_p;
    $T_p := b_p :=$ entier $(B_{p+1}/M_p - 0.5) - $ sb;
    sb := sb + b_p;
    if $b_p >$ mb **then**
    begin ib := p; nb := nb − mb; mb := bp **end**
    else nb := nb − b_p **end**;
$T_{ib} :=$ nb;
I := 1;
for p := 1 **step** 1 **until** k **do**
    I := I × $(a_p - b_p + 1)$;
comment Having counted the number of possible iterations
and established a set of trial values for the T_n's, a trial m is
found;
D := 1; E := F := 0;
newm: **for** p := 1 **step** 1 **until** k **do**
    begin D := D × h_{p,T_p};
        u := 1;
        if p ≠ 1 **then for** q := 1 **step** 1 **until** p − 1
        do u := u × h_{q,T_q};
        E := E + S_{p,T_p} × u;
        v := 0;
        if p ≠ 1 **then for** r := 1 **step** 1 **until** p
        do v := v + X_r;
        F := F + u × v **end**;
m := (L × D + F)/E;
    comment Now find out whether m is good enough
    q := 1; F := D := 0;

for p := 1 **step** 1 **until** t **do**
begin get F: F := (D + m − E)/(1 + i_q);
    if $B_{q+1} \geqq$ F **then** D := F **else** q := q + 1;
    if D ≠ F **go to** get F **end**;
    if abs (D − L) ≦ tol **then go to** exit;
comment If not within tolerance, adjust T_n's and try
  again;
p := 0;
redo: p := p + 1;
    if p ≠ ib **then**
    begin if $T_p \geqq a_p$ **then**
        begin $T_{ib} := T_{ib} + T_p − b_p$
            $T_p := b_p$ **end end**
    else begin
        $T_p := T_p + 1$;
        $T_{ib} := T_{ib} − 1$;
        p := k **end**;
    if p = k **then** I := I − 1 **else go to** redo;
    go to if I > 0 **then** newm **else** goof;
exit: **end** monpay;

CERTIFICATION OF ALGORITHM 45
INTEREST [Peter Z. Ingerman, *Comm. ACM* Apr. 1961
and Oct. 1960]
CARL B. WRIGHT
Dartmouth College, Hanover, N. H.

INTEREST was translated into Dartmouth College Computation Center's "Self Contained ALGOL Processor" for the Royal-
McBee LGP-30. When using SCALP, memory capacity is severely
limited and thus it was necessary to run this program in two
blocks. Block I ended with the computation of I, and Block II
started with the "*newm*" loop. After making the changes listed
below, test problems using up to three interest rates and up to 18
time periods were used with the following results:

Loan	Periods	Interest Rates	Payments	Final Balance*	Tolerance
$100.00	1	0.05	$105.00	$0.00	$0.25
1800.00	10	0.03	211.01	0.05	4.50
875.65	8	0.08 to 500.00			
		0.05 over 500.00	139.78	−1.49	2.19
14750.00	18	0.06 to 5000.00			
		0.05 to 10,000.00			
		0.04 over 10,000.00	1201.70	10.3ʊ	36.88

* Hand calculation.

It is noted that in each case the final balance is within the prescribed tolerance (0.0025 of the loan).

In the following corrections bracketed subscripts replace
ordinary subscripts and exponentiation is represented by ↑
rather than superscript.

The following corrections should be made in the Note on Interest in the October, 1960, issue of *Comm. ACM*:

1. Definition of *B[n]*: Replace "*minimum*" by "*maximum*".
Replace "*j[n]*" by "*j[n−1]*".

2. Define *B[k+1]* ≡ *L*.

3. Definition of *K[n]*: Replace "*B[n]*" by "*B[n+1]*".

The following corrections were found necessary in the procedure:

1. The upper limit of the vector B is $k+1$, not k. It is not necessary to change the upper limit of the I-vector. (See correction 4 below.)

2. D, E, F, u, v were not declared and must be declared as **real**.

3. In the **array** declaration replace "$M[1:k]$" by "$M[1:k+1]$".

4. As j approaches 0, i approaches 1 and lim $(h/S) = 1/t$. Thus for $j[k+1] = 0$, $i[k+1] = 1$, and $M[k+1] = L/t$. Thus after

$M[p] := L \times (h[p,t]/S[p,t])$ **end**;

`insert

$M[k+1] := L/t$; $B[k+1] := L$;

5. In the conditional statement following computation of $b[p]$, replace "$>$" by "$\geq$".

6. In same conditional statement, next line, "$mb := bp$" should read "$mb := b[p]$".

7. $D := 1$; $E := F := 0$;
 newm: for $p := 1$ **step** 1 **until** k **do**

should be changed to
 newm: $D := 1$; $E := F := 0$;
 for $p := 1$ **step** 1 **until** k **do**

8. **begin** *get F*: $F := (D+m-E)/(1+i[q])$;
   if $B[q+1] \geq F$ **then** $D := F$ **else** $q := q + 1$;
   if $D \neq F$ **go to** *get F* **end**;

should be changed to read as follows:
   begin *get F*: $F := (D+m)/i[q]$;
   if $B[q+1] \geq F$ **then** $D := F$ **else**
   begin if $q < k$ **then** $q := q + 1$ **else** $D := F$ **end**;
   if $D \neq F$ **then go to** *get F* **end**;

Note that the "then" in the last line was omitted from the original procedure.

9. In the "*redo*" loop insert a semicolon after the statement
   $T[ib] := T[ib] + T[p] - b[p]$;

10. In the "*redo*" loop, next line, omit the second "*end*".

11. In the "*redo*" loop,

$$p := k \text{ end};$$

should be changed to

$$p := k \text{ end end};$$

ALGORITHM 46
EXPONENTIAL OF A COMPLEX NUMBER
John R. Herndon
Stanford Research Institute, Menlo Park, California

```
procedure  EXPC (a, b, c, d);  value a, b;  real a, b, c, d;
comment  This procedure computes the number, c+di, which
   is equal to e^(a+bi);
begin   c := exp (a);
        d := c × sin (b);
        c := c × cos (b)
end EXPC;
```


CERTIFICATION OF ALGORITHM 46
EXPONENTIAL OF A COMPLEX NUMBER (J. R.
   Herndon, *Comm. ACM 4* (Apr., 1961), 178)
A. P. Relph
Atomic Power Div., The English Electric Co., Whetstone,
   England

Algorithm 46 was translated using the Deuce Algol compiler,
no corrections being required, and gave satisfactory results.

ALGORITHM 47

ASSOCIATED LEGENDRE FUNCTIONS OF THE FIRST KIND FOR REAL OR IMAGINARY ARGUMENTS

JOHN R. HERNDON
Stanford Research Institute, Menlo Park, California

```
procedure  LEGENDREA (m, n, x, r);  value m, n, x, r;
            integer m, n;  real x, r;
comment  This procedure computes any Pₙᵐ(x) or Pₙᵐ(ix) for
n an integer less than 20 and m an integer no larger than n.
The upper limit of 20 was taken because (42)! is larger than
10⁴⁹. Using a modification of this procedure values up to n=35
have been calculated. If Pₙᵐ(x) is desired, r is set to zero. If
r is nonzero, Pₙᵐ(ix) is computed;

begin
            integer i, j;  array Gamma [1:41];
            real p, z, w, y;
            if n = 0 then
                begin p := 1;
                go to gate end;
            if n < m then
                begin p := 0;
                go to gate end;
            z := 1; w := z;
            if n=m then go to main;
            for i := 1 step 1 until n−m do
                z := x × z;
    main:   Gamma [1] := 1;
            for i := 2 step 1 until n+n+1 do
                begin Gamma [i] := w × Gamma [i−1];
                w := w+1 end;
            w := 1;  y := w/(x × x);
            if r=0 then
                begin y := −y;
                w := −w end;
            if x=0 then
                begin i := (n−m)/2;
                if (i+i) ≠ (n−m) then
                    begin p := 0;
                    go to gate end;
                p := Gamma [m+n+1]/(Gamma [i+1] × Gamma
                    [m+i+1]);
                go to last end;
            j := 3;  p := 0;
            for i := 1 step 1 until 12 do
                begin if (n−m+2)/2 < i then go to last end;
                p := p + Gamma [n+n−i−i+3] × z/(Gamma
                    [i] × Gamma [n−i+2] × Gamma [n−i−i−
                    m+j]);
                z := z × y end;
    last:   z := 1;
            for i := step 1 until n do
                z := z+z;
            p := p/z;
            if r ≠ 0 then
                begin i := n−n/4;
                if 1 < i then
                    p := −p end;
```

```
            if m = 0 then go to gate;
            j := m/2;  z := abs(w+x × x);
            if m ≠ (j+j) then
                begin z := sqrt (z);
                j := m end;
            for i := step 1 until j do
                p := p × z;
    gate:   LEGENDREA := p
end LEGENDREA;
```

CERTIFICATION OF ALGORITHM 47
ASSOCIATED LEGENDRE FUNCTIONS OF THE FIRST KIND FOR REAL OR IMAGINARY ARGUMENTS [John R. Herndon, *Comm. ACM*, Apr. 1961]
RICHARD GEORGE*
Argonne National Laboratory, Argonne, Ill.

* Work supported by United States Atomic Energy Commission.

This procedure was programmed in FORTRAM for the IBM 1620 and was tested with a number of real arguments. A few errors were detected:

1. In the following sequence the *end* must be removed:

$$\textbf{begin if } (n - m + 2)/2 < i \textbf{ then go to last end};$$

2. In these, the lower bound of 1 is needed:

$$\textbf{for } i := \textbf{step 1 until } n \textbf{ do}$$
$$\textbf{for } i := \textbf{step 1 until } j \textbf{ do}$$

3. There are four places where integer arithmetic is clearly intended and we must substitute the symbol ÷ for the symbol /.

In addition, it might be mentioned that the statement

$$\textbf{if } n = m \textbf{ then go to main};$$

could be omitted from the ALGOL program without harm, though the FORTRAN version requires it. Here, and elsewhere in the procedure, one might make an equivalent but more succinct statement. With change in style, the variable j could be eliminated.

CERTIFICATION OF ALGORITHM 47 [S16]
ASSOCIATED LEGENDRE FUNCTIONS OF THE FIRST KIND FOR REAL OR IMAGINARY ARGUMENTS [John R. Herndon, *Comm. ACM 4* (Apr. 1961), 178]

S. M. COBB (Recd. 6 Feb. 1969, 12 May 1969 and 9 July 1969)
The Plessey Co. Ltd., Roke Manor, Romsey, Hants, England

KEYWORDS AND PHRASES: Legendre function, associated Legendre function, real or imaginary arguments
CR CATEGORIES: 5.12

This procedure was tested and run on the I.C.T. Atlas computer.

In addition to the errors mentioned in the certification of August 1963 [2] the following points were noted.

1. The requirement that when $n < m$ $p := 0$ must take precedence over $p := 1$ when $n = 0$. Hence the order of the first two **if** statements must be interchanged.

2. Most computers fail on division by zero. Hence the statement beginning **if** $x = 0$ **then** and ending with **go to** *last* **end;** should be inserted between $w := 1;$ and $y := w/(x \times x)$.

3. When $x = 0$, if the argument of the Legendre function is to be considered as real p must be multiplied by $(-1)^i$. This is achieved by inserting after the statement beginning $p := Gamma$ $[m+n+1]$ the **if** statement

if r **then** $p := p \times (-1) \uparrow i;$

(For a change in the meaning of r see item 5 below.)

4. After the label *last* in the compound statement beginning **if** $r \neq 0$ the statement $i := n - n \div 4;$ is wrong. This should read

$i := n - 4 \times (n \div 4);$

5. Since r is used only as an indicator it is better that it be declared as **Boolean**. It can then be given the value **true** if the argument of the Legendre function is x and **false** if it is ix. The following program changes are then necessary. The statement beginning

if $r = 0$ **then**

becomes

if r **then**

The statement beginning

if $r \neq 0$ **then**

becomes

if $\neg r$ **then**

6. Computing time can be saved in several ways. First we should declare another integer k and set it equal to $n - m$. The first statement of the procedure is then

$k := n - m;$

The next statement will begin

if $k < 0$ **then**

(This replaces **if** $n < m$ **then** whose position has been changed in accordance with item 1 above.)
$n - m$ is then replaced by k in the lines
for $i := 1$ **step** 1 **until** $n - m$ **do**

and

if $(i+1) \neq (n-m)$ **then**

Removing j as suggested in the previous certification leaves it free to be set to $k \div 2$. This requires the following modification: instead of the unnecessary statement **if** $n = m$ **then go to** *main* put

$j := k \div 2;$

In the statement beginning **if** $x = 0$ **then** replace the line

begin $i := (n-m) \div 2;$

by

begin $i := j;$

In the **for** loop beginning **for** $i := 1$ **step** 1 **until** 12 **do** a further small saving in computer time could be achieved by setting k to $n - i$. The loop thus becomes

for $i := 1$ **step** 1 **until** 12 **do**
begin if $j + 1 < i$ **then go to** *last;*
    $k := n - i;$

$p := p + Gamma[2 \times k+3] \times z/Gamma[i] \times Gamma[k+2] \times Gamma[k-i-m+3]);$
$z := z \times y$
end

For real argument the program was tested as follows.

(i) $x = 0(0.1)1,\ m = 0(1)3,\ n = 0(1)3$
(ii) $x = 1.2(0.2)2.8,\ m = 0(1)2,\ n = 0(1)2$
(iii) $m = 0,\ n = 9,\ x = 0(0.2)1,\ 2(2)10.$

For imaginary argument we used

$x = 0(0.2)2,\ m = 0(1)2,\ n = 0(1)2.$

Checking for real argument was carried out where possible using [1], agreement being obtained in all cases to the maximum number of figures available, which varied between 6 and 8. For all other cases [3] had to be used, giving only a 5 figure check.

REFERENCES:
1. ABRAMOWITZ, M., AND STEGUN, I. A. Handbook of mathematical functions. AMS 55, Nat. Bur. Stand. US Govt. Printing Off., Washington, D.C., 1964.
2. GEORGE, R. Certification of Algorithm 47. *Comm. ACM 6* (Aug. 1963), 446.
3. MORSE, P. M., AND FESBACH, H. *Methods of Theoretical Physics Pt. II*. McGraw Hill, New York, 1953.

ALGORITHM 48
LOGARITHM OF A COMPLEX NUMBER
JOHN R. HERNDON
Stanford Research Institute, Menlo Park, California

procedure LOGC(a, b, c, d); **value** a, b; **real** a, b, c, d;
comment This procedure computes the number, c+di, which
is equal to $\log_e(a+bi)$;
begin c := sqrt (a $\times$ a + b $\times$ b);
d := arctan (b/a);
c := log (c);
if a < 0 **then** d := d+3.1415927
end LOGC;

CERTIFICATION OF ALGORITHM 48
LOGARITHM OF A COMPLEX NUMBER (J. R.
Herndon, *Comm. ACM 4* (Apr., 1961), 179)
A. P. RELPH
Atomic Power Div., The English Electric Co., Whetstone,
England

Algorithm 48 was translated using the DEUCE ALGOL compiler,
after certain modifications had been incorporated, and then gave
satisfactory results.

The original version will fail if $a = 0$ when the procedure for
arctan is entered. It also assumes that $-\pi/2 < d < 3\pi/2$, whereas the
principal value for logarithm of a complex number assumes
$-\pi < d \leqq \pi$.

Incidentally, the ALGOL 60 identifier for natural logarithm is ln,
not log.

The modified procedure is as follows:

procedure LOGC (a,b,c,d); **value** a,b; **real** a,b,c,d;
comment This procedure computes the number $c + di$ which is
equal to the principal value of $\log_e (a + bi)$. If $a = 0$ then c is
put equal to $-{}_{10}47$ which is used to represent "$-$ infinity";
begin integer m,n
m := sign (a); n := sign (b);
if a = 0 **then begin** c := $-{}_{10}47$;
d := 1.5707963 $\times$ n;
go to k
end;
c := sq rt(a $\times$ a + b $\times$ b);
c := ln (c);
d := 1.5707963 $\times$ (1$-$m) $\times$ (1+n$-$n$\times$n) + arctan (b/a);
k: **end** LOGC;

REMARK ON ALGORITHM 48
LOGARITHM OF A COMPLEX NUMBER [John R.
Herndon, *Comm. ACM 4* (Apr. 1961)]
MARGARET L. JOHNSON AND WARD SANGREN
Computer Applications, Inc., San Diego, Calif.

Considerable care must be taken in using the arctan function.
In Algorithm 48 two such difficulties are ignored. First it is
necessary, because of a resulting division by zero, to deal separately with the case where the real part of the complex number
is zero. Second, if the real part of the complex number is negative

and the argument of the logarithm is to have a value between
$-\pi$ and π then the action depends upon the sign of the imaginary
part of the complex number. For clarity the following procedure
exhibits in sequence the alternatives:
procedure LOGC (a, b, c, d); **value** a, b; **real** a, b, c, d;
comment This procedure computes the number $c+di$ which is
equal to $\log_e (a+bi)$. It is assumed that the arctan has a value
between $-\pi/2$ and $\pi/2$.
begin if a>0 **then begin** THETA := 0; **go to** SOL **end**;
if a<0$\wedge$b$\geqq$0 **then begin** THETA := 3.1415927;
go to SOL **end**;
if a<0$\wedge$b<0 **then begin** THETA := $-$3.1415927;
go to SOL **end**;
if a=0$\wedge$b=0 **then begin** c := d := 0;
go to RETURN **end**;
if a=0$\wedge$b>0 **then begin** c := ln(b); d := 1.570963;
go to RETURN **end**;
if a=0$\wedge$b<0 **then begin** c := ln(abs(b));
d := 1.570963; **go to** RETURN **end**;
SOL: d := arctan (b/a) + THETA;
c := sqrt(a$\times$a+b$\times$b);
c := ln(c);
RETURN: **end** LOGC

REMARK ON REMARKS ON ALGORITHM 48 [B3]
LOGARITHM OF A COMPLEX NUMBER [John R.
Herndon, *Comm. ACM 4* (Apr. 1961), 179; *5* (Jun. 62),
347; *5* (Jul. 62), 391]
DAVID S. COLLENS (Recd. 24 Jan. 1964 and 1 Jun. 1964)
Computer Laboratory, The University, Liverpool 3,
England

This procedure was designed to compute $\log_e(a+bi)$, namely
$c+di$, and although some very necessary precautions about its
use have already been stated, some points seem to have escaped
notice. In particular, A. P. Relph [*Comm. ACM*, June 1962, 347]
remarked that if $a = 0$, then c becomes '$-$infinity', but this is only
the case if $b = 0$ also. Margaret L. Johnson and Ward Sangren
[*Comm. ACM*, July 1962, 391] conceded that $a = b = 0$ was a special
case, but wrongly gave zero as the result. The only reasonable way
of dealing with this case is to exit to some nonlocal label and to
let the user decide whether to terminate his program or to assign
particular values to c and d. The obvious values to use here are, for
c, a negative number, larger than the largest which would be given
by the procedure, and possibly zero for d. (In an implementation
where 2^{-129} is the smallest representable nonzero number, the
largest negative value of c possible is $-$89.416.) Finally, in the
Johnson-Sangren version of the procedure, the last conditional
statement should read

if $a = 0 \wedge b < 0$ then begin $c := ln(abs(b))$;

$d := -1.570963$; **go to** *RETURN* **end**;

the omission of the minus sign in the original being probably
typographical in origin.

ALGORITHM 49
SPHERICAL NEUMANN FUNCTION
JOHN R. HERNDON
Stanford Research Institute, Menlo Park, California

real procedure SPHBEN (r,x); **value** r,x; **real** r,x;
comment This procedure computes the spherical Neumann
function $(\pi/2x)^{\frac{1}{2}}N_{r+1/2}(x)$. Infinity is represented by 10^{47};
begin real z, g, t;
    if x=0 **then**
      begin s := 10 ↑ 47;
        go to gate

      end;
s := −cos (x)/x;
if r = 0 **then**
    go to gate;
t := sin (x)/x;
for g := 1 **step** 1 **until** r **do**
    begin z := s;
      s := s × (g+g−1)/(x−t):
      t := z
    end;
gate: SPHBEN := s
end SPHBEN;

ACM Transactions on Mathematical Software, Vol. 4, No. 3, September 1978, Page 295.

REMARK ON ALGORITHM 49

Spherical Neumann Function
[J.R. Herndon, *Comm. ACM 4*, 4 (April 1961), 179]

John P. Coleman [Recd 17 February 1978]
Department of Mathematics, University of Durham, Durham, England

There is a typographical error in this algorithm. The line

$$s := s \times (g + g - 1)/(x - t);$$

should read

$$s := s \times (g + g - 1)/x - t;$$

The algorithm provides overflow protection only when $x = 0$. Overflow will still occur for x very close to zero. The range of values of x for which overflow occurs will depend both on the value of r and on the largest number the machine can hold.

ALGORITHM 50
INVERSE OF A FINITE SEGMENT OF THE
  HILBERT MATRIX
JOHN R. HERNDON
Stanford Research Institute, Menlo Park, California

```
procedure  INVHILBERT (n,S);  value n;  real n;
             real array S;
comment  This procedure computes the elements of the inverse
  of an n × n finite segment of the Hilbert matrix and stores them
  in the array S;
begin      real i, j, k;
           S[1, 1] = n × n;
           for i := 2 step 1 until n do
             begin
               S[i, i] := (n+i−1) × (n−i+1)/((i−1) × (i−1));
               S[i, i] := S[i−1, i−1] × S[i, i] × S[i, i]
             end;
           for i := 1 step 1 until n−1 do
             begin
               for j := i+1 step 1 until n do
                 begin
                   k := j−1;
                   S[i, j] := −S[i, k] × (n+k) × (n−k)/(k × k)
                 end
             end;
           for i := 2 step 1 until n do
             begin S[i, i] := S[i, i]/(i+i−1);
               for j := 1 step 1 until i−1 do
             begin S[j, i] := S[j, 1]/(i+j−1);
                   S[i, j] := S[j, i]
             end
           end
end INVHILBERT;
```

CERTIFICATION OF ALGORITHM 50
INVERSE OF A FINITE SEGMENT OF THE HIL-
  BERT MATRIX [J. R. Herndon, *Comm. ACM 4*
  (Apr. 1961)]
B. RANDELL
Atomic Power Division, The English Electric Co., Whet-
  stone, England

INVHILBERT was translated using the DEUCE ALGOL compiler and the following corrections being needed.
1.   S[1, 1] = n × n, replaced by S[1, 1] := n × n;
2.   S[j, i] := S(j, 1]/(i + j − 1)
replaced by   S[j, i] := S[j, i]/(i + j − 1)
The compiled program, which used a 20 bit mantissa floating point
notation then produced the following 4 × 4 segment

16.0	−120.0	240.0002	−140.0
−120.0	1200.0	−2700.0	1680.0019
240.0	−2700.0	6480.0	−4200.0
−140.0	1680.0019	−4200.0	2800.0039

REMARKS ON AND CERTIFICATION OF AL-
  GORTHM 50
INVERSE OF A FINITE SEGMENT OF THE
  HILBERT MATRIX [J. R. Herndon, *Comm. ACM*,
  Apr. 1961]
P. NAUR
Regnecentralen, Copenhagen, Denmark

In addition to inserting the corrections indicated by B. Randell
[*Comm. ACM 5* (Jan. 1962), 50], we have modified and simplified
the algorithm as follows:

1. The types of n, i, j and k have been changed to **integer**.
This saves roundoff operations in subscripts.

2. Explicit multiplications have been replaced by squaring.
This saves code length and execution time, at least in a compiler
like ours for the GIER.

3. Repeated references to subscripted variables have been
eliminated, partly with the aid of an additional simple working
variable, w, partly by using simultaneous assignments.

4. An unnecessary **begin end** pair has been removed.

In total, these changes, in addition to reducing the code length,
have increased speed by a factor of 1.6.

The resulting algorithm is as follows:

```
procedure INVHILBERT(n,S);
value n;  integer n;  real array S;
comment  ALG. 50: This procedure computes the elements of
  the inverse of an n × n finite segment of the Hilbert matrix and
  stores them in the array S. The Hilbert matrix has the elements
  HILBERT[i,j] = 1/(i+j−1). The segments of this are known
  to be increasingly ill-conditioned with increasing size;
begin integer i, j, k;  real w;
w := S[1,1] := n↑2;
for i := 2 step 1 until n do w := S[i,i] := w × ((n+i−1) ×
    (n−i+1)/(i−1)↑2)↑2;
for i := 1 step 1 until n−1 do for j := i+1 step 1 until n do
  begin
  k := j−1;
  S[i,j] := −S[i,k] × (n+k) × (n−k)/k↑2
  end;
for i := 2 step 1 until n do for j := 1 step 1 until i do
  S[i,j] := S[j,i] := S[j,i]/(i+j−1)
end INVHILBERT;
```

Both the original version and the above improved one have
been run successfully on the GIER ALGOL system (30-bit mantissa). The test program included:

(a) Output of the 4 × 4 matrix, to be compared with the results
of Randell [loc. cit.]. Results:

16.000000	−120.000000	240.000000	−140.000000
−120.000000	1200.000000	−2700.000000	1680.000000
240.000000	−2700.000000	6480.000000	−4200.000000
−140.000000	1680.000000	−4200.000000	2799.999977

(b) For $n := 1$ **step 1 until** 15, the inverse of the segment was
calculated by INVHILBERT and multiplied by the segment of
the Hilbert matrix, and the result was compared with the unit
matrix. The maximum error was divided by the largest element of
the inverse to form a relative error. Some of the results, which
were entirely satisfactory throughout, are given below:

Order	Element of max error	abs (max error)	Largest element of INVHILBERT	Relative error
3	$S[3,3]$	$2.38_{10}-7$	$1.92_{10}2$	$1.24_{10}-9$
6	$S[2,4]$	$4.39_{10}-3$	$4.41_{10}6$	$9.96_{10}-10$
9	$S[2,8]$	$1.24_{10}2$	$1.22_{10}11$	$1.01_{10}-9$
12	$S[5,9]$	$1.54_{10}6$	$3.66_{10}15$	$4.21_{10}-10$
15	$S[1,12]$	$1.06_{10}11$	$1.15_{10}20$	$9.22_{10}-10$

(c) The time for a call of the revised INVHILBERT was found as follows:

n	
5	0.2 seconds
10	0.6 "
15	1.3 "

ALGORITHM 51

ADJUST INVERSE OF A MATRIX WHEN AN
ELEMENT IS PERTURBED

JOHN R. HERNDON

Stanford Research Institute, Menlo Park, California

```
procedure  ADJUST (n, d, i, j, A, B);  value i, j, n, d;
             integer i, j, n;  real d;  real array A, B;
comment   If the n × n matrix A=M⁻¹ and a change, d, is made
  in the i, j-th element of M this procedure will calculate the
  corrected matrix for M⁻¹ by adjusting matrix A. The adjusted
  matrix is stored in B;
begin      integer r, s;
           real t;
           t := d/(A[j, i] × d+1);
           for r := 1 step 1 until n do
             begin for s := 1 step 1 until n do
               B[r, s] = A[r, s] − t × A[r, i] × Aj, s] end
end ADJUST
```

CERTIFICATION OF ALGORITHM 51

ADJUST INVERSE OF A MATRIX WHEN AN
ELEMENT IS PERTURBED [John R. Herndon,
Comm. ACM 4 (Apr. 1961)]

RICHARD GEORGE*

Argonne National Laboratory, Argonne, Ill.

This procedure was programmed in FORTRAN and reduced to
machine code mechanically. It was run on the Argonne-built
computing machine, GEORGE. A floating-point routine was used
which allows maximum accuracy to 31 bits.

The procedure was tested for matrices with n ranging from
2 to 10. For each value of n, there were 20 successive trials; each
trial consisted of a random perturbation of a randomly selected
element of the matrix M, followed by a use of ADJUST, followed
by the matrix multiplication N := B·M. For each trial, the
adjustment was evaluated by computing

$$\text{sum} := \left\{ \sum_{i=1}^{n} \sum_{j=1}^{n} N[i, j] \right\} - n.$$

For random perturbations between −1.0 and +1.0, the value
of sum never exceeded $2.0_{10}-8$.

There are two typographical errors present:

```
B[r,s]=A[r,s]−t×A[r,i]×Aj,s] end
```

should be

```
B[r,s] := A[r,s]−t×A[r,i]×A[j,s] end
```

---

* Work supported by the U. S. Atomic Energy Commission.

ALGORITHM 52
A SET OF TEST MATRICES
JOHN R. HERNDON
Stanford Research Institute, Menlo Park, California

```
procedure  TESTMATRIX (n,A);  value n;  integer n;
           real array A;
comment  This procedure places in A an n × n matrix whose
  inverse and eigenvalues are known. The n-th row and the n-th
  column of the inverse are the set: 1, 2, 3, ... , n. The matrix
  formed by deleting the n-th row and the n-th column of the
  inverse is the identity matrix of order n−1;
begin      integer i, j;
           real t, c, d, f;
           c := t × (t+1) × (t+t−5)/6;
           d := 1/c;
           A[n, n] := −d;
           for i := 1 step 1 until n−1 do
             begin f := i;
               A[i, n] := d × f;
               A[n, i] := A[i, n];
               A[i, i] := d × (c−f × f);
               for j := 1 step 1 until i−1 do
                 begin t := j;
                   A[i, j] := −d × f × t;
                   A[j, i] := A[i, j]
                 end
             end
end TESTMATRIX;
```

CERTIFICATION OF ALGORITHM 52
A SET OF TEST MATRICES (J. R. Herndon, *Comm.
  ACM*, Apr. 1961)
H. E. GILBERT
University of California at San Diego, La Jolla, Calif.

The statement

    c := t × (t+1) × (t+t−5)/6;

was changed to

    c := n × (n+1) × (n+n−5)/6;

to make the inverse have the form described in the algorithm. The
algorithm was translated to FORTRAN and tested with a matrix
eigenvalue program on the CDC 1604 computer at UCSD.

The eigenvalues for the 20 × 20 test matrix are:

```
 1.    1.000000
 2.    1.000000
 :     :
19.     .01636693
20.    −.02493833
```

REMARK ON ALGORITHM 52
A SET OF TEST MATRICES (John R. Herndon, *Comm.
  ACM*, Apr. 1961)
G. H. DUBAY
University of St. Thomas, Houston, Tex.

In the assignment statement

    c := t×(t + 1)×(t + t − 5)/6;   (a)

the t is undefined. A suitable definition would be provided by
preceding (a) with t := n;

REMARKS ON AND CERTIFICATION OF
  ALGORITHM 52
A SET OF TEST MATRICES [J. R. Herndon, *Comm.
  ACM*, Apr. 1961]
P. NAUR
Regnecentralen, Copenhagen, Denmark

In addition to inserting the correction indicated by H. E.
Gilbert [*Comm. ACM* (Aug. 1961), 339] the algorithm was simplified by using the simultaneous assignment and by eliminating the
local variables t and f. The resulting algorithm is as follows:

```
procedure TESTMATRIX(n,A);
value n;  integer n;  real array A;
comment ALG. 52: This procedure places in A an n × n matrix
  whose inverse and eigenvalues are known. The nth row and the
  nth column of the inverse are the set: 1, 2, 3, ..., n. The matrix
  formed by deleting the nth row and the nth column of the in-
  verse is the identity matrix of order n−1;
begin integer i,j;  real c,d;
c := n × (n+1) × (n+n−5)/6;
d := 1/c;
A[n,n] := −d;
for i := 1 step 1 until n−1 do
  begin
  A[i,n] := A[n,i] := d × i;
  A[i,i] := d × (c−i↑2);
  for j := 1 step 1 until i−1 do A[i,j] := A[j,i] := −d × i × j
  end
end TESTMATRIX;
```

This version of the algorithm was successfully run in the GIER
ALGOL system together with the inversion procedures INVER-
SION II and gjr (see Certification of Algorithm 120 below). From
the figures produced by INVERSION II it looks as if the determinant of these matrices is given by $6/(n(n+1)(5−2n))$, which is
also the value of the element $A[n,n]$. For $n > 3$ the absolutely
greatest element is $A[1,1] = 1 + A[n,n]$.

CERTIFICATION OF ALGORITHM 52
A SET OF TEST MATRICES [J. R. Herndon, *Comm.
  ACM*, Apr. 1961]
J. S. HILLMORE
Elliott Bros. (London) Ltd., Borehamwood, Herts.,
England

The algorithm was corrected as recommended by H. E. Gilbert
in his certification [*Comm. ACM*, Aug. 1961] and then successfully
run using the Elliott ALGOL translator on the National-Elliott 803.
The matrices so generated were used to test the matrix inversion
procedure GJR given by H. R. Schwarz in his article "An Introduction to ALGOL" [Comm. ACM, Feb. 1962].

ADDITIONAL REMARKS ON ALGORITHM 52
A SET OF TEST MATRICES [J. R. Herndon, *Comm. ACM* (Apr. 1961), 180]

P. NAUR

Regnecentralen, Copenhagen, Denmark

From an inspection of the results of eigenvalue-finding algorithms I conclude that all but two of the eigenvalues of TEST-MATRIX are unity while the two remaining are given by the expressions $6/(p \times (n+1))$ and $p/(n \times (5-2 \times n))$ where

$$p = 3 + \text{sqrt} ((4 \times n - 3) \times (n-1) \times 3/(n+1)).$$

These expressions have been used for the determination of absolute errors of the eigenvalues calculated by JACOBI, Algorithm 85, and Householder Tridiagonalisation, etc. as reported below. They were also used to calculate the following table (using GIER ALGOL, with 29 significant bits):

n	Determinant	Eigenvalues Differing from unity	
3	− .500 000 00	.224 744 87	−2.224 744 9
4	− .100 000 00	.153 112 89	− .653 112 89
5	− .040 000 000	.113 238 08	− .353 238 08
6	− .020 408 163	.088 290 570	− .231 147 71
7	− .011 904 762	.071 428 571	− .166 666 67
8	− .007 575 757 6	.059 386 081	− .127 567 90
9	− .005 128 205 2	.050 422 549	− .101 704 60
10	− .003 636 363 6	.043 532 383	− .083 532 383
11	− .002 673 796 8	.038 097 478	− .070 183 039
12	− .002 024 291 5	.033 718 770	− .060 034 559
13	− .001 569 858 7	.030 128 103	− .052 106 125
14	− .001 242 236 0	.027 139 206	− .045 772 747
15	− .001 000 000 0	.024 619 013	− .040 619 013
16	− .000 816 993 47	.022 470 157	− .036 359 046
17	− .000 676 132 52	.020 619 902	− .032 790 288
18	− .000 565 930 96	.019 012 916	− .029 765 605
19	− .000 478 468 90	.017 606 429	− .027 175 807
20	− .000 408 163 27	.016 366 903	− .024 938 332

The figures for $n = 20$ agree very well with the results quoted by H. E. Gilbert in his certification [*Comm. ACM 4* (Aug. 1961), 339].

ALGORITHM 53
Nth ROOTS OF A COMPLEX NUMBER
John R. Herndon
Stanford Research Institute, Menlo Park, California

```
procedure  NTHROOT (n, r, u, REAL, UNREAL);  value
           n, r, u;  integer n;
           real r, u;  real array REAL, UNREAL;
comment  This procedure computes the n roots of the equation
  xⁿ = r+ui. The real parts of the roots are stored in the vector
  REAL [  ]. The imaginary parts are stored in the corresponding
  locations in the vector UNREAL [  ];
begin      integer n1, n2;  real en, th, s, th 1;
           REAL [n] := 0;
           en := 1/n;
           if u=0 then
             begin s := (abs(r)) ↑ en;
             th := 0,
             go to main end;
           if r=0 then
             begin s := (abs(u)) ↑ en;
             th := 1.5707963;
             if u < 0 then
               th := −th
             go to main end;
             s := (r × r+u × u) ↑ (en/2);
             th := arctan (u/r);
main:      if r < 0 then
             th := th + 3.1415926;
           th := en × th;
           th1 := 6.2831853 × en;
           for n2 := 1 step 1 until n do
             begin REAL [n2] := s × cos (th);
             UNREAL [n2] := s × sin (th);
             th = th+th 1 end
end NTHROOT;
```

REMARK ON ALGORITHM 53
Nth ROOTS OF A COMPLEX NUMBER (John R.
  Herndon, *Comm. ACM 4*, Apr. 1961)
C. W. Nestor, Jr.
Oak Ridge National Laboratory, Oak Ridge, Tennessee

A considerable saving of machine time for $N \geq 3$ would result
from the use of the recursion formulas for the sine and cosine in
place of an entry into a sine-cosine subroutine in the do loop
associated with the Nth roots of a complex number. That is, one
could use

$$\sin (n + 1)\theta = \sin n\theta \cos\theta + \cos n\theta \sin\theta$$
$$\cos (n + 1)\theta = \cos n\theta \cos\theta - \sin n\theta \sin\theta,$$

at the cost of some additional storage.

We have found this procedure to be very efficient in problems
dealing with Fourier analysis, as suggested by G. Goerzel in
chapter 24 of *Mathematical Methods for Digital Computers*.

ALGORITHM 54
GAMMA FUNCTION FOR RANGE 1 TO 2
JOHN R. HERNDON
Stanford Research Institute, Menlo Park, California

real procedure Q(x); **value** x; **real** x,
comment This procedure computes $\Gamma(x)$ for $1 \leq x \leq 2$. This is
a reference procedure for the more general gamma function
procedure. $\Gamma(x) = Q(x-1)$;
begin   Q := $((((((0.035868343 \times x - 0.19352782) \times x$
           $+ 0.48219939) \times x - 0.75670408) \times x$
           $+ 0.91820686) \times x - 0.89705694) \times x$
           $+ 0.98820589) \times x - 0.57719165) \times x + 1.0$
end   Q;

REMARKS ON:
ALGORITHM 34 [S14]
GAMMA FUNCTION
   [M. F. Lipp, *Comm. ACM 4* (Feb. 1961), 106]
ALGORITHM 54 [S14]
GAMMA FUNCTION FOR RANGE 1 TO 2
   [John R. Herndon, *Comm. ACM 4* (Apr. 1961), 180]
ALGORITHM 80 [S14]
RECIPROCAL GAMMA FUNCTION OF REAL
ARGUMENT
   [William Holsten, *Comm. ACM 5* (Mar. 1962), 166]
ALGORITHM 221 [S14]
GAMMA FUNCTION
   [Walter Gautschi, *Comm. ACM 7* (Mar. 1964), 143]
ALGORITHM 291 [S14]
LOGARITHM OF GAMMA FUNCTION
   [M. C. Pike and I. D. Hill, *Comm. ACM 9* (Sept. 1966),
   684]
M. C. PIKE AND I. D. HILL (Recd. 12 Jan. 1966)
Medical Research Council's Statistical Research Unit,
University College Hospital Medical School,
London, England

Algorithms 34 and 54 both use the same Hastings approximation, accurate to about 7 decimal places. Of these two, Algorithm 54 is to be preferred on grounds of speed.

Algorithm 80 has the following errors:
(1) *RGAM* should be in the parameter list of *RGR*.
(2) The lines
   if $x = 0$ **then begin** $RGR := 0$; **go to** *EXIT* **end**
and
   if $x = 1$ **then begin** $RGR := 1$; **go to** *EXIT* **end**
should each be followed either by a semicolon or preferably by an
else.
(3) The lines
   if $x = 1$ **then begin** $RGR := 1/y$; **go to** *EXIT* **end**

and
   if $x < -1$ **then begin** $y := y \times x$; **go to** *CC* **end**
should each be followed by a semicolon.
(4) The lines
   BB: **if** $x = -1$ **then begin** $RGR := 0$; **go to** *EXIT* **end**
and
   if $x > -1$ **then begin** $RGR := RGAM(x)$: **go to** *EXIT* **end**
should be separated either by **else** or by a semicolon and this
second line needs terminating with a semicolon.
(5) The declarations of **integer** i and **real array** $B[0:13]$ in *RGAM*
are in the wrong place; they should come immediately after
   begin real z;

With these modifications (and the replacement of the array B
in *RGAM* by the obvious nested multiplication) Algorithm 80 ran
successfully on the ICT Atlas computer with the ICT Atlas
ALGOL compiler and gave answers correct to 10 significant digits.

Algorithms 80, 221 and 291 all work to an accuracy of about 10
decimal places and to evaluate the gamma function it is therefore
on grounds of speed that a choice should be made between them.
Algorithms 80 and 221 take virtually the same amount of computing time, being twice as fast as 291 at $x = 1$, but this advantage
decreases steadily with increasing x so that at $x = 7$ the speeds are
about equal and then from this point on 291 is faster—taking only
about a third of the time at $x = 25$ and about a tenth of the time
at $x = 78$. These timings include taking the exponential of *loggamma*.

For many applications a ratio of gamma functions is required
(e.g. binomial coefficients, incomplete beta function ratio) and the
use of algorithm 291 allows such a ratio to be calculated for much
larger arguments without overflow difficulties.

ALGORITHM 55
COMPLETE ELLIPTIC INTEGRAL OF THE FIRST
    KIND

JOHN R. HERNDON
Stanford Research Institute, Menlo Park, California

real procedure ELLIPTIC 1(k); **value** k; **real** k;
comment This procedure computes the elliptic integral of the
    first kind K(k, $\pi/2$);
begin   **real** t;
                t := 1−k × k;
                ELLIPTIC 1 := (((0.032024666 × t +
                    0.054555509) × t
                    + 0.097932891) × t + 1.3862944)
                    − (((0.010944912 × t + 0.060118519) × t
                    + 0.12475074) × t + 0.5) × log (t)
end   ELLIPTIC 1;

CERTIFICATION OF ALGORITHM 55
COMPLETE ELLIPTIC INTEGRAL OF THE FIRST
    KIND [John R. Herndon, *Comm. ACM*, Apr. 1961]
    and
CERTIFICATION OF ALGORITHM 149
COMPLETE ELLIPTIC INTEGRAL [J. N. Merner,
    Comm. ACM, Dec. 1962]
HENRY C. THACHER, JR.*
Reactor Eng. Div., Argonne National Laboratory,
Argonne, Ill.
    * Work supported by the U.S. Atomic Energy Commission.

The bodies of Algorithm 55 and of the second procedure of
Algorithm 149 were tested on the LGP-30 computer using SCALP,
the Dartmouth "LOAD-AND-GO" translator for a substantial subset of ALGOL 60. The floating-point arithmetic for this translator
carries 7+ significant digits.

In addition to modifications required because of the limitations
of the SCALP subset, the following need correction:

In Algorithm 55:
    1. The constant 0.054555509 should be 0.054544409.
    2. The function *log* should be *ln*.

In procedure ELIP 2 of Algorithm 149, the statement $a := c$
should be $a := C$.

The parameters of Algorithm 149 are related to the complete
elliptic integral of the first kind by: $K = a \times ELIP(a, b)$ where
the parameter $m = k^2 = 1 - b/a$.

The maximum approximation error in Algorithm 55 is given by
Hastings as about $0.6_{10}-6$. In addition there is the possibility of
serious cancellation error in forming the complementary parameter $t = 1 - k \times k$. For k near 1, errors as great as 4 significant
digits were sustained. In these regions, the complementary
parameter itself is a far more satisfactory parameter.

The accuracy obtainable with Algorithm 149 is limited only by
the arithmetic accuracy and the amount of effort which it is
desired to expend. Six-figure accuracy was obtained with 5 applications of the arithmetic-geometric mean for $a = 1000$, $b = 2$,
and with one application for $a = 500$, $b = 500$.

Neither algorithm is satisfactory for $k = 1$. The behavior for
Algorithm 55 will be governed by the error exit from the logarithm
procedure. Under these circumstances, Algorithm 149 goes into an
endless loop. Algorithm 149 may also go into an endless loop of the
terminating constant ($_{10}-8$ in the published algorithm) is too
small for the arithmetic being used. For the SCALP arithmetic it
was found necessary to increase this tolerance to $5.0_{10}-7$. The
resulting values of the elliptic integrals were, however, accurate
to within 2 in the 7th significant digit (6th decimal).

The relative efficiency of the two algorithms will depend
strongly on the efficiency of the square-root and logarithm subroutines. With most systems, Algorithm 55 will provide sufficient
accuracy, and will be more efficient. If a square-root operation or
a highly efficient square-root subroutine is available, Algorithm
149 may well be the better method.

ALGORITHM 56
COMPLETE ELLIPTIC INTEGRAL OF THE
SECOND KIND
JOHN R. HERNDON
Stanford Research Institute, Menlo Park, California

```
real procedure  ELLIPTIC 2(k);  value k;  real k:
comment  This procedure computes the elliptic integral of the ·
    second kind E(k, π/2);
begin        real t;
             t := 1 − k × k;
             ELLIPTIC 2 := ((((0.040905094 × t +
                 0.085099193) × t
                 + 0.44479204) × t + 1.0 − (((0.01382999 × t
                 + 0.08150224) × t + 0.24969795) × t) × log (t)
end          ELLIPTIC 2;
```

CERTIFICATION OF ALGORITHM 56 [S21]
COMPLETE ELLIPTIC INTEGRAL OF THE
SECOND KIND
[J. R. Herndon, *Comm. ACM 4*, (Apr. 1961), 180]
GERHARD MEIDELL LARSSEN (Recd. 9 Aug. 1965)
Institut für Statik und Dynamik der Luft- und Raum-
    fahrtkonstruktionen mit Rechengruppe der Luftfahrt,
    Technische Hochschule, Stuttgart, Germany

Algorithm 56 was run on a UNIVAC 1107 using the UNIVAC 1107
ALGOL 60 compiler (dated January 25, 1965). The single-precision
floating-point arithmetic of this translator carries eight significant
digits.

Two syntactical errors were removed from the algorithm:
1. The line

$$ELLIPTIC\ 2 := ((((0.040905094 \times t +$$

was changed to

$$ELLIPTIC\ 2 := (((0.040905094 \times t +$$

2. The function *log* was changed to *ln*.
In addition, the statement

$$t := 1 - k \times k$$

was removed from the algorithm and the complementary parameter itself used as input to the procedure:

real procedure *ELLIPTIC* 2 (*t*); **value** *t*; **real** *t*;

to avoid cancellation error for values of k near 1. [While the use
of t as input parameter is good computationally, the name of the
procedure is then slightly misleading.—J.G.H.]
Several values of the complete elliptic integral of the second
kind were computed for $1 \geq t > 0$. The maximum error was found
to be about $7_{10}-7$, compared with A. M. Legendre, Tafeln der
Elliptischen Normalintegrale, Stuttgart, 1931. For $t = 0$ an error
exit from the *ln* routine takes place.

ALGORITHM 57
BER OR BEI FUNCTION
JOHN R. HERNDON
Stanford Research Institute, Menlo Park, California

```
real procedure  BERBEI (r, z);  value r, z;  real r, z;
comment  This procedure computes ber(z) if r is set equal to
   zero. bei(z) is produced if r equals 1.0;
begin
                real s, k, c, f, t;
                if r = 0 then
                  s := 1
                else
                  s := (z × z)/4;
                k := s;
                f := z × z;
                f := f × f;
                for c := 2 step 2 until 100 do
                  begin
                    if s = s + k then
                      go to gate;
                    t := (c+r) × (c+r−1);
                    k := −0.0625 × k × f/(t × t);
                    s := s+k end;
gate:  BERBEI := s
end BERBEI;
```

CERTIFICATION OF ALGORITHM 57
BER OR BEI FUNCTION [J. R. Herndon, *Comm. ACM*
   4 (Apr. 1961)]
A. P. RELPH
The English Electric Co. Whetstone, England

Algorithm 57 was translated using the DEUCE ALGOL compiler.
No corrections were required, and the results were satisfactory.

CERTIFICATION OF ALGORITHM 57
BER OR BEI FUNCTION [John R. Herndon, *Comm.
   ACM*, Apr. 1961]
HENRY C. THACHER, JR.*
Reactor Engineering Div., Argonne National Lab.,
   Argonne, Ill.

* Work supported by the U. S. Atomic Energy Commission.

The body of Algorithm 57 was tested on the LGP-30 using the
ALGOL 60 translator developed by the Dartmouth College Computer Center. No syntactical errors were found. For $z = 0.1(0.1)1.0$,
with a 7+ significant decimal arithmetic routine, the program
gave results with errors less than 5 (and for $z = 1(1)5$ less than 12)
in the seventh digit. For large values of z, serious cancellation
errors may occur. For example, for $z = 20$, more than 2 decimals
of significance can be lost in this way.

ALGORITHM 58
MATRIX INVERSION
DONALD COHEN
Burroughs Corporation, Pasadena, Calif.

```
procedure invert (n) array: (a);
comment   matrix inversion by Gauss-Jordan elimination;
    value n;
    array a;   integer n;
begin
    array b, c [1:n];   integer i, j, k, ℓ, p;
    integer array z [1:n];
        for j := 1 step 1 until n do z[j] := j;
        for i := 1 step 1 until n do begin
        k := i;   y := a[i, i];   ℓ := i − 1;   p := i + 1;
        for j := p step 1 until n do begin
        w := a[i, j];   if abs(w) > abs(y) then begin
        k := j;   y := w end end;
        for j := 1 step 1 until n do begin
        c[j] := a[j, k];   a[j, k] := a[j, i];
        a[j, i] := −c[j]/y;   b[j] := a[i, j] := a[i, j]/y end   ;
        a[i, i] := 1/y;   j := z[i];   z[i] := z[k];   z[k] := j   ;
        for k := 1 step 1 until ℓ, p step 1 until n do
        for j := 1 step 1 until ℓ, p step 1 until n do
        a[k, j] := a[k, i] − b[j] × c[k] end;   ℓ := 0   ;
 back:   ℓ := ℓ + 1;   k := z[ℓ];   if ℓ ≤ n then begin
        for j := ℓ while k ≠ j do begin
        for i := 1 step 1 until n do begin
        w := a[j, i];   a[j, i] := a[k, i];   a[k, i] := w end   ;
        go to back end
        end invert.
```

CERTIFICATION OF ALGORITHM 58
MATRIX INVERSION (Donald Cohen, *Comm. ACM 4,*
    May 1961)
RICHARD A. CONGER
Yalem Computer Center, St. Louis University, St.
Louis, Mo.

Invert was hand-coded in FORTRAN for the IBM 1620. The
following corrections were found necessary:
The statement $a_{k,j} := a_{k,i} − b_j × c_k$ *should be*

$$a_{k,j} := a_{k,j} − b_j × c_k$$

The statement   **go to** back   *should be changed to*

$$i := z_k ;   z_k := z_j ;   z_j := i;   \textbf{go to } back$$

After these corrections were made, the program was checked by
inverting a 6 × 6 matrix and then inverting the result. The second
result was equal to the original matrix within round-off.

CERTIFICATION OF ALGORITHM 58
MATRIX INVERSION [Donald Cohen, *Comm. ACM,*
    May, 1961]
RICHARD GEORGE*
Particle Accelerator Div., Argonne National Lab.,
Argonne, Ill.

   * Work supported by the U. S. Atomic Energy Commission.

This procedure was programmed in FORTRAN and reduced to
machine code mechanically. It was run on the Argonne-built computing machine, GEORGE. A floating-point routine was used which
allows maximum accuracy to 31 bits.
   There are a number of errors of various types:
(1) There are eight **begin**'s and only seven **end**'s.
(2) The line

$$a[k, j] := a[k, i] − b[j] × c[k]   \textbf{end};$$

   should be

$$a[k, j] := a[k, j] − b[j] × c[k]   \textbf{end};$$

(3) The permutation of rows of the inverted matrix and permutation of elements of the integer array z must be carried out simultaneously. This algorithm fails to do this, and consequently the
matrix at the time of exit from the procedure is left in a permuted
condition.
(4) The algorithm permits the statement

$$k := z[l];$$

to be executed even though the declarations place an upper limit
of n on the integer array z, and the test for $l \leq n$ has not yet been
made. Obviously, Mr. Cohen's compiling system would allow an
out-of-bounds array look-up. One could easily incorporate into an
ALGOL compiler a guard against such illicit array references, and
therefore the published algorithm might be considered machine
dependent.
(5) This algorithm requires $3n^2$ divisions, most of which are unnecessary. By inserting the statement

$$y := 1.0/y;$$

at the proper place, one may accomplish the obvious economy
of reducing this to only n divisions plus $2n^2$ multiplications.
(6) If a matrix should be singular (or nearly so), some pivot
element will be zero (or very small), and a test should be made,
with provision for a jump to *ALARM*, a non-local label.
(7) The identifiers w and y should be declared within this procedure, to avoid trouble.
(8) This algorithm omits calculation of the determinant of the
matrix. This could be computed with very little extra effort.
   The revised algorithm was then tested on the LGP-30 computer, using ALGOL-30, a small subset of ALGOL. Within the restrictions of this subset, the program worked satisfactorily on test
matrices.

REMARK ON ALGORITHM 58
MATRIX INVERSION [Donald Cohen, *Comm. ACM*,
  May, 1961]

GEORGE STRUBLE
University of Oregon, Eugene, Oregon

For the last seven lines, beginning with $a[k, j] := a[k, i]$, substitute:

```
a[k, j] := a[k, j] − b[j] × c[k] end;
l := 0;
back:   l := l+1;
again:  k := z[l];
if k ≠ l then
  begin for i := 1 step 1 until n do
    begin w := a[l, i];
          a[l, i] := a[k, i];
          a[k, i] := w end;
    z[l] := z[k];
    z[k] := k;
    go to again end;
  else if l ≠ n go to back
end invert
```

REMARK ON ALGORITHM 58
MATRIX INVERSION [D. Cohen, *Comm. ACM*,
  May 61]

PETER G. BEHRENZ
Matematikmaskinnmänden, Box 6131, Stockholm 6,
  Sweden

invert was run on FACIT EDB using FACIT-ALGOL 1. Some changes in the procedure had to be made:

1. y and w had to be declared in the procedure-body as **real** y, w;

2. The last part of the procedure starting with $l := 0$; which should interchange the matrix rows did not work correctly, even with the corrections proposed by R. A. Conger [*Comm. ACM*, June 62]. We propose the following code:

```
for l := 1 step 1 until n do begin
k := z[l];   for j := l while k ≠ j do begin
for i := 1 step 1 until n do begin
w := a[j, i];  a[j, i] := a[k, i];  a[k, i] := w end;
i := z[k];  z[k] := z[j];  k := z[j] := i end end end invert
```

If the matrix a is singular, the value of the pivot element y will once be zero or very nearly zero and division by zero would occur in the course of the calculation. It would therefore be advantageous to introduce an empirical tolerance parameter *epsilon* into the procedure.

To calculate the determinant of the matrix a it is only necessary to put three more statements into the code. With these augmentations *invert* should read:

```
procedure invert (n, a, epsilon, determinant);
value n, epsilon;   real epsilon, determinant;
array a;   integer n;
begin real y, w;   integer i, j, k, l, p;
array b, c[1:n];   integer array z[1:n];
determinant := 1;
```

followed by the same code as before until:

```
y := w end end;
determinant := y × determinant;
if k ≠ i then determinant := −determinant;
if abs (y) < epsilon then go to singular;
```

followed by the same code as before with the changes mentioned in the certification by R. A. Conger [*Comm. ACM*, June 62] and

the changes given above. *singular* should be a nonlocal label in the main program.

ALGORITHM 59
ZEROS OF A REAL POLYNOMIAL BY RESULTANT
  PROCEDURE
E. H. Bareiss and M. A. Fisherkeller
Argonne National Laboratory, Argonne, Ill.

```
procedure RES (n, c, alpha, mu, re, im, rt, gc)  ;  value n,
         c, alpha  ;  integer n, alpha  ;  integer array
         mu  ;  array c, re, im, rt, gc  ;
comment  RES finds simultaneously all zeros of a polynomial of
    degree n with real coefficients, cj (j = 0, ...  n), where cn
    is the constant term. The real part, rei , and imaginary part,
    imi , of each zero, with corresponding multiplicity, mui , and
    remainder term, rti , (i = 1, ..., n), are found and a poly-
    nomial with coefficients gcj (j = 0, ..., n), is generated
    from these zeros. Alpha provides an option for local or nonlocal
    selection of M, the number of root-squaring iterations, and
    delta and epsilon, acceptance criteria. If alpha = 1, these
    parameters are assigned locally. If alpha = 2, M, delta and
    epsilon are set equal to the global parameters Mp, deltap,
    and epsilonp, respectively. In cases where zeros may be found
    more than once, the superfluous ones are eliminated by fac-
    torization. The method has been described by E. H. Bareiss
    (J. ACM 7, Oct. 1960, pp. 346-386).  ;
    begin integer M  ;  real delta, epsilon  ;  switch U :=
    U1, U2  ;
    go to U [alpha];
U1:          M := 10  ;  delta := 0.2  ;  epsilon := 10⁻⁸  ;
             go to START  ;
U2:          M := Mp  ;  delta := deltap  ;  epsilon :=
             epsilonp  ;
START:       begin integer CT, nu, nuc, beta, m, j, jc, k,
             i, p  ;  Boolean ROOT  ;
             real X, Y, GX, rp  ;  array a, ac [0:n, 0:M],
             R, Rc, t [0:n],
                 s [-1:n], ag [-2:n], rh, q, G, F [1:2×n]  ;
             switch S := S1, S2  ;  switch T := T1, T2  ;
             switch V := V1, V2  ;
             real procedure min (u,v)  ;  real u,v  ;
                 min := if u ≤ v then u else v  ;
             real procedure SYND (W, Q, I, T)  ;
             integer I  ;  real W, Q  ;
                 array T  ;
SYNTHETIC    begin s [-1] := 0  ;  s [0] := T [0]  ;  for
DIV:         m := 1 step 1 until I do
                 s [m] := T [m] - W*s [m - 1] - Q×s
             [m - 2]  ;
             if Q = 0 then SYND := abs (s[I]) else
                 SYND := abs (W/2×s [I - 1] + s[I])
             end SYND  ;
             CT := beta := 1  ;  for j := 0 step 1 until
             n do a [j,0] := c[j]  ;
SQUARING     begin integer el  ;  real h  ;  for m :=
OPERATION:   1 step 1 until M do
             begin for j := 1 step 1 until n do
             begin h := 0  ;  for el := 1 step 1 until
             min (n - j, j) do
                 h :=   + (-1) ↑ el × a [j - el,m - 1] × a
             (j + el     - 1]  ;
             a [j,m] := (-1) ↑ j × (a [j, m - 1] ↑
             2 + 2×h) end end end  ;
             for j := 0 step 1 until n do R [j] := (-1) ↑
             j×a [j, M - 1] ↑ 2/a [j,M]  ;
                 j := 0  ;  nu := 1  ;
RD:          if (1 - delta ≤ R [j]) ∧ (R [j] ≤ 1 + delta)
             then
             begin rp := (a [j,M]/a [j - nu, M]) ↑ (1/(2 ↑
             M×nu))  ;
             go to T [beta] end  ;
             nu := nu + 1  ;
1:
2:           j := j + 1  ;  if j = n then go to S [beta]
             else go to RD  ;
3:           nu := 1  ;  go to 2  ;
T1:          rh [CT] := rp  ;  X := rp + epsilon × rp  ;
             Y := X + epsilon × rp  ;
             for k := 0 step 1 until n do t [k] := abs (c[k])  ;
             F [CT] := SYND (Y,0,n,t) - SYND
             (X,0,n,t)  ;
             G [CT] := SYND (rh [CT],0,n,c)  ;  if
             F [CT] > G [CT] then
             begin ROOT := true  ;  q [CT] := 0  ;
             CT := CT + 1  ;  F [CT] := F[CT - 1] end  ;
             rh [CT] := -rp  ;  G [CT] := SYND (rh
             [CT],0,n,c)  ;
             if F [CT] > G [CT] then begin ROOT :=
             true  ;  q [CT] := 0  ;  CT := CT + 1  ;
             F [CT] := F [CT - 1] end  ;  if nu = 1 then
             go to 2  ;
             q [CT] := rp ↑ 2  ;  nuc := nu  ;  jc := j  ;
             for j := 0 step 1 until n do
             begin Rc [j] := R [j]  ;  ac [j,M] := a [j,M]
             end  ;
RESULTANT:   begin real h  ;  array b [-1:n + 1,
             -1:n + 1], A [1:n],
                 r [0:n, 0:n], CB [-1:n + 1]  ;
             b [-1,0] := CB [-1] := CB [n + 1] := 0  ;
             for j := 0 step 1 until n do
             CB [j] := c[j]  ;  b [0,0] := 1  ;  for k :=
             1 step 1 until n do
             begin b [k,-1] := 0  ;  for j := 0 step 1
             until k do
                 b [k + 1,j] := b [k,j - 1] - q [CT] × b
             [k - 1,j]  ;
                 b [k + 1, k + 1] := h := 0  ;  for j :=
             n - k step -1 until 0 do
                 h := h + (CB [j]×CB [k + j] - CB [j - 1]
             ×CB[k +j + 1]) × q [CT] ↑ (n - k - j)  ;
                 A [k] := (-1 ↑ k×h  ;  for j := 0 step
             1 until k - 1 do
             begin r [0,j] := 0  ;  r [k,j] := r [k - 1,j] +
             A [k] × b [k,j] end  ,
                 r [k,k] := A[k] end  ;  beta := 2  ;  for
             j := 0 step 1 until n do
                 a [j,0] := r [n,j] end  ;  go to SQUAR-
             ING OPERATION  ;
T2:          if (rp/2) ↑ 2 ≥ q [CT] then go to 3  ;  rh
             [CT] := rp  ;
             G [CT] := SYND (rh [CT], q [CT], n,c)  ;
```

```
        if F [CT] > G [CT] then
        begin CT := CT + 1 ; F [CT] := F
        [CT − 1] ; q [CT] := q [CT − 1] end ;
        rh [CT] := −rp ; G [CT] := SYND [rh [CT],
        q [CT], n,c) ;
        if F [CT] > G [CT] then begin CT := CT
        + 1 ; F [CT] := F [CT − 1] ;
        q [CT] := q [CT − 1] end ; go to 3 ;
S2:     for j := 0 step 1 until n do begin a [j,M] :=
        ac [j,M] ;
        R [j] := Rc [j] end ; j := jc ; beta := 1 ;
        if ROOT then go to 3 else
            nu := nuc ; go to 1 ;
S1:     ag [−2] := ag [−1] := 0 ; ag [0] := 1 ;
        for j := 1 step 1 until n do
        ag [j] := 0 ; k := 1 ; i := n ; m := 1 ;
        for j := 0 step 1 until n do
            t [j] := c [j] ;
MULT:   mu [m] := 0 ; p := if q [k] = 0 then 1
        else 2 ;
IT:     GX := SYND (rh [k], q [k],i,t) ; if F [k]
        > GX then
        begin for j := 1 step 1 until n do
            ag [j] := ag [j] − rh [k] × ag [j − 1] + q
            [k] × ag [j − 2] ;
            mu [m] := mu [m] + p ; i := i − p ;
            for j := 0 step 1 until i do
            t [j] := s [j] ; go to IT end else if
            mu [m] ≠ 0 then begin
            rt [m] := G [k] ; go to V [p] end else
            go to D ;
V1:     re [m] := rh [k] ; im [m] := 0 ; go to E ;
V2:     re [m] := rh [k]/2 ; im [m] := sqrt (q [k] −
        re [m] ↑ 2) ;
E:      m := m + 1 ;
D:      k := k + 1 ; if k ≦ CT ∧ m ≦ n then go to
        MULT ;
        for j := 0 step 1 until n do gc [j] := ag [j] end
        end RES
```

ALGORITHM 60
ROMBERG INTEGRATION
F. L. BAUER
Gutenberg University, Mainz, Germany

```
real procedure rombergintegr (fct, ℓgr, rgr, ord)  ;
  value ℓgr, rgr, ord  ;
  real ℓgr, rgr;  integer ord  ;  real procedure fct  ;
  comment rombergintegr is the value of the integral of the
    function fct between the limits ℓgr and rgr, calculated by the
    algorithm of Romberg with an error term of the order
    2×ord+2, ord≧0 Computation time will roughly be doubled
    when ord is increased by 1;
  begin
    real array t[1 : ord+1];
    real  ℓ,  u,  m  ;
    integer f, h, j, n  ;
    ℓ  :=  rgr−ℓgr  ;
    t[1]  :=  (fct(ℓgr)+fct(rgr))/2  ;
    n := 1  ;
    for h := 1 step 1 until ord do
      begin u  :=  0  ;
        m  :=  ℓ/(2×n)  ;
        for j := 1 step 2 until 2×n−1 do
          u := u+fct(ℓgr+j×m)  ;
        t[h+1]  :=  (u/n+t[h])/2  ;
        f  :=  1  ;
        for j := h step − 1 until 1 do
          begin f  :=  4×f  ;
            t[j]  :=  t[j+1]+(t[j+1]−t[j])/(f−1)
          end  ;
        n := 2×n
      end  ;
    rombergintegr := t[1]×ℓ
  end
```

CERTIFICATION OF ALGORITHM 60
ROMBERG INTEGRATION (F. L. Bauer, *Comm. ACM*, June, 1961)
HENRY C. THACHER, JR.*
Argonne National Laboratory, Argonne, Ill.

* Work supported by the U. S. Atomic Energy Commission.

This procedure was translated to the ACT III compiler language for the Royal Precision LGP-30 computer. This system provides 7+ significant decimal digits. The program was used to integrate x^n between the limits 0.01 and 1.1, and between the limits 1.1 and 0.01. The results in Table I were obtained. The pole at 0 for negative n affords a test of the reliability of the method when the higher derivatives of the integrand are large. The agreement between integrations in the forward and backward directions is an indication of the effects of round-off error.

It is apparent that the procedure gives results well within the noise level for the positive powers, and that even the effect of a closely adjacent singularity for the negative powers can be overcome.

The flexibility of the algorithm would be improved by adding to the formal parameters a procedure, check, to decide if sufficient

TABLE I. INTEGRATION OF $\int_{0.01}^{1.1} x^n dx$ AND $\int_{1.1}^{0.01} x^n dx$

n	0	+12	+12	−1
True Value	1.0900000	.26555932	− .26555932	4.7004831
Order 1	1.0899997	.57076812	− .57076842	19.641113
Order 2	1.0899997	.30614608	− .30614626	10.656923
Order 5	1.0899991	.26555693	− .26555818	4.9017590
Order 10				4.7002345

n	−1	−5	−5
True Value	−4.7004831	.25000000×10⁸	−18.166667 ×10⁸
Order 1	−19.641125	18.166655 ×10⁸	− .25000000×10⁸
Order 2	−10.656929	8.4777719 ×10⁸	−8.4777766 ×10⁸
Order 5	−4.9017805	1.0408634 ×10⁸	−1.0408640 ×10⁸
Order 10	−4.7004402	.25000715×10⁸	− .25000727×10⁸
Order 12		.24999291×10⁸	− .25001311×10⁸

accuracy had been obtained without carrying through the entire iteration. A possible form for this procedure would be:

```
procedure check (t1, t2, f, exit);
  real t1, t2;
  label exit;
  integer f;
begin if abs ((t2 − t1) × f) / t1 < tolerance ∧ f > minimum order
  then go to exit end.
```

The global variables tolerance, which is the maximum relative difference between approximations of increasing order, and the minimum acceptable order should be selected by the programmer for the exigencies of the problem. A check of this sort is clearly not as sound as an a priori estimate of the necessary order, but is frequently an acceptable expedient.

The Romberg quadrature algorithm is analyzed in the following references:

Romberg, W. Vereinfachte numerische Integration. *Det Kongelinge Norske Videnskaber Selskab Forhandlinger 28*, (1955), 30–36.

Stiefel, E., and Rutishauser, H. Remarques concernant l'integration numerique. *Comptes Rendus Acad. Scil (Paris) 252*, (1961), 1899–1900.

CERTIFICATION OF ALGORITHM 60
ROMBERG INTEGRATION (F. L. Bauer, *Comm. ACM*, June 1961)
KARL HEINZ BUCHNER
Lurgi Gesellschaft fur Mineraloltechnik m.b.H., Frankfurt, Germany

Since August 1961, the Rombert Integration has been successfully applied in FORTRAN language to various problems on an IBM 1620. Due to its elegant method and the memory saving features, the Romberg Integration has succeeded other methods in our program library, e.g., the Newton-Cotes integration of order 10.

Reference is made to Stiefel, *Numerische Mathermatik* (Teubner Verlag. Stuttgart). Stiefel discusses in his book various methods of numerical integration including the Romberg algorithm.

CERTIFICATION OF ALGORITHM 60
ROMBERG INTEGRATION (F. L. Bauer, *Comm. ACM*, June 1961)

KARL HEINZ BUCHNER

Lurgi Gesellschaft fur Mineraloltechnik m.b.H., Frankfurt, Germany

Since August 1961, the Rombert Integration has been successfully applied in FORTRAN language to various problems on an IBM 1620. Due to its elegant method and the memory saving features, the Romberg Integration has succeeded other methods in our program library, e.g., the Newton-Cotes integration of order 10.

Reference is made to Stiefel, *Numerische Mathermatik* (Teubner Verlag. Stuttgart). Stiefel discusses in his book various methods of numerical integration including the Romberg algorithm.

REMARK ON ALGORITHM 60 [D1]
ROMBERG INTEGRATION [F. L. Bauer, *Comm. ACM 4* (June 1961) 255; *5* (Mar. 1962), 168; *5* (May 1962), 281]

HENRY C. THACHER, JR.* (Recd. 20 Feb. 1964 and 23 Mar. 1964)

Argonne National Laboratory, Argonne, Ill.

* Work supported by the U. S. Atomic Energy Commission.

The Romberg integration algorithm has been used with great success by many groups [1, 2], and appears to be among the most generally reliable quadrature methods available. It is, therefore, worth pointing out that it is not entirely foolproof, and that a significant class of integrands exists for which the extrapolated values are poorer estimates of the integral than the corresponding trapezoidal sums.

The validity of the Romberg procedure depends upon the possibility of expanding the error of the trapezoidal rule in powers of h^2, where h is the stepsize. One expansion of this type is the Euler-Maclaurin sum formula. An alternative expression may be obtained from the Fourier series expansion. The coefficients of h^{2r} in the Euler Maclaurin formula are proportional to the difference of the values of the $(2r+1)$-th derivative at the two ends of the range. Thus, any integral for which the odd derivatives of the integrand either vanish or are equal at the limits will not be improved by Romberg extrapolation. Among the common examples of such integrals are integrals of periodic functions over a period and integrals for which the derivatives vanish at both limits. An example of the last type is the integral approximation to the modified Hankel function [3], $e^x K_p(x) = \int_0^L e^{x(1-\cosh t)} \cosh (pt) dt$, where L is taken so large that the contribution of the integral from L to ∞ may be neglected. Several other examples are given under the heading "Exceptional cases" by Bauer, Rutishauser and Stiefele [7]. This paper is among the most extensive discussions of the Romberg method in English.

The algorithm also fails when the expansion of the error term contains other powers of h along with the even ones. Rutishauser [4] discusses estimating integrals of the form $\int_0^a f(x) \, dx = \int_0^a (\varphi(x)/\sqrt{x}) \, dx$. If such integrals are estimated by the trapezoidal rule, assigning the value 0 to $f(0)$, the error may be expressed in the form $\sum c_k h^{2k} + \sqrt{h} \sum d_k h^k$. Although the standard Romberg extrapolation fails when applied to this sequence of estimates, Rutishauser presents a modified procedure which is effective.

The extrapolation is also invalid when the integrand is discontinuous, although this exception is trivial from the computational standpoint.

It has also been pointed out [5, 6] that the Romberg procedure may amplify round-off errors. The losses, while significant, do not appear prohibitive for most applications.

REFERENCES:
1. THACHER, H. C., JR. Certification of algorithm 60. *Comm. ACM 5* (Mar. 1962), 168.
2. BUCHNER, K. H. Certification of algorithm 60. *Comm. ACM 5* (May, 1962), 281.
3. FETTIS, H. E. Algorithm 163, modified Hankel function. *Comm. ACM 6* (Apr. 1963), 161-2; *6* (Sep. 1963), 522.
4. RUTISHAUSER, H. Ausdehnung des Rombergschen Prinzips. *Numer. Math. 5* (1963), 48-54.
5. MCKEEMAN, W. M. Personal communication, Sept. 1963.
6. ENGELI, M. Personal communication, Jan. 1964.
7. BAUER, F. L., RUTISHAUSER, H., AND STIEFELE, E. New aspects in numerical quadrature. Proc. Symp. Appl. Math 15, 1963, 199–218.

ALGORITHM 61

PROCEDURES FOR RANGE ARITHMETIC

ALLAN GIBB*

University of Alberta, Calgary, Alberta, Canada

```
begin
  procedure  RANGESUM (a, b, c, d, e, f);
    real  a, b, c, d, e, f;
  comment   The term "range number" was used by P. S. Dwyer,
```
Linear Computations (Wiley, 1951). Machine procedures for range arithmetic were developed about 1958 by Ramon Moore, "Automatic Error Analysis in Digital Computation," LMSD Report 48421, 28 Jan. 1959, Lockheed Missiles and Space Division, Palo Alto, California, 59 pp. If $a \leq x \leq b$ and $c \leq y \leq d$, then RANGESUM yields an interval $[e, f]$ such that $e \leq (x + y) \leq f$. Because of machine operation (truncation or rounding) the machine sums $a + c$ and $b + d$ may not provide safe end-points of the output interval. Thus RANGESUM requires a non-local real procedure ADJUSTSUM which will compensate for the machine arithmetic. The body of ADJUSTSUM will be dependent upon the type of machine for which it is written and so is not given here. (An example, however, appears below.) It is assumed that ADJUSTSUM has as parameters real v and w, and integer i, and is accompanied by a non-local real procedure CORRECTION which gives an upper bound to the magnitude of the error involved in the machine representation of a number. The output ADJUSTSUM provides the left end-point of the output interval of RANGESUM when ADJUSTSUM is called with $i = -1$, and the right end-point when called with $i = 1$. The procedures RANGESUB, RANGEMPY, and RANGEDVD provide for the remaining fundamental operations in range arithmetic. RANGESQR gives an interval within which the square of a range number must lie. RNGSUMC, RNGSUBC, RNGMPYC and RNGDVDC provide for range arithmetic with complex range arguments, i.e. the real and imaginary parts are range numbers;

```
begin
    e := ADJUSTSUM (a, c, -1);
    f := ADJUSTSUM (b, d, 1)
end  RANGESUM;
procedure  RANGESUB (a, b, c, d, e, f);
    real  a, b, c, d, e, f;
comment   RANGESUM is a non-local procedure;
begin
    RANGESUM (a, b, -d, -c, e, f)
end  RANGESUB;
procedure  RANGEMPY (a, b, c, d, e, f);
    real  a, b, c, d, e, f;
comment   ADJUSTPROD, which appears at the end of this
procedure, is analogous to ADJUSTSUM above and is a non-
local real procedure. MAX and MIN find the maximum and
minimum of a set of real numbers and are non-local;
begin
    real  v, w;
    if  a < 0 ∧ c ≧ 0 then
1:  begin
        v := c;  c := a;  a := v;  w := d;  d := b;  b := w
    end  1;
    if  a ≧ 0  then
```

```
2: begin
      if c ≧ 0 then
3:begin
        e := a × c; f := b × d; go to 8
      end 3;
      e := b × c;
      if d ≧ 0 then
4:    begin
        f := b × d;  go to 8
      end 4;
      f := a × d;  go to 8
5:  end 2;
    if b > 0 then
6:  begin
      if d > 0 then
      begin
        e := MIN(a × d, b × c);
        f := MAX(a × c, b × d);  go to 8
      end 6;
      e := b × c;  f := a × c;  go to 8
    end 5;
    f := a × c;
    if d ≦ 0 then
7:  begin
      e := b × d;  go to 8
    end 7;
    e := a × d;
8:  e := ADJUSTPROD (e, -1);
    f := ADJUSTPROD (f, 1)
end  RANGEMPY;
procedure  RANGEDVD (a, b, c, d, e, f);
  real  a, b, c, d, e, f;
comment   If the range divisor includes zero the program
```
exists to a non-local label "zerodvsr". RANGEDVD assumes a non-local real procedure ADJUSTQUOT which is analogous (possibly identical) to ADJUSTPROD;
```
begin
    if c ≦ 0 ∧ d ≧ 0 then go to zerodvsr;
    if c < 0 then
1:  begin
      if b > 0 then
2:    begin
        e := b/d;  go to 3
      end 2;
      e := b/c;
3:    if a ≧ 0 then
4:    begin
        f := a/c;  go to 8
      end 4;
      f := a/d; go to 8
    end 1;
    if a < 0 then
5:  begin
      e := a/c;  go to 6
    end 5;
    e := a/d;
6:  if b > 0 then
7:  begin
      f := b/c;  go to 8
    end 7;
    f := b/d;
```

8: e := ADJUSTQUOT (e, −1); f := ADJUSTQUOT (f,1)
end RANGEDVD;
procedure RANGESQR (a, b, e, f);
  real a, b, e, f;
comment ADJUSTPROD is a non-local procedure;
begin
        if a < 0 **then**
1:    **begin**
        if b < 0 **then**
2:      **begin**
          e := b × b; f := a × a; **go to** 3
          end 2;
          e := 0; m := MAX (-a,b); f := m × m; **go to** 3
        end 1;
        e := a × a; f := b × b;
3:    ADJUSTPROD (e, −1);
      ADJUSTPROD (f, 1)
    end RANGESQR;
procedure RNGSUMC (aL, aR, bL, bU, cL, cR, dL, dU, eL,
eR, fL, fU);
  real aL, aR, bL, bU, cL, cR, dL, dU, eL, eR, fL, fU;
comment Rangesum is a non-local procedure;
begin
  RANGESUM (aL, aR, cL, cR, eL, eR);
  RANGESUM (bL, bU, dL, dU, fL, fU)
end RNGSUMC;
procedure RNGSUBC (aL, aR, bL, bU, cL, cR, dL, dU, eL,
eR, fL, fU);
  real aL, aR, bL, bU, cL, cR, dL, dU, eL, eR, fL, fU;
comment RNGSUMC is a non-local procedure;
begin
  RNGSUMC (aL, aR, bL, bR, −cR, −cL, −dU, −dL, eL, eR,
  fL, fU)
end RNGSUBC;
procedure RNGMPYC (aL, aR, bL, bU, cL, cR, dL, dU, eL,
eR, fL, fU);
  real aL, aR, bL, bU, cL, cR, dL, dU, eL, eR, fL, fU;
comment RANGEMPY, RANGESUB, and RANGESUM are
non-local procedures;
begin
  real L1, R1, L2, R2, L3, R3, L4, R4;
  RANGEMPY (aL, aR, cL, cR, L1, R1);
  RANGEMPY (bL, bU, dL, dU, L2, R2);
  RANGESUB (L1, R1, L2, R2, eL, eR);
  RANGEMPY (aL, aR, dL, dU, L3, R3);
  RANGEMPY (bL, bU, cL, cR, L4, R4);
  RANGESUM (L3, R3, L4, R4, fL, fU);
end RNGMPYC;
procedure RNGDVDC (aL, aR, bL, bU, cL, cR, dL, dU, eL,
eR, fL, fU);
  real aL, aR, bL, bU, cL, cR, dL, dU, eL, eR, fL, fU;
comment RNGMPYC, RANGESQR, RANGESUM, and
RANGEDVD are non-local procedures;
begin
  real L1, R1, L2, R2, L3, R3, L4, R4, L5, R5;
  RNGMPYC (aL, aR, bL, bU, cL, cR, −dU, −dL, L1, R1, L2,
  R2);
  RANGESQR (cL, cR, L3, R3);
  RANGESQR (dL, dU, L4, R4);
  RANGESUM (L3, R3, L4, R4, L5, R5);
  RANGEDVD (L1, R1, L5, R5, eL, eR);
  RANGEDVD (L2, R2, L5, R5, fL, fU)
  end RNGDVDC
end

EXAMPLE

real procedure CORRECTION (p); **real** p;
comment CORRECTION and the procedures below are intended for use with single-precision normalized floating-point
arithmetic for machines in which the mantissa of a floating-point
number is expressible to s significant figures, base b. Limitations
of the machine or requirements of the user will limit the range of
p to $b^m \leqq |p| < b^{n+1}$ for some integers m and n. Appropriate
integers must replace s, b, m and n below. Signal is a non-local
label. The procedures of the example would be included in the
same block as the range procedures above;
begin
  integer w;
  for w := m **step** 1 **until** n **do**
1: **begin**
    if (b ↑ w ≦ abs (p)) ∧ (abs (p) < b ↑ (w + 1)) **then**
2:    **begin**
      CORRECTION := b ↑ (w+1−s); **go to** exit
      end 2
    end 1;
    go to signal;
exit: **end** CORRECTION;
real procedure ADJUSTSUM (w, v, i); **integer** i;
  real w, v;
comment ADJUSTSUM exemplifies a possible procedure for use
with machines which, when operating in floating point addition,
simply shift out any lower order digits that may not be used. No
attempt is made here to examine the possibility that every digit
that is dropped is zero. CORRECTION is a non-local real procedure which gives an upper bound to the magnitude of the error
involved in the machine representation of a number;
begin
  real r, cw, cv, cr;
  r := w + v;
  if w = 0 ∨ v = 0 **then go to** 1;
  cw := CORRECTION (w);
  cv := CORRECTION (v);
  cr := CORRECTION (r);
  if cw = cv ∧ cr ≦ cw **then go to** 1;
  if sign (i × sign (w) × sign (v) × sign (r)) = −1 **then go to** 1;
  ADJUSTSUM := r + i × MAX (cw, cv, cr); **go to** exit;
1: ADJUSTSUM := r;
exit: **end** ADJUSTSUM;
real procedure ADJUSTPROD (p, i); **real** p; **integer** i;
comment ADJUSTPROD is for machines which truncate when
lower order digits are dropped. CORRECTION is a non-local real
procedure;
begin
  if p × i ≦ 0 **then**
1: **begin**
    ADJUSTPROD := p; **go to** out
    end 1;
    ADJUSTPROD := p + i × CORRECTION (p);
out: **end** ADJUSTPROD;
comment Although ordinarily rounded arithmetic is preferable
to truncated (chopped) arithmetic, for these range procedures
truncated arithmetic leads to closer bounds than rounding does.

———

* These procedures were written and tested in the Burroughs
220 version of the ALGOL language in the summer of 1960 at
Stanford University. The typing and editorial work were done
under Office of Naval Research Contract Nonr-225(37). The author
wishes to thank Professor George E. Forsythe for encouraging
this work and for assistance with the syntax of ALGOL 60.

ALGORITHM 62
A SET OF ASSOCIATE LEGENDRE POLYNOMIALS
OF THE SECOND KIND*

JOHN R. HERNDON
Stanford Research Institute, Menlo Park, California

comment This procedure places a set of values of $Q_n^m(x)$ in the array $Q[\ ]$ for values of n from 0 to nmax for a particular value of m and a value of x which is real if ri is 0 and is purely imaginary, ix, ortherwise. $R[\ ]$ will contain the set of ratios of successive values of Q. These ratios may be especially valuable when the $Q_n^m(x)$ of the smallest size is so small as to underflow the machine representation (e.g. 10^{-60} if 10^{-51} were the smallest representable number). 9.9×10^{45} is used to represent infinity. Imaginary values of x may not be negative and real values of x may not be smaller than 1.

Values of $Q_n^m(x)$ may be calculated easily by hypergeometric series if x is not too small nor (n − m) too large. $Q_n^m(x)$ can be computed from an appropriate set of values of $P_n^m(x)$ if x is near 1.0 or ix is near 0. Loss of significant digits occurs for x as small as 1.1 if n is larger than 10. Loss of significant digits is a major difficulty in using finite polynomial representations also if n is larger than m. However, QLEG has been tested in regions of x and n both large and small;

```
procedure  QLEG(m, nmax, x, ri, R, Q);  value m, nmax, x, ri;
              real m, nmax, x, ri;  real array R, Q;
begin   real t, i, n, q0, s;
        n := 20;
        if  nmax > 13 then
            n := nmax + 7;
        if  ri = 0 then
            begin  if m = 0 then
                        Q[0] := 0.5 × log((x + 1)/(x − 1))
                    else
                begin  t := −1.0/sqrt(x × x − 1);
                       q0 := 0;
                       Q[0] := t;
                       for i := 1 step 1 until m do
                         begin  s   :=  (x+x)×(i−1)×t
                            ×Q[0]+(3i−i×i−2)×q0;
                            q0 := Q[0];
                            Q[0] := s end end;
            if        x = 1 then
                      Q[0] := 9.9 ↑ 45;
            R[n + 1] := x − sqrt(x × x − 1);
            for       i := n step −1 until 1 do
                      R[i] := (i + m)/((i + i + 1) × x
                      +(m − i − 1) × R[i + 1]);
            go to the end;
        if m = 0 then
            begin if x < 0.5 then
                      Q[0] := arctan(x) − 1.5707963 else
                      Q[0] := − arctan(1/x)end else
            begin  t := 1/sqrt(x × x + 1);
               q0 := 0;
               Q[0] := t;
               for       i := 2 step 1 until m do
                  begin s := (x + x) × (i − 1) × t × Q[0]
                  +(3i + i × i − 2) × q0;
```

```
               q0 := Q[0];
               Q[0] := s end end;
        R[n + 1] := x − sqrt(x × x + 1);
        for  i := n step − 1 until 1 do
             R[i] := (i + m)/((i − m + 1) × R[i + 1]
             −(i + i − 1) × x);
        for  i := 1 step 2 until nmax do
             R[i] := − R[i];
the:  for i := 1 step 1 until nmax do
             Q[i] := Q[i − 1] × R[i]
end   QLEG;
```

---

* This procedure was developed in part under the sponsorship of the Air Force Cambridge Research Center.

REMARK ON ALGORITHM 62
A SET OF ASSOCIATE LEGENDRE POLYNOMIALS
OF THE SECOND KIND (John R. Herndon, *Comm. ACM 4* (July, 1961))

JOHN R. HERNDON
Stanford Research Institute, Menlo Park, California

In regard to Algorithm 62 in *Communications of the ACM*, two errors were found:

The 14th line of the procedure
```
    for i := 1 step 1 until m do
```
should read
```
    for i := 2 step 1 until m do
```

The 35th line
```
    +(3i − iXi − 2)Xq0
```
should read
```
    +(3i − i × i − 2) × q0
```

The procedure *QLEC* was developed from the standard recurrence formula

$$(n + m - 1)Q_{n-2}^m = (2n - 1) \cdot x \cdot Q_{n-1}^m - (n - m)Q_n^m.$$

Invert and multiply by $(n + m - 1)Q_{n-1}^m$.

$$\frac{Q_{n-1}^m}{Q_{n-2}^m} = \frac{(n + m - 1)}{(2n - 1) \cdot x - (n - m)Q_n^m/Q_{n-1}^m},$$

or

$$R_{n-1}^m = \frac{(n + m - 1)}{(2n - 1) \cdot x - (n - m)R_n^m}.$$

Analysis (and testing) shows that, for n large, this infinite continued fraction need only be carried to about eight terms for eight-digit accuracy if the final term is evaluated with the asymptotic value derived by setting

$$R_{n-1}^m = R_n^m, \quad \lim_{n \to \infty} R_n^m = x \pm \sqrt{x^2 - 1},$$

the minus sign being chosen since in general $Q_n^m < Q_{n-1}^m$. The formulas pertaining to purely imaginary parameters follow readily. The value of

$$Q_0^0(x) = \frac{1}{2} \log_e \frac{x + 1}{x - 1},$$

while

$$Q_1^0(x) = x \cdot Q_0^0(x) - 1,$$

and

$$Q_0^1(x) = \frac{-1}{\sqrt{x^2 - 1}}.$$

Other values are derived using the ratios $R_n^m(x)$ and/or the recurrence formula

$$Q_n^m = -\frac{2(m-1)x}{\sqrt{x^2-1}} Q_n^{m-1} + (n - m + 2)(n + m - 2)Q_n^{m-2}.$$

The derivation of the expression for $Q_0^0(ix)$ is not trivial and proceeds as follows:

$$i \cdot Q_0^0(ix) = \frac{1}{2}\log_e \frac{ix + 1}{ix - 1} = \frac{1}{2}\log_e\left[-\frac{x^2 - 1}{x^2 + 1} + \frac{2x}{x^2 + 1}\right]$$

$$e^{a+ib} = e^a \cdot e^{ib} = e^a \cos b + i \sin b.$$

Thus

$$\tan b = \frac{-2x}{1 - x^2}$$

and

$$Q_0^0(ix) = (\arctan x - \pi/2)i.$$

ALGORITHM 63
PARTITION
C. A. R. HOARE
Elliott Brothers Ltd., Borehamwood, Hertfordshire, Eng.

procedure partition (A,M,N,I,J); **value** M,N;
      array A; **integer** M,N,I,J;
comment I and J are output variables, and A is the array (with subscript bounds M:N) which is operated upon by this procedure. Partition takes the value X of a random element of the array A, and rearranges the values of the elements of the array in such a way that there exist integers I and J with the following properties:

$$M \leqq J < I \leqq N \text{ provided } M < N$$
$$A[R] \leqq X \text{ for } M \leqq R \leqq J$$
$$A[R] = X \text{ for } J < R < I$$
$$A[R] \geqq X \text{ for } I \leqq R \leqq N$$

The procedure uses an integer procedure random (M,N) which chooses equiprobably a random integer F between M and N, and also a procedure exchange, which exchanges the values of its two parameters;

```
begin      real X;  integer F;
           F := random (M,N);  X := A[F];
           I := M;  J := N;
up:        for I := I step 1 until N do
                   if X < A [I] then go to down;
           I := N;
down:      for J := J step −1 until M do
                   if A[J]<X then go to change;
           J := M;
change:  if I < J then begin exchange (A[I], A[J]);
                          I := I + 1; J := J − 1;
                          go to up
                     end
else       if I < F then begin exchange (A[I], A[F]);
                          I := I + 1
                     end
else       if F < J then  begin exchange (A[F], A[J]);
                          J := J − 1
                     end;
end        partition
```

CERTIFICATION OF ALGORITHMS 63, 64, 65 PARTITION, QUICKSORT, FIND [C. A. R. Hoare, *Comm. ACM*, July 1961]

J. S. HILLMORE
Elliott Bros. (London) Ltd., Borehamwood, Herts., England

The body of the procedure find was corrected to read:

```
begin integer I, J;
if M < N then begin partition (A, M, N, I, J);
               if K ≤ I then find (A, M, J, K)
               else if J ≤ K then find (A, I, N, K)
               end
end find
```

and the trio of procedures was then successfully run using the Elliott ALGOL translator on the National-Elliott 803.

The author's estimate of $\frac{1}{2}(N-M)1n(N-M)$ for the number of exchanges required to sort a random set was found to be correct. However, the number of comparisons was generally less than $2(N-M)1n(N-M)$ even without the modification mentioned below.

The efficiency of the procedure quicksort was increased by changing its body to read:

```
begin integer I, J;
if M < N−1 then begin partition (A, M, N, I, J);
                  quicksort (A, M, J);
                  quicksort (A, I, N)
                  end
else if N−M = 1 then begin if A[N] < A[M] then
                         exchange (A[M], A[N])
                  end
end quicksort
```

This alteration reduced the number of comparisons involved in sorting a set of random numbers by 4–5 percent, and the number of entries to the procedure partition by 25–30 percent.

CERTIFICATION OF ALGORITHMS 63, 64 AND 65, PARTITION, QUICKSORT, AND FIND, [*Comm. ACM*, July 1961]

B. RANDELL AND L. J. RUSSELL
The English Electric Company Ltd., Whetstone, England

Algorithms 63, 64, and 65 have been tested using the Pegasus ALGOL 60 Compiler developed at the De Havilland Aircraft Company Ltd., Hatfield, England.

No changes were necessary to Algorithms 63 and 64 (Partition and Quicksort) which worked satisfactorily. However, the comment that Quicksort will sort an array without the need for any extra storage space is incorrect, as space is needed for the organization of the sequence of recursive procedure activations, or, if implemented without using recursive procedures, for storing information which records the progress of the partitioning and sorting.

A misprint ('if' for '**if**' on the line starting '**else if** $J \leqq K$ **then** $\cdots$') was corrected in Algorithm 65 (Find), but it was found that in certain cases the sequence of recursive activations of *Find* would not terminate successfully. Since *Partition* produces as output two integers J and I such that elements of the array $A[M:N]$ which lie between $A[J]$ and $A[I]$ are in the positions that they will occupy when the sorting of the array is completed, *Find* should cease to make further recursive activations of itself if K fulfills the condition $J < K < I$.

Therefore the conditional statement in the body of *Find* was changed to read

```
if K ≤ J then find (A,M,J,K)
else if I ≤ K then find (A,I,N,K)
```

With this change the procedure worked satisfactorily.

ALGORITHM 64
QUICKSORT
C. A. R. HOARE
Elliott Brothers Ltd., Borehamwood, Hertfordshire, Eng.

procedure quicksort (A,M,N); **value** M,N;
       array A; **integer** M,N;
comment Quicksort is a very fast and convenient method of
sorting an array in the random-access store of a computer. The
entire contents of the store may be sorted, since no extra space is
required. The average number of comparisons made is 2(M−N) ln
(N−M), and the average number of exchanges is one sixth this
amount. Suitable refinements of this method will be desirable for
its implementation on any actual computer;
begin   **integer** I,J;
      if M < N **then begin** partition (A,M,N,I,J);
                 quicksort (A,M,J);
                 quicksort (A, I, N)
              end
end   quicksort

CERTIFICATION OF ALGORITHMS 63, 64, 65
PARTITION, QUICKSORT, FIND [C. A. R. Hoare,
Comm. ACM, July 1961]
J. S. HILLMORE
Elliott Bros. (London) Ltd., Borehamwood, Herts.,
England

The body of the procedure find was corrected to read:
begin integer I, J;
if $M < N$ **then begin** *partition* (A, M, N, I, J);
      if $K \leq I$ **then** *find* (A, M, J, K)
      else if $J \leq K$ **then** *find* (A, I, N, K)
      end
end find
and the trio of procedures was then successfully run using the
Elliott ALGOL translator on the National-Elliott 803.

The author's estimate of $\frac{1}{3}(N-M)1n(N-M)$ for the number of
exchanges required to sort a random set was found to be correct.
However, the number of comparisons was generally less than
$2(N-M)1n(N-M)$ even without the modification mentioned
below.

The efficiency of the procedure quicksort was increased by
changing its body to read:
begin integer I, J;
if $M < N-1$ **then begin** *partition* (A, M, N, I, J);
      quicksort (A, M, J);
      quicksort (A, I, N)
      end
else if $N-M = 1$ **then begin if** $A[N] < A[M]$ **then**
          exchange $(A[M], A[N])$
      end
end quicksort
This alteration reduced the number of comparisons involved in
sorting a set of random numbers by 4–5 percent, and the number
of entries to the procedure partition by 25–30 percent.

CERTIFICATION OF ALGORITHMS 63, 64 AND 65,
PARTITION, QUICKSORT, AND FIND, [*Comm. ACM*,
July 1961]
B. RANDELL AND L. J. RUSSELL
The English Electric Company Ltd., Whetstone, England

Algorithms 63, 64, and 65 have been tested using the Pegasus
ALGOL 60 Compiler developed at the De Havilland Aircraft Company Ltd., Hatfield, England.

No changes were necessary to Algorithms 63 and 64 (Partition
and Quicksort) which worked satisfactorily. However, the comment that Quicksort will sort an array without the need for any
extra storage space is incorrect, as space is needed for the organization of the sequence of recursive procedure activations, or, if
implemented without using recursive procedures, for storing information which records the progress of the partitioning and
sorting.

A misprint ('if' for '**if**' on the line starting '**else** if $J \leq K$ **then**
···') was corrected in Algorithm 65 (Find), but it was found that
in certain cases the sequence of recursive activations of *Find*
would not terminate successfully. Since *Partition* produces as
output two integers J and I such that elements of the array
$A[M:N]$ which lie between $A[J]$ and $A[I]$ are in the positions that
they will occupy when the sorting of the array is completed, *Find*
should cease to make further recursive activations of itself if K
fulfills the condition $J < K < I$.

Therefore the conditional statement in the body of *Find* was
changed to read

      if $K \leq J$ **then** find (A,M,J,K)
      else if $I \leq K$ **then** find (A,I,N,K)

With this change the procedure worked satisfactorily.

ALGORITHM 65
FIND
C. A. R. Hoare
Elliott Brothers Ltd., Borehamwood, Hertfordshire, Eng.

```
procedure  find (A,M,N,K);  value M,N,K;
            array A;  integer M,N,K;
```
comment Find will assign to A [K] the value which it would
have if the array A [M:N] had been sorted. The array A will be
partly sorted, and subsequent entries will be faster than the first;
```
begin        integer I,J;
             if M < N then begin partition (A, M, N, I, J);
                           if K≦I then find (A,M,I,K)
                           else if J≦K then find (A,J,N,K)
                           end
end          find
```


CERTIFICATION OF ALGORITHMS 63, 64, 65
PARTITION, QUICKSORT, FIND [C. A. R. Hoare,
   Comm. ACM, July 1961]
J. S. Hillmore
Elliott Bros. (London) Ltd., Borehamwood, Herts.,
   England

The body of the procedure find was corrected to read:
```
begin integer I, J;
if M < N then begin partition (A, M, N, I, J);
              if K ≦ I then find (A, M, J, K)
              else if J ≦ K then find (A, I, N, K)
              end
end find
```
and the trio of procedures was then successfully run using the
Elliott Algol translator on the National-Elliott 803.

The author's estimate of $\frac{1}{3}(N-M)1n(N-M)$ for the number of
exchanges required to sort a random set was found to be correct.
However, the number of comparisons was generally less than
$2(N-M)1n(N-M)$ even without the modification mentioned
below.

The efficiency of the procedure quicksort was increased by
changing its body to read:
```
begin integer I, J;
if M < N−1 then begin partition (A, M, N, I, J);
               quicksort (A, M, J);
               quicksort (A, I, N)
               end
else if N−M = 1 then begin if A[N] < A[M] then
                       exchange (A[M], A[N])
                    end
end quicksort
```
This alteration reduced the number of comparisons involved in
sorting a set of random numbers by 4–5 percent, and the number
of entries to the procedure partition by 25–30 percent.

CERTIFICATION OF ALGORITHMS 63, 64 AND 65,
PARTITION, QUICKSORT, AND FIND, [*Comm. ACM*,
July 1961]
B. Randell and L. J. Russell
The English Electric Company Ltd., Whetstone, England

Algorithms 63, 64, and 65 have been tested using the Pegasus
Algol 60 Compiler developed at the De Havilland Aircraft Company Ltd., Hatfield, England.

No changes were necessary to Algorithms 63 and 64 (Partition
and Quicksort) which worked satisfactorily. However, the comment that Quicksort will sort an array without the need for any
extra storage space is incorrect, as space is needed for the organization of the sequence of recursive procedure activations, or, if
implemented without using recursive procedures, for storing information which records the progress of the partitioning and
sorting.

A misprint ('if' for '**if**' on the line starting '**else if** $J \leq K$ **then**
···') was corrected in Algorithm 65 (Find), but it was found that
in certain cases the sequence of recursive activations of *Find*
would not terminate successfully. Since *Partition* produces as
output two integers J and I such that elements of the array
$A[M:N]$ which lie between $A[J]$ and $A[I]$ are in the positions that
they will occupy when the sorting of the array is completed, *Find*
should cease to make further recursive activations of itself if K
fulfills the condition $J < K < I$.

Therefore the conditional statement in the body of *Find* was
changed to read

```
if K ≤ J then find (A,M,J,K)
else if I ≤ K then find (A,I,N,K)
```

With this change the procedure worked satisfactorily.

ALGORITHM 66
INVRS
JOHN CAFFREY
Director of Research, Palo Alto Unified School District,
Palo Alto, California

procedure Invrs (t) size : (n); **value** n; **real array** t; **integer** n;
comment Inverts a positive definite symmetric matrix t, of order n, by a simplified variant of the square root method. Replaces the $n(n+1)/2$ diagonal and superdiagonal elements of t with elements of t^{-1}, leaving subdiagonal elements unchanged. Advantages: only n temporary storage registers are required, no identity matrix is used, no square roots are computed, only n divisions are performed, and, as n becomes large, the number of multiplications approaches $n^3/2$;
begin integer i, j, s; **real array** v[1:n−1]; **real** y, pivot;
for s := 0 **step** 1 **until** n−1 **do**
begin pivot := 1.0/t[1,1];
   begin pivot := 1.0/t[1,1];
         comment If t[1,1] $\leq$ 0, t is not positive definite;
     for i := 2 **step** 1 **until** n **do** v[i−1] := t[1, i];
     for i := 1 **step** 1 **until** n−1 **do**
       begin t[i,n] := y := −v[i] $\times$ pivot;
       for j := i **step** 1 **until** n−1 **do**
       t[i, j] := t[i + 1, j +1] + v[j] $\times$ y
       end;
     t[n,n] := −pivot
  end;
      comment At this point, elements of t^{-1} occupy the original array space but with signs reversed, and the following statements effect a simple reflection;
  for i := 1 **step** 1 **until** n **do**
  for j := i **step** 1 **until** n **do** t[i,j] := −t[i,j]
end Invrs

CERTIFICATION OF ALGORITHM 66
INVRS (J. Caffrey, Comm. ACM, July 1961)
B. RANDELL, C. G. BROYDEN.
Atomic Power Division, The English Electric Company,
Whetstone, England.

INVRS was translated using the DEUCE ALGOL Compiler, and needed the following correction.
The repeat of the line,
        begin pivot := 1.0/t[1, 1];
was deleted.
The compiled program, which used a 20 bit mantissa floating point notation, was tested using Wilson's matrix

5	7	6	5
7	10	8	7
6	8	10	9
5	7	9	10

and gave results

67.9982	−40.9991	−16.9995	9.9997
−40.9991	24.9995	9.9997	−5.9998
−16.9995	9.9997	4.9998	−2.9999
9.9997	−5.9998	−2.9999	1.9999

(The output routine completed the symmetric matrix)

INVRS will in fact invert non-positive symmetric matrices, the only restriction appearing to be that the leading minors of the matrix must be non-zero. The variable T[1, 1] takes as its successive values ratios of the $(r + 1)$th to the r th leadng minors of the matrix, and if it becomes zero the variable 'pivot' cannot be computed.

The following matrix, for which the successive values of T[1, 1] were $+2, -2, -1, -0.6, +5$ gave results correct to one unit in the fifth significant figure.

2	−3	1	−1	4
−3	2	−4	3	−2
1	−4	−3	2	4
−1	3	2	−2	−3
4	−2	4	−3	2

CERTIFICATION OF ALGORITHM 66
INVRS (J. Caffrey, *Comm. ACM*, July 1961)
JOHN CAFFREY
Palo Alto Unified School District, Palo Alto, California

INVRS was translated using the Burroughs 220 Algebraic Computer (BALCOM) at Stanford University, using 8-digit floating-point arithmetic. The misprint noted by Randell and Broyden (*Comm. ACM*, Jan. 1962, p. 50) was corrected, and the same example (Wilson's 4 $\times$ 4 matrix) was used as a test case. The resulting inverse was:

68.0000	−41.0000	−17.0000	10.0000
	25.0000	10.0000	6.0000
		5.0000	−3.0000
			2.0000

It may also be useful to note that the determinant of the matrix may be obtained as the successive product of the pivots. That is, if $t_i \ (= T(1, 1))$ is the ith pivot of a matrix of order n,
$$\text{determinant} = \prod_i^n t_i .$$
For the above *input* example,
$$\text{determinant} = 1.0$$
Randell and Broyden's observation concerning the *apparent* limitation of INVRS to positive definite cases is correct: That is, any nonsingular real symmetric matrix (positive, indefinite, or negative) may be inverted using this algorithm. The original INVRS should therefore be modified as follows:
     if pivot = 0 **then go to** singular;
Randell and Broyden's second example (of order 5) was also used as a test case, with the resulting inverse:

−.0000	.9999	.0000	.0000	.9999
	1.5333	−.7333	−.1333	.7999
		−.8666	−1.0666	−.5999
			−1.4666	−.1999
				.2000

$$\text{determinant} = -14.999999$$

An attempt to invert the *inverse* of the 4×4 segment of the
Hilbert matrix, as presented by Randell (*Comm. ACM*, Jan.
1962, p. 50), yielded the following results:

.9999	.4999	.3333	.2499
	.3333	.2499	.1999
		.1999	.1666
			.1428

determinant = 6048020.6

ALGORITHM 67
CRAM
JOHN CAFFREY
Director of Research, Palo Alto Unified School District,
Palo Alto, California

procedure CRAM (n, r, a) Result: (f); **value** n, r; **integer**
n, r; **real array** a, f;
comment CRAM stores, via an unspecified input procedure
READ, the diagonal and superdiagonal elements of a square symmetric matrix e, of order n, as a pseudo-array of dimension
$1:n(n + 1)/2$. READ (u) puts one number into u. Elements $e[i, j]$
are addressable as $a[c + j]$, where $c = (2n - i)(i - 1)/2$ and $c[i + 1]$
may be found as $c[i] + n - i$. Since $c[1] = 0$, it is simpler to develop
a table of the $c[i]$ by recursion, as shown in the sequence labelled
"table". Further manipulation of the elements so stored is illustrated by premultiplying a rectangular matrix f, of order n, r, by
the matrix e, replacing the elements of f with the new values, requiring a temporary storage array v of dimension $1:n$;
begin integer i, j, k, m; **real array** $v[1:n]$; **real** s;
  integer array $c[1:n]$;
table: $j := - n$; $k := n + 1$;   **for** $i := 1$ **step** 1 **until** n **do**
      begin
      $j := j + k - i$; $c[i] := j$ **end**;
load: **for** $i := 1$ **step** 1 **until** n **do**
      begin for $j := i$ **step** 1 **until** n **do** READ $(v[j])$; $m :=$
      $c[i]$;
      for $k := i$ **step** 1 **until** n **do** $a[m + k] := v[k]$ **end**;
premult: **for** $j := 1$ **step** 1 **until** r **do**
      begin for $i := 1$ **step** 1 **until** n **do**
          begin $s := 0.0$;
          for $k := 1$ **step** 1 **until** i **do**
          begin $m := c[k]$; $s := s + a[m + i]$
          $\times f[k, j]$ **end**;
          for $k := i + 1$ **step** 1 **until** n **do**
             $s := s + a[m + k] \times f[k, j]$; $v[i] = s$
          end;
          for $k := 1$ **step** 1 **until** n **do** $f[k, j] = v[k]$
      end
end CRAM

CERTIFICATION OF ALGORITHM 67
CRAM (J. Caffrey, *Comm. ACM 4* (July 1961), 322)
A. P. RELPH
Atomic Power Div., The English Electric Co., Whetstone,
England

CRAM was translated using the DEUCE ALGOL compiler with
the following corrections:
      $V[i] = S$ *was changed to* $V[i] := S$
      $f[k,j] = V[k]$ *was changed to* $f[k,j] := V[k]$
  It is quicker not to use the table of the $C[i]$ in the "load"
sequence and instead use the following sequence:
    load: $m := n \times (n+1)/2$;
        for $i := 1$ **step** 1 **until** m **do** READ $(a[i])$;

ALGORITHM 68
AUGMENTATION
H. G. RICE
Computer Sciences Corp., Palos Verdes, Calif.

real procedure Aug(x,y); **value** x,y; **integer** x,y;
comment This algorithm makes use of the implicitly defined recursive properties of ALGOL procedures to compute the augment of x by y, using the basic technique of incrementation by unit step size;
begin Aug := **if** x = 0 **then** (**if** y > x **then** (Aug(y − 1, x) + 1)
  else y)
else Aug(x − 1, y + 1) **end** Aug

CERTIFICATION OF ALGORITHM 68
AUGMENTATION (H. G. Rice, *Comm. ACM*, Aug.
  1961)
L. M. BREED
Stanford University, Stanford, Calif.

AUGMENTATION was transliterated into BALGOL for the Burroughs 220, and proved successful in a number of test cases. However, the following algorithm has exactly the same effect and is considerably simpler:
real procedure Aug(x, y); **value** x, y; **integer** x, y;
begin if x<0 **then** L : **go to** L **else** Aug := x+y **end** Aug

ALGORITHM 69
CHAIN TRACING
Brian H. Mayoh
Regnecentralen, Gl. Carlsbergvet. 2, Copenhagen.

```
procedure CHAIN tracing (iteration counter, number of
        identifiers, number of identifier links, final linkage
        matrix, couples);
    Boolean array final linkage matrix;
    integer array couples;
    integer iteration counter, number of identifiers, number of
        identifier links;
begin comment  This procedure is given a list of pairs of inte-
        gers, the second being related to the first in some way. It finds
        those pairs of integers which are related to each other if the
        relation is transitive. It is supplied with,
    couples  a matrix whose bound pairlist is [1:2, 1:number of
        identifier links] where couples [2, i] is related to couples
        [1, i] in some way.
    final linkage matrix  a matrix whose bound pair list is
        [1:number of identifiers, 1:number of identifiers] and into
        which the procedure puts true if the second subscript
        expression is an integer which is related to the integer
        corresponding to the first subscript expression, if the
        relation is irreflexive then the diagonal entries of this
        matrix are false.
    iteration counter  a place for the procedure to put the
        length of the longest chain it finds. CHAIN tracing can be
        applied to any system which can be represented by a Turing
        machine by letting the integers in couples correspond to
        the Turing machine states. Two integers j, k are related if
        there is an input symbol which causes state j to change to
        state k. If the Turing machine always stops whatever the
        sequence of input symbols, then its final linkage matrix
        will have false for all leading diagonal entries;
    integer i, j;
    Boolean array working linkage matrix [1:number of identi-
        fiers, 1:number of identifiers];
    Boolean procedure PROGRESS;
        begin PROGRESS := false;
        for i := 1 step 1 until number of identifiers
            do for j := 1 step 1 until number of identifiers
                do begin if Working linkage matrix [i, j] = ¬ Final
                        linkage matrix [i, j] then PROGRESS := true;
                        Final linkage matrix [i, j] := Working linkage
                        matrix [i, j]
                    end of comparison
        end of PROGRESS;
BEGIN OF PROGRAM:
    for iteration counter := −1, 0, iteration counter + 1 while
        PROGRESS
        do for i := 1 step 1 until number of identifier links
        do for j := 1 step 1 until number of identifiers
            do begin if iteration number = −1
                thenFinal linkage Matrix [couples [1, i], j]
                    := Working linkage Matrix [couples [1, i], j]
                    := couples [2, i] = j
                else Working linkage Matrix [couples [1, i], j]
                    := Working linkage Matrix [couples [1, i], j]
                    ∨ Working linkage Matrix [couples [2, i], j];
```

end of setting one linkage
end of CHAIN tracing;

ALGORITHM 70
INTERPOLATION BY AITKEN
CHARLES J. MIFSUD
General Electric Co., Bethesda, Md.

```
procedure  AITKEN (x, f, n, X, F);  real array x, f;
              integer n;  real X, F;
comment  If given x₀,x₁, . . . xₙ,  n+1 abscissas and also given
    f(x₀), f(x₁), . . . f(xₙ),  n+1 functional values, this procedure
    generates a Lagrange polynomial, F(X) of the nth degree so that
    F(xᵢ) = f(xᵢ). Hence, for any given value X, a functional value
    F(X) is generated. The procedure is good for either equal or
    unequal intervals of the xᵢ. Aitken's interative scheme is used
    in the generation of F(X). Since the f array is used for tem-
    porary storage, as the calculation proceeds its original values
    are destroyed;
begin integer i, j, t;
    for j := 0 step 1 until n−1 do
      begin t := j+1
        for i := t step 1 until n do
          f[i] := ((X−x [j]) × f [i] − (X−x [i]) × f[j])/
                (x[i] −x[j])     end
            F := f [n]
end
```

CERTIFICATION OF ALGORITHM 70
INTERPOLATION BY AITKEN [C. J. Mifsud, *Comm.
ACM 4* (Nov. 1961)]
A. P. RELPH
The English Electric Co., Whetstone, England

Algorithm 70 was translated using the DEUCE ALGOL compiler
and gave satisfactory results after semicolons had been added to

$$t := j+1 \quad \textit{to make it} \quad t := j+1;$$

and $\quad$ (x[i]−x[j]) **end** *to make it* (x[i]−x[j]) **end**;

The identifier *t* can be eliminated and the algorithm shortened
by the following changes:

Replace **begin integer** i, j, t; $\quad$ *by* $\quad$ **begin integer** i, j;
Replace $\quad$ t := j+1; $\qquad$ *by* $\quad$ **for** i := j+1 **step** 1 **until**
$\qquad$ **for** i := t **step** 1 **until** $\qquad$ n **do**
$\qquad\qquad$ n **do**

ALGORITHM 71
PERMUTATION
R. R. Coveyou and J. G. Sullivan
Oak Ridge National Laboratory, Oak Ridge, Tenn.

```
procedure  PERMUTATION (I, P, N);
    value I, N;  integer N;  integer array P;  boolean I;
    comment  This procedure produces all permutations of the
      integers from 0 thru N. Upon entry with I = false the pro-
      cedure initializes itself producing no permutation. Upon each
      successive entry into the procedure with I = true a new
      permutation is stored in P[0] thru P[N]. When the process has
      been exhausted a sentinel is set:
      P[0] : −1,
      N ≥ 0;
    begin
      integer i;  own integer array x[0:N];
      if ¬ I then
      begin for i := 0 step 1 until N−1 do x[i] := 0;  x[N] := −1;
        go to E end;
      for i := N step −1 until 0 do begin if x[i]≠i then go to A;
        x[i] := 0 end;
      P[0] := −1;  go to E;
A:    x[i] := x[i]+1;  P[0] := 0;
      for i := 1 step 1 until N do
        begin P[i] := P[i−x[i]];  P[i−x[i]] := i end;
E:  end PERMUTATION
```

CERTIFICATION OF ALGORITHM 71
PERMUTATION (R. R. Coveyou and J. G. Sullivan,
  Comm. ACM, Nov. 1961)
P. J. Brown
University of North Carolina, Chapel Hill, N. C.

PERMUTATION was transliterated into GAT for the Univac 1105 and successfully run for a number of cases.

CERTIFICATION OF ALGORITHM 71
PERMUTATION (R. R. Coveyou and J. G. Sullivan,
  Comm. ACM, Nov. 1961)
J. E. L. Peck and G. F. Schrack
University of Alberta, Calgary, Alberta, Canada

PERMUTATION was translated into Fortran for the IBM
1620 and it performed satisfactorily. The **own integer array**
x[0:n] may be shortened to x[1:n], provided corresponding corrections are made in the first two **for** statements.

However, PERMUTE (Algorithm 86) is superior to PERMU
TATION in two respects.

(1) PERMUTATION, using storage of order 2n, is designed to
permute the specific vector 0, 1, 2, $\cdots$, n − 1 rather than an
arbitrary vector. Thus storage of order 3n is required to permute
an arbitrary vector. PERMUTE, in contrast, only needs storage
of order 2n to permute an arbitrary vector.

(2) PERMUTE is built up from cyclic permutations. The
number of permutations actually executed internally (the redundant ones are suppressed) by PERMUTE is asymptotic to

(e − 1)n! rather than n!. In spite of this, PERMUTE is dis-
tinctly faster (1316 against 2823 seconds for n = 8) than PERMU-
TATION. If t_n is the time taken for all permutations of a vector
with n components, and if $r_n = t_n/nt_{n-1}$, then one would expect
r_n to be close to 1. Experiment with small values of n gave the
following results for r_n .

n	6	7	8
PERMUTE	0.96	0.99	1.00
PERMUTATION	1.10	1.13	1.12

Is there yet a faster way to do it?

See also: C. Tompkins, "Machine Attacks on Problems whose
Variables are Permutations", Proceedings of Symposia in Applied
Mathematics, Vol. VI: *Numerical Analysis* (N. Y., McGraw-Hill,
1956).

CERTIFICATION OF ALGORITHM 71
PERMUTATION [R. R. Coveyou and J. G. Sullivan,
  Comm. ACM, Nov. 1961]
J. S. Hillmore
Elliott Bros. (London) Ltd., Borehamwood, Herts.,
England

The algorithm was successfully run using the Elliott Algol
translator on the National-Elliott 803. The integer array x was
made a parameter of the procedure in order to avoid having an
own array with variable bounds.

ALGORITHM 72
COMPOSITION GENERATOR
L. HELLERMAN AND S. OGDEN
IBM-Product Development Laboratory, Poughkeepsie,
N. Y.

```
procedure comp (c, k);  value k;  integer array c;
   integer k;
comment Given a k-part composition c of the positive integer n,
   comp generates a consequent composition if there is one. If
   comp operates on each consequent composition after it is found,
   all compositions will be generated, provided that 1, 1, . . . , 1,
   n−k+1 is the initial c. If c is of the form n−k+1, 1, 1, . . . , 1,
   there is no consequent, and c will be replaced by a k vector of
   0's. Reference: John Riordan, An Introduction to Combi-
   natorial Analysis, John Wiley and Sons, Inc., New York, 1958,
   Chapter 6;
begin integer j;  integer array d [1:k];
   if k = 1 then go to last;
   for j := 1 step 1 until k do d [j] := c [j] − 1;
test:  if d[j]>0 then go to set;
       j := j−1;
       go to if j = 1 then last else test;
set:   d [j] := 0;
       d [j − 1] := d [j − 1] + 1;
       d [k] := c [j] − 2;
       for j := 1 step 1 until k do c [j] := d[j] + 1;
       go to exit;
last:  for j := 1 step 1 until k do c [j] := 0;
exit:  end comp
```


CERTIFICATION OF ALGORITHM 72
COMPOSITION GENERATOR [L. Hellerman and S.
   Ogden, Comm. ACM, Nov. 1961]
D. M. COLLISON
Elliott Bros. (London) Ltd., Borehamwood, Herts.,
   England

After
```
        for j := 1 step 1 until k do d[j] := c[j]−1;
```
the statement
```
                j := k;
```
should be inserted (see ALGOL 60 report, para 4.6.5). With this
alteration, the algorithm was successfully run using the Elliott
ALGOL translator on the National-Elliott 803.

ALGORITHM 73
INCOMPLETE ELLIPTIC INTEGRALS
DAVID K. JEFFERSON
U. S. Naval Weapons Laboratory, Dahlgren, Virginia

```
procedure ellint (k, phi, E, F);
value k, phi;
real phi, F, k, E;
comment   ellint computes the value of the incomplete elliptic
    integrals of the first and second kinds, F(phi, k) and E(phi, k),
    where phi is in radians. If | k | > 1 or | phi | > π/2, E and F
    will be set equal to 100,000,000, otherwise they will contain the
    computed integrals. For the formulation of this procedure, see
    DiDonato, A. R., and Hershey, A. V., "New Formulae for
    Computing Incomplete Elliptic Integrals of the First and
    Second Kind", J. ACM 6, 4 (Oct. 1959);
begin real kp, sinphi, n, cosphi;
    real array H [1:2], A [1:2], sigma [1:4], L [1:2], M [1:2],
        N [1:2], T [1:2], del [1:4];
    sigma [1] := sigma [2] := sigma [3] := sigma [4] := 0;
    H [1] := 1;
    n := 0;
    sinphi := sin(phi);
if abs (k × sinphi) ≤ tanh (1) then go to small else if abs (k) ≤
    1 ∧ abs(phi) ≤ π/2 then go to large;
    E := F := 100000000;
    go to stop;
small:  A [1] := phi;
step 1:  n := n + 1;
    cosphi := cos (phi);
    E := (2 × n−1) /(2 × N);
    H [2] := E × k ↑ 2 × H [1];
    A [2] := E × A [1] − sinphi ↑ (2 × n−1) × cosphi /(2 × n);
    del [1] := H [2] × A [2];
    del [2] := −k ↑ 2 × H [1] × A [2] /(2 × n);
    sigma [1] := sigma [1] + del [1];
    sigma [2] := sigma [2] + del [2];
    H [1] := H [2];
    A [1] := A [2];
if abs ((sigma [1] + del [1]) − sigma [1]) > 0 ∧ phi × sinphi
    ↑ (2 × n) ≥ A [2] then go to step 1;
    F := phi + sigma [1];
    E := phi + sigma [2];
    go to stop;
large:  kp := sqrt (1−k↑2);
    A [1] := 1;
    L [1] := M [1] := N [1] := 0;
step 2:  n := n+1;
    E := (2 × n−1) /(2 × n);
    F := abs (k) × sqrt (1−sinphi ↑2) × (1−k↑2 × sinphi
        ↑2) ↑ ((2 × n−1) /(2 × n));
    H [2] := E × H [1];
    A [2] := E↑2 × kp↑2 × A [1];
    L [2] := L [1] + 1 /(n × 2 × n−1));
M [2] := (M [1] − F × H [2]) × ((2 × n+1) /(2 × n+2)) ↑2 ×
    kp↑2;
N [2] := (N[1] − F × H [1]) × E × (2 × n+1) × kp ↑ 2 /(2 ×
    n+2);
del [1] := M [2] − A [2] × L [2];
del [2] := N [2] − E × kp ↑ 2 × A [1] × L [2] + kp ↑ 2 × A [1]
    /((2 × n) ↑ 2);
del [3] := A [2];
del [4] := (2 × n+1) × A [2] /(2 × n+2);
sigma [1] := sigma [1] + del [1];
sigma [2] := sigma [2] + del [2];
sigma [3] := sigma [3] + del [3];
sigma [4] := sigma [4] + del [4];
H [1] := H [2];
A [1] := A [2];
L [1] := L [2];
M [1] := M [2];
N [1] := N [2];
if abs ((sigma [1] + del [1]) − sigma [1]) > 0 then go to step 2;
T [1] := ln (4 /(sqrt (1 − k ↑ 2 × sinphi ↑ 2) + abs (k) × sqrt(1−
    sinphi ↑ 2)));
T [2] := abs (k) × sqrt ((1−sinphi ↑ 2) /(1−k↑2 × sinphi ↑2));
F := T [1] × (1+sigma [3]) + T [2] × ln (.5 + .5 × abs (k ×
    sinphi)) + sigma [1];
E := (.5 + sigma [4]) × kp↑2 × T [1] + 1−T [2] × (1−abs
    (k × sinphi)) + sigma [2];
stop:  end
```

CERTIFICATION OF ALGORITHM 73
INCOMPLETE ELLIPTIC INTEGRALS (David K.
Jefferson, *Comm. ACM*, Dec. 1961)
DEAN C. KRIEBEL
U. S. Naval Weapons Laboratory, Dahlgren, Virginia

This algorithm was originally coded in NORC machine language
and K. Pearson's incomplete elliptic integral tables of the first
and second kind generated. (See DiDonato, A. R., and Hershey,
A. V., "New Formulae for Computing Incomplete Elliptic Integrals of the First and Second Kind", *J.ACM 6*, 4 (Oct. 1959)).

The algorithm was coded for the MAD Compiler exactly as
written in ALGOL and run on an IBM 7090. Forty cases were computed with K ranging from 0° to 90° and PHI ranging from 0° to
90°. The results contained eight significant digits which agreed
with the DiDonato and Hershey tables to within 0 to 2 units in the
8th digit. (This may be attributed to the decimal to binary, binary
to decimal input-output conversion used with a binary computer
as compared to straight decimal computation on the NORC.)

CERTIFICATION OF ALGORITHM 73
INCOMPLETE ELLIPTIC INTEGRALS [David K.
Jefferson, *Comm. ACM 4*, Dec. 1961]
NOELLE A. MEYER
E. I. du Pont de Nemours & Co., Wilmington, Del.

Ellint was hand-coded in FORTRAN for the IBM 7070. The following corrections were made

The statement

$$E := (2 \times n - 1)/(2 \times N);$$

should be

$$E := (2 \times n - 1)/(2 \times n);$$

The statement

$$F := abs(k) \times sqrt(1 - sinphi \uparrow 2) \times (1 - k \uparrow 2 \times sinphi \uparrow 2) \uparrow$$
$$((2 \times n - 1)/(2 \times n));$$

should be

$$F := (abs(k) \times sqrt(1 - sinphi \uparrow 2) \times$$
$$(1 - k \uparrow 2 \times sinphi \uparrow 2) \uparrow (n - .5))/(2 \times n)$$

The statement

$$L[2] := L[1] + 1/(n \times 2 \times n - 1));$$

should be

$$L[2] := L[1] + (1/(n \times (2 \times n - 1));$$

In order to accommodate negative ϕ the following changes were made:

The statement

if $abs((sigma[1] + del[1]) - sigma[1]) > 0 \wedge phi \times sinphi \uparrow$
$$(2 \times n) \geq A[2] \textbf{ then go to } step\ 1;$$

was changed to

if $abs((sigma[1] + del[1]) - sigma[1]) > 0 \wedge abs(phi \times sinphi \uparrow (2 \times n))$
$$\geq abs(A[2]) \textbf{ then go to } step\ 1;$$

Also the following was inserted before the last statement (*stop*: **end**)

$$\textbf{if } phi < 0 \textbf{ then go to } wait \textbf{ else go to } stop;$$
$$wait:\ \ F := -F;$$
$$E := -E;$$

The revised algorithm yielded satisfactory answers when compared with the DiDonato and Hershey tables. Differences occurred in the eighth significant digit as shown in the following difference tables.

DIFFERENCE TABLES

F-TABLE

ϕ (in degrees)	θ (in degrees)			
	0	30	60	90
0	0.	0.	0.	0.
30	-1×10^{-8}	-1×10^{-8}	-1×10^{-8}	-3×10^{-8}
60	1×10^{-8}	1×10^{-8}	2×10^{-8}	-3×10^{-8}
90	0.	2×10^{-8}	6×10^{-8}	0.

E-TABLE

	0	30	60	90
0	0.	0.	0.	0.
30	-1×10^{-8}	-1×10^{-8}	-1×10^{-8}	-1×10^{-8}
60	1×10^{-8}	1×10^{-8}	-7×10^{-8}	3×10^{-8}
90	0.	0.	1×10^{-8}	0.

CERTIFICATION OF ALGORITHM 73
INCOMPLETE ELLIPTIC INTEGRALS [David K Jefferson, *Comm. ACM* Dec. 1961]

R. P. VAN DE RIET
Mathematical Centre, Amsterdam

The algorithm contained three misprints:
The 26th line of the procedure

$$E := (2 \times n - 1)/(2 \times N);$$

should read

$$E := (2 \times n - 1)/(2 \times n);$$

The 46th line of the procedure

$$\uparrow 2) \uparrow ((2 \times n - 1)/(2 \times n));$$

should read

$$\uparrow 2) \uparrow ((2 \times n - 1)/2)/(2 \times n);$$

The 49th line of the procedure

$$L[2] := L[4] + 1/(n \times 2 \times n - 1));$$

should read

$$L[2] := L[1] + 1/(n \times (2 \times n - 1));$$

The program was run on the X1 computer of the Mathematical Centre. For $phi = 45°$, $k = sin(10°(10°)180°)$, E and F were calculated. The result contained 12 significant digits.

Comparison with a 12-decimal table of Legendre-Emde (1931) showed that the 12th digit was affected with an error, at most 4 units large. After about 10 minutes of calculation (i.e. more than 100 cycles) no results were obtained for $k = sin\ 89°$, $phi = 1°$ and the calculation was discontinued.

REMARKS. As phi is unchanged during the calculation, we placed the statement $cos\ phi := cos\ (phi)$ in the beginning of the program, to be certain that the cosine was not calculated 30 or more times. Moreover, in the expression for $T[1]$ and $T[2]$, $sqrt(1 - sin\ phi \uparrow 2)$ was replaced by $cos\ phi$, so that loss of significant figures does not occur.

The expression $2 \times n$ was changed in a new variable, to obtain a more rapid program.

REMARK ON ALGORITHM 73
INCOMPLETE ELLIPTIC INTEGRALS [David K. Jefferson, *Comm. ACM* (Dec. 1961)]

DAVID K. JEFFERSON
U. S. Naval Weapons Laboratory, Dahlgren, Virginia

In regard to Algorithm 73, two errors were found:
The 34th line of the procedure

$$F := abs(k) \times sqrt(1 - sinphi \uparrow 2)$$
$$\times (1 - k \uparrow 2 \times sinphi \uparrow 2) \uparrow ((2 \times n - 1)/(2 \times n));$$

should read

$$F := abs(k) \times sqrt(1 - sinphi \uparrow 2)$$
$$\times (1 - k \uparrow 2 \times sinphi \uparrow 2) \uparrow ((2 \times n - 1)/2)/(2 \times n);$$

The 37th line

$$L[2] := L[1] + 1/(n \times 2 \times n - 1));$$

should read

$$L[2] := L[1] + 1/(n \times (2 \times n - 1));$$

In addition, efficiency is improved by interchanging lines 13 and 14:

$$Step\ 1:\ \ n := n + 1;$$
$$cosphi := cos(phi);$$

can be replaced by

$$cosphi := cos(phi);$$
$$Step\ 1:\ \ n := n + 1;$$

ALGORITHM 74
CURVE FITTING WITH CONSTRAINTS
J. E. L. PECK,
University of Alberta, Calgary, Alberta, Canada

procedure Curve fitting (k,a,b,m,x,y,w,n,alpha,beta,s,sgmsq,x0, gamma,c,z,r);
comment This procedure finds, by the method of least squares, the polynomial of degree n, $k \leq n < k+m$, whose graph contains $(a_1, b_1), \cdots, (a_k b_k)$ and approximates $(x_1, y_1), \cdots, (x_m, y_m)$, where w_i is the weight attached to the point (x_i, y_i). The details will be found in the reference cited below, where a similar notation is used. A nonlocal label "error" is assumed;
value a, x, y, w; **integer** k, m, n, r; **real** x0, gamma; **array**
a, b, x, y, w, alpha, beta, s, sgmsq, c, z;
begin integer i, j; **array** w1[1:k]; **real** p, f, lambda;
comment We shall first define several procedures to be used in the main program, which begins at the label START;

procedure Evalue (x, nu);
comment This procedure evaluates $f = s_0p_0 + s_1p_1 + \cdots + s_\nu p_\nu$, where $p_{-1}(x) = 0$, $p_0(x) = 1$, $\beta_0 = 0$ and $p_{i+1}(x) = (x - \alpha_i)p_i(x) - \beta_i p_{i-1}(x)$, $i = 0, 1, \cdots, \nu-1$. The value of $p_\nu(x)$ remains in p;
real x; **integer** nu;
begin real p0, temp; **integer** i; p0 := 0; p := 1; f := s[0];
for i := 0 **step** 1 **until** nu−1 **do**
  begin temp := p;
  p := (x−alpha[i]) × p−beta[i] × p0;
  p0 := temp; f := f + p × s[i+1] **end** i
end Evalue;

procedure Coda (n, c);
comment This procedure finds the c's when $c_0 + c_1x + \cdots + c_nx^n = s_0p_0(x) + \cdots + s_np_n(x)$;
integer n; **array** c;
begin integer i,r; **real** t1,t2; **array** pm,p[0:n];
for r := 1 **step** 1 **until** n **do**
  c[r] := pm[r] := p[r] := 0;
pm[0] := 0; p[0] := 1; c[0] := s[0];
for i := 0 **step** 1 **until** n−1 **do**
  begin t2 := 0;
  for r := 0 **step** 1 **until** i+1 **do**
    begin t1 := (t2−alpha[i]× p[r]−beta[i] × pm[r])/lambda;
    t2 := pm[r] := p[r]; p[r] := t1;
    c[r] := c[r] + t1 × s[i+1]**end** r
  end i
end Coda;

procedure GEFYT (n,n0,x,y,w,m);
comment This is the heart of the main program. It computes the $\alpha_i, \beta_i, s_i, \sigma_i^2$, using the method of orthogonal polynomials, as described in the reference;
integer n,n0,m; **array** x,y,w;
begin real dsq,wpp,wpp0,wxpp,wyp,temp;
integer i,j,freedom; **array** p,p0[1:m]; **boolean** exact;
if n−n0 > m $\lor$ n < n0 **then go to** error;
beta[n0] := dsq := wpp := 0; exact := n−n0 $\geq$ m−1;
for j := 1 **step** 1 **until** m **do**
  begin p[j] := 1; p0[j] := 0; wpp := wpp + w[j];
  if ¬ exact **then** dsq := dsq + w[j] × y[j] × y[j] **end** initialise;

for i := n0 **step** 1 **until** n **do**
  begin freedom := m−1−(i−n0); wyp := wxpp := 0;
  for j := 1 **step** 1 **until** m **do**
    begin temp := w[j] × p[j];
    if i < n **then** wxpp := wxpp + temp × x[j] × p[j];
    if freedom $\geq$ 0 **then** wyp := wyp + temp × y[j] **end** j;
  if freedom $\geq$ 0 **then** s[i] := wyp/wpp;
  if ¬ exact **then begin** dsq := dsq − s[i] × s[i] × wpp;
  sgmsq[i] := dsq/freedom **end if**;
  if i < n **then begin** alpha[i] := wxpp/wpp; wpp0 := wpp;
  wpp := 0;
  for j := 1 **step** 1 **until** m **do**
    begin temp := (x[j]−alpha[i]) × p[j] − beta[i] × p0[j];
    wpp := wpp + w[j] × temp × temp;
    p0[j] := p[j]; p[j] := temp **end** j;
  beta[i+1] := wpp/wpp0 **end if**
  end i
end GEFYT;
  START: **for** j := 1 **step** 1 **until** k **do**
begin w1[j] := 1; a[j] = (a[j]−x0)/gamma **end** j;
GEFYT (k,0,a,b,w1,k);
comment This finds the polynomial of degree k−1 whose graph contains $(a_1,b_1),\cdots,(a_k,b_k)$ supplying the α_i,β_i,s_i, $0 \leq i \leq k$;
begin real rho; rho := 0;
for j := 1 **step** 1 **until** m **do**
  begin rho := rho + w[j];
  x[j] := (x[j] − x0)/gamma **end** j; rho := m/rho;
comment The factor ρ is used to normalize the weights. We shall now put $s_k = 0$ in order to evaluate $p_k(x)$ and the polynomial of degree k−1 simultaneously;
s[k] := 0;
for j := 1 **step** 1 **until** m **do**
  begin Evalue (x[j],k);
  if p = 0 **then go to** error;
  y[j] := (y[j] − f)/p;
  w[j] := w[j] × p × p × rho **end** j
end rho;
comment We have now normalized the weights and adjusted the weights and ordinates ready for the least squares approximation;
GEFYT (n,k,x,y,w,m);
comment The coefficients α_i,β_i, $0 \leq i < n$, and s_i, $0 \leq i \leq n$ are now ready. The polynomial may be evaluated for x = $z_1,z_2, \cdots,z_r$, but the variable must be adjusted first. Note that we may evaluate the best polynomial of lower degree by decreasing n;
begin real x;
for j := 1 **step** 1 **until** r **do**
  begin x := (z[j]−x0)/gamma;
  Evalue (x,n); **comment** the values of z_j and f should now be printed; **end** j;
comment We may now adjust the coefficients for scale and then find the coefficients of the power series $c_0 + c_1x + \cdots + c_nx^n = s_0p_0(x) + \cdots + s_np_n(x)$;
for i := 0 **step** 1 **until** n−1 **do**
  begin alpha[i] := alpha[i] × gamma + x0;
  beta[i] := beta[i] × gamma **end** i; lambda := gamma;
Coda (n,c);
comment We may now re-evaluate the polynomial from the power series;
for j := 1 **step** 1 **until** r **do**

```
begin x := z[j];  f := c[n];
 for i := n−1 step −1 until 0 do
   f := f × x + c[i];
 comment   the values of x and f should now be printed;  end j
 end x
end Curve fitting
```

REFERENCE: PECK, J. E. L. Polynomial curve fitting with
constraint, *Soc. Indust. Appl. Math. Rev.* (1961).

CERTIFICATION OF ALGORITHM 74
CURVE FITTING WITH CONSTRAINTS [J. E.
  Peck, *Comm. ACM*, Jan. 62]
KAZUO ISODA
Japan Atomic Energy Research Institute, Tokai, Ibaraki,
  Japan

Algorithm 74 was hand-compiled into SOAP IIa for the IBM
650 and run successfully with no corrections except the case in
which the origin $(0, 0)$ are given as both a constraint and a sample.

ALGORITHM 75
FACTORS
J. E. L. Peck,
University of Alberta, Calgary, Alberta, Canada

```
procedure  factors (n,a,u,v,r,c);
comment  This procedure finds all the rational linear factors of
    the polynomial a₀xⁿ + a₁xⁿ⁻¹ + · · · + aₙ₋₁x + aₙ, with integral
    coefficients. An absolute value procedure abs is assumed;
value n,a;  integer r,n,c;  integer array a,u,v;
begin comment  We find whether p divides a₀,  1 ≤ p ≤ |a₀| and
    q divides aₙ,  0 ≤ q ≤ |aₙ|. If this is the case we try (px ± q);
integer p,q,a0,an;
r := 0;  c := 1;  comment r will be the number of linear factors
    and c the common constant factor;
TRY AGAIN:  a0 := a[0];  an := a[n];
for p := 1 step 1 until abs(a0) do
  begin if (a0 ÷ p) × p = a0 then
    begin comment p divides a₀;
    for q := 0 step 1 until abs(an) do
      begin if q = 0 ∨ (an ÷ q) × q = an then
        begin comment  q divides aₙ (or q = 0). If p = q we
          may have a common constant factor, therefore;  if q
          > 1 ∧ p = 1 then
        begin integer j;
        for j := 1 step 1 until n−1 do
          if (a[j] ÷ q) × q ≠ a[j] then go to NO CONSTANT;
        for j := 0 step 1 until n do
          a[j] := a[j]/q;
        c := c × q;  go to TRY AGAIN
        end the search for a common constant factor;
NO CONSTANT:
        begin comment  try (px − q) as a factor;
        integer f,g,i;  f := a0;  g := 1;
        comment  we try x = q/p;
        for i := 1 step 1 until n do
          begin g := g × p;  f := f × q + a[i] × g
          end evaluation;
        if f = 0 then
          begin comment  we have found the factor (px − q);
          r := r + 1;  u[r] := p;  v[r] := q;
          comment  there are now r linear factors;
            begin comment  we divide by (px − q);
            integer i,t;  t := 0;
            for i := 0 step 1 until n do
              begin a[i] := t := (a[i] + t)/p;  t := t × q
              end i;
            n := n − 1
            end reduction of polynomial. Therefore;
          go to if n = 0 then REDUCED else TRY AGAIN
          end discovery of px − q as a factor. But
        if we got this far it was not a factor so try px + q;
        q := −q;  if q < 0 then go to NO CONSTANT
        end trial of px ± q,
      end q divides aₙ and
    end of q loop.
  end p divides a₀, also
```

```
  end p loop, which means;
REDUCED:  if n = 0 then
    begin c := c × a0;  a0 := 1
    end if n = 0
end factors procedure.
```
There are now r (r > 0) rational linear
factors (uᵢx − vᵢ), 1 < i < r, and the reduced polynomial of
reduced degree n replaces the original. The common constant
factor is c. Acknowledgments to Clay Perry.

CERTIFICATION OF ALGORITHM 75
FACTORS [J. E. L. Peck, *Comm. ACM 5* (Jan. 1962)]
A. P. Relph
The English Electric Co., Whetstone, England

Algorithm 75 was translated using the Deuce Algol compiler
and gave satisfactory results after the following corrections had
been made:
```
    begin if q=0∨(an÷q)×q=an then
      begin if q>1∧p=1 then
```
was changed to
```
    begin if q≦1 then go to NO CONSTANT;
      if (an÷q)×q=an then
      begin if p=q then
```
———————
```
    begin c := c×a0;  a0 := 1
    end
```
was changed to
```
    begin c := c×a[0];  a[0] := 1;
    end
```
———————

There are now r (r>0) rational linear factors (u ᵢ x−vᵢ),
    1<i<r,
was changed to
If r>0 there are now r rational linear factors (uᵢx−vᵢ), 1≦i≦ r,
———————

CERTIFICATION OF ALGORITHM 75
FACTORS [J. E. L. Peck, *Comm. ACM*, Jan. 1962]
J. S. Hillmore
Elliott Bros. (London) Ltd., Borehamwood, Herts.,
England

The following changes had to be made to the algorithm:
(1) *For*   if q > 1 ∧ p = 1 then
    put   if q > 1 ∧ p = q then
(2) *For*   begin c := c × a0; a0 := 1 end
    put   begin c := c × a[0]; a[0] := 1 end
(3) *For*   if q = 0 ∨ (an ÷ q) × q = an then
    put   if (if q = 0 then true else (an ÷ q) × q = an) then
This change is necessary to ensure that the term (an ÷ q) is not
evaluated when q = 0.

The algorithm, thus modified, was successfully run using the
Elliott Algol translator on the National-Elliott 803.

To return to the state $(p=1, q=0)$ after every factor or constant is found is inefficient. This can be avoided by substituting a[0] and a[n] for the identifiers a0 and an respectively. The procedure then becomes:

```
procedure factors (n, a, u, v, r, c);  value n, a;
   integer array a, u, v;
integer r, n, c;
begin integer p, q;
  r := 0;  c := 1;
ZERO:  if a[n]=0 then
          begin r := r+1;  u[r] := 1;  v[r] := 0;  n := n−1;
            go to ZERO
          end;
          for p := 1 step 1 until abs (a[0]) do
          begin if (a[0]÷p)×p=a[0] then
            begin for q := 1 step 1 until abs (a[n]) do
              begin if q=1 then go to NO CONSTANT;
          TRY AGAIN:  if (a[n]÷q)×q=a[n] then
                      begin integer j;
                        for j := 0 step 1 until n−1 do
                          if (a[j]÷q)×q≠a[j] then go to
                          NO CONSTANT;
                        for j := 0 step 1 until n do
                          a[j] := a[j]/q;
                        c := c×q;  go to TRY AGAIN
                      end;
          NO CONSTANT:  begin integer f, g, i;  f := a[0];
                        g := 1;
                        for i := 1 step 1 until n do
                        begin g := g×p;
                          f := f×q+a[i]×g
                        end;
                        if f=0 then
                        begin r := r+1;  u[r] := p;
                          v[r] := q;
                          begin integer i, t;  t := 0;
                            for i := 0 step 1 until n do
                            begin a[i] := t := (a[i]+t)/p;
                              t := t×q
                            end;
                            n := n−1
                          end
                          go to if n=0 then REDUCED
                          else NO CONSTANT
                        end;
                        q := −q;  if q<0 then go to NO
                        CONSTANT
                      end
              end
            end
          end;
          REDUCED:  if n=0 then
            begin c := c×a[0];  a[0] := 1
            end
end
```

ALGORITHM 76
SORTING PROCEDURES
IVAN FLORES
Private Consultant, Norwalk, Connecticut

comment The following ALGOL 60 algorithms are procedures for the sorting of records stored within the memory of the computer. These procedures are described in detail, flow-charted, compared, and contrasted in "Analysis of Internal Computer Sorting" by Ivan Flores [*J. ACM 8* (Jan. 1961)]. Although sorting is usually a business computer application, it can be described completely in ALGOL if we stretch our imagination a little. Sorting is ordering with respect to a key contained within the record. If the key *is* the active record, the sorting is trivial. A means is required to extract the key from the record. This is essentially string manipulation, for which no provision, as yet, has been made in ALGOL. We circumambulate this difficulty by defining an **integer procedure** K(I) which "creates" a key from the record, I. ALGOL *does* provide for machine language code substitutions, which is one way to think of K(I). This could be more accurately represented by using the string notation proposed by Julien Green ["Remarks on ALGOL and Symbol Manipulation," *Comm. ACM 2* (Sept. 1959), 25–27]. The function **sub** ($,i,g) represents the procedure, K(I). $ corresponds to the record I, i corresponds to the starting position of the key and g corresponds to the length of the key. Both i and g are **values** which must be specified when the sort procedure is called for as a statement instead of a declaration.

Another factor, which might vex some, is that the key might be alphabetic instead of numeric. Then, of course, K(I) would not be integer. It would, however, be string when such is defined eventually. Note, also, that keys are frequently compared. This is done using the ordering relations ">" for "greater than," etc. These are not really defined in the ALGOL statement [NAUR, PETER, ET AL. "Report on the Algorithmic Language ALGOL 60". *Comm. ACM 3* (May 1960), 294–314]. They can simply be defined so that $Z > Y > \cdots > A > 9 > \cdots > 1 > 0$. Also the assignment X[i] := z should be interpreted as "Assign the key 'z' which is larger than any other key." For any sort procedure (I,N,S), "I" is the set of unsorted records, "N" is their number, and "S" the sorted set of records.

Caution, these algorithms were developed purely for the love of it: No one was available with the combined knowledge of sorting and ALGOL to check this work. Hence each algorithm should be carefully checked before use. I will be glad to answer any questions which may arise;

```
Sort insert (I,N,S);  value N;   array I[1:N], S[1:N];
   integer procedure K(I);   integer N;
   begin integer i, j, k;
     S[1] := I[1];
     for i := 2 step 1 until N do begin
       for j := i − 1,  j − 1 while K(I[i]) > K(S[j]) do
       for k := i step − 1 until j + 1 do
         S[k] := S[K − 1];
       S[j + 1] := I[i] end end

Sort count (I,N,S);  value N;   array I[1:N], S[1:N];
   integer procedure K(I);   integer N;
   begin integer array C[1:N];   integer i,j;
     for i := 1 step 1 until N do C[i] := 0;
     for i := 2 step 1 until N do
       for j := 1 step 1 until i − 1 do
         if K(I[i]) > K(I[j]) then C[i] := C[i] + 1
         else C[j] := C[j] + 1;
     for i := 1 step 1 until N do
       S[C[i]] := I[i] end

Sort select (I,N,S); value N;   array I[1:N], S[1:N];
   integer procedure K(I);   integer N;
   begin integer i,j,A,h;
     for i := 1 step 1 until N do begin
     h := K(I[1]);
     for j := 2 step 1 until N do
     if h > K(I[j]) then begin h := K(I[j]);   A := j end;
     S[i] := I[A];
     I[A] := z end end

Sort select exchange (I,N);   value N;   array I[1:N];
   integer procedure K(I);   integer N;
   begin integer h,i,j,H;   real T;
     for i := 1 step 1 until N do begin
       H := K(I[i]);   h := i;
       for j := i + 1 step 1 until N do
       if K(I[j]) < H then begin
       H := K(I[j]);   h := j end
       T := I[i];   I[i] := I[h];   I[A] := T end
   end

Sort binary insert (I,N,S);  value N;   array I[1:N], S[1:N];
   integer procedure K(I);   integer N;
   begin integer i,k,j,l;
     if K(I[1]) < K(I[2]) then begin
       S[1] := I[1];   S[2] := I[2] end
     else begin S[1] := I[2];   S[2] := I[1] end;
start:   for    i := 3 step 1 until N do begin
             j := (i + 1) ÷ 2;
find spot:     for k := (i + 1) ÷ 2, (k + 1) ÷ 2 while k > 1 do
               if K(I[i]) < K(S[j]) then j := j − k;
               else j := j + k;
               if K(I[i]) ≥ K(S[j]) then j := j − 1;
move items:    for l := i step − 1 until j do
               S[l + 1] := S[l];
enter this
one:           S[j] := I[i] end end

Sort address calculation (I,N,S,F);   value N;
   array S[1:M], I[1:N];   integer procedure F(K), K(I);
   integer N,M;
             begin integer i,j,G,H,F,M;
             M := entier(2.5 × N)
             for i := 1 step 1 until M do S[i] = 0;
Address:     for i := 1 step 1 until N do begin
             F := F(K(I[i]));
             if S[F] = 0 then begin S[F] := I[i];
               go to NEXT end
             else if K(S[F]) > K(I[i]) then go to SMALLER;
LARGER:      for H := F, H + 1 while K(S[H]) < K(I[i]) do
             for G := H, G + 1 while K(S[G]) ≠ 0 do
             for j := G step −1 until H + 1 do
               S[j] := S[j − 1];
             S[H] := I[i];   go to NEXT;
SMALLER:     for H := F, H − 1 while K(S[H]) > K(I[i]) do
```

```
                        for G := H, G − 1 while K(S[G]) ≠ 0 do
                        for j := G step 1 until H − 1 do
                          S[j] = S[j + 1];
                          S[H] := I[i];
NEXT:        end end

Sort quadratic select (I,N,S);  value N;  array I[1:N], S[1:N];
  integer procedure K(I);  integer N;
                begin integer i,j,k,C,D,J,M;
                  integer array C[1:M], D[1:M];
                  array I[1:M, 1:M];
  Divide inputs:    M := entier (sqrt (N)) + 1;  j := k := 1;
                    for i := 1 step 1 until N do begin
                    I[j,k] := I[i];  k := k + 1;
                    if k > M then begin k := 1;
                      j := j + 1 end end
  Fill up inputs:   I[j,k] := z;  k := k + 1;
                    if k > M then begin k := 1;  j := j + 1 end
                    if j ≤ M then go to Fill up inputs;
  Set controls:     for j := 1 step 1 until M do begin
                    C[j] := K(I[j, 1]);  D[j] := 1;
                    for k = 2 step 1 until M do
                      if C[j] > K(I[j,k]) then begin
                      C[j] := K(I[j,k]);  D[j] := k end end;
                    i := 1;
  Find least:       C := C[1];  D := D[1];  J := 1;
                    for j := 2 step 1 until M do
                    if C > C[j] then begin C := C[j];
                      D := D[j];  J := j end;
  Fill file:        S[i] := I[J,D];  i := i + 1;  I[J,D] := z;
                    if i = N + 1 go to STOP;
  Reset controls:   for j := J do begin
                    C[j] := K(I[j, 1]);  D[j] := 1;
                    for k := 2 step 1 until M do
                      if C[j] > K(I[j,k]) then begin C[j] :=
                      K(I[j,k];  D[j] := k end end;
                    go to Find least;
  STOP:        end

Presort quadratic selection (I,N,S);  value N;
  array I[1:N], S[1:N];  integer procedure K(I);  integer N;
                begin integer i,j,k,C,J,M;
                  integer array C[1:M], D[1:M];
                  array I[1:M,1:M];
  Divide inputs:    M := entier (sqrt(N)) + 1;  j := k := 1;
                    for i := 1 step 1 until N do begin
                    I[j,k] := I[i];  k := k + 1;
                    if k > M then begin k := 1;
                      j := j + 1 end end
  Fill up inputs:   I[j,k] := z;  k := k + 1;
                    if k > M then begin k := 1;  j = j + 1 end
                    if j ≤ M then go to Fill up inputs;
  First sort:       for j := 1 step 1 until M do
                    sort select exchange (I[j,k],M);
  Set controls:     for j := 1 step 1 until M do begin
                    C[j] := K(I[j,1]);  D[j] := 1 end
                    i := 1;
  Find least:       C := C[1];  J := 1;
                    for j := 1 step 1 until M do
                    if C > C[j] then begin C := C[j];
                      J := j end;
  Fill file:        S[i] := I[J,D[J]];  i := i + 1;
                    if i = N + 1 go to STOP
  Reset control:    for j := J do begin
                    D[j] := D[j] + 1;
                    if D[j] > M then C[j] := z else C[j] :=
                      K(I[j, D[j]]) end
                    go to Find least;
  STOP:        end
```

```
Sort binary merge (I,N,S);  value N; array I[1:N];
  integer procedure K(I);  integer N;
  begin real array S[1:N];
  integer array A[0:1, 0:J[a]], B[0:1, 0:K[b]], Aloc[0:1, 0:J[a]],
    Bloc[0:1, 0:K[b]], J[0:1], K[0:1], j[0:1], k[0:1];
    integer a,b,i,j,k;
  distribute:       a := b := j[0] := j[1] := 1;
                    for i := 1 step 1 until N do begin
                    if K(I[i]) < K(I[i−1] then
                      if a = 1 then a := 0 else a := 1;
                    A[a, j[a]] := K(I[i]);  Aloc[a, j[a]] := i;
                    j[a] := j[a] + 1 end;
                    J[0] := j[0];  J[1] := j[1];
  next sort:        begin a := b := j[0] := j[1] := k[0] :=
                    k[1] := 1;
  two inputs:       if A[1, j[1]] ≤ A[0, j[0]] then a := 1 else
                    a := 0;
                    B[b, k[b]] := A[a, j[a]];
                    Bloc[b, k[b]] := Aloc[a, j[a]];
                    j[a] := j[a] + 1;  k[b] := k[b] + 1;
                    if A[a, j[a]] ≥ A[a, j[a] − 1] then go to two
                      inputs else
                    if a = 1 then a := 0 else a := 1;
  single step:      B[b, k[b]] := A[a, j[a]];
                    Bloc[b, k[b]] := Aloc[a, j[a]];
                    j[a] := j[a] +1;  k[b] := k[b] + 1;
                    if A[a, j[a]] ≥ A[a, j[a] − 1] then go to
                      single step;
  switch file:      if b = 1 then b := 0 else b := 1;
  check rollout:    for a := 0, 1 do
                    if j[a] = J[a] then go to rollout;
                    go to two inputs;
  rollout:          B[b, k[b]] := A[a, j[a]];
                    Bloc[b, k[b]] := Aloc [a, j[a]];
                    k[b] := k[b] + 1;  j[a] := j[a] + 1;
                    if j[a] = J[a] then go to interchange files;
                    if A[a, j[a]] < A[a, j[a] − 1] then
                      if b = 1 then b := 0 else b := 1;
                    go to rollout;
  interchange files: K[0] := k[0];  K[1] := k[1];
                    if K[0] = 1 then go to output end
                    for b := 1, 0 do begin
                      for k[b] := 1 step 1 until K[b] do begin
                      A[b, k[b]] := B[b, k[b]];
                        Aloc[b, k[b]] := Bloc[b, k[b]];
                      J[b] := K[b] end end
                    go to next sort;
  output:           for i := 1 step 1 until N do
                      S[i] := I[Bloc[0, i]];
                    end
```

REMARK ON ALGORITHM 76

SORTING PROCEDURES (Ivan Flores, *Comm. ACM*
5, Jan. 1962)

B. Randell

Atomic Power Div., The English Electric Co., Whetstone,
England

The following types of errors have been found in the Sorting

Procedures:

1. Procedure declarations not starting with **procedure**.

2. Bound pair list given with array specification.

3. = used instead of := , in assignment statements, and in a **for** clause.

4. A large number of semicolons missing (usually after **end**).

5. Expressions in bound pair lists in array declarations depending on local variables.

6. Right parentheses missing in some procedure statements.

7. Conditional statement following a **then.**

8. No declarations for A, or z, which is presumably a misprint.

9. In several procedures attempt is made to use the same identifier for two different quantities, and sometimes to declare an identifier twice in the same block head.

10. In the Presort quadratic selection procedure an array, declared as having two dimensions, is used by a subscripted variable with only one subscript.

11. At one point a subscripted variable is given as an actual parameter corresponding to a formal parameter specified as an array.

12. In several of the procedures, identifiers used as formal parameters are redeclared, and still assumed to be available as parameters.

13. In every procedure K is given in the specification part, with a parameter, whilst not given in the formal parameter list.

No attempt has been made to translate, or even to understand the logic of these procedures. Indeed it is felt that such a grossly inaccurate attempt at ALGOL should never have appeared as an algorithm in the *Communications*.

ALGORITHM 77
INTERPOLATION, DIFFERENTIATION, AND IN-
TEGRATION
PAUL E. HENNION
Grumman Aircraft Engineering Corporation, Bethpage,
L. I., New York

real procedure AVINT (nop, jt, xarg, xlo, xup, xa, ya);
        value nop, jt, xarg, xlo, xup; **real** xarg, xlo, xup;
        integer nop, jt; **real array** xa, ya;
comment This procedure will perform interpolation, differentiation, or integration operating upon functions of one variable which over part or all of the interval of interest are ade-
quately described by a di-parabolic fit.

The routine was originally programmed as an open subroutine for the IBM 704 in FORTRAN II and occupied 323 memory
locations. It is based upon a Lagrange interpolation scheme
specialized for averaged second order parabolas. The technique finds the slope of a function numerically defined at
points 1, 2, 3 and 4 by fitting a parabola through the points
1, 2, 3, and another parabola through the points 2, 3, and 4.
The slope then, at point 2, is the average analytical derivative
of the two parabolas, i.e. the coefficients of the parabola
through points 1, 2 and 3 $(a_1x^2+b_1x_2+c_1)$ and the coefficients
of the parabola through points 2, 3, and 4 $(a_2x^2+b_2x_2+c_2)$
are determined by applying Lagrange's equations as shown below. The arithmetic mean of these coefficients $a = (a_1+a_2)/2$,
$b = (b_1+b_2)/2$, $c = (c_1+c_2)/2$ are used to supply the slope
in the interval from 2 to 3, namely $(2ax + b)$.

The interpolation is calculated in similar fashion, except the
final formula is that a parabola $(ax^2 + bx + c)$.

The integration is performed likewise by a curve fitting
process, e.g. the integral between any two points say 2 and 3
is the average integral of the two parabolas between the independent coordinate limits for points 2 and 3. The averaging
process is done for each interval along the abscissa as the
results obtained are accumulated to evaluate the definite
integral.

Applying Lagrange's equations, the coefficients a, b, and c
may be found by defining: $T_j = y_j/\prod_{i=1,\ i\neq j}^{n} (X_j - X_i)$ where
$y = f(x)$, $n = 3$, $j = 1, 2, \cdots, n$, then $a = \sum_{i=1}^{n} T_i$,
$b = \sum_{i=1}^{n} T_i \sum_{j=1,\ j\neq i}^{n} X_j$, $c = \sum_{i=1}^{n} T_i \prod_{j=1,\ j\neq i}^{n} X_j$;

begin real ca, cb, cc, a, b, c, syl, syu, term1, term2, term3, da,
        dif, sum;
integer   jm, js, jul, ia, ib;
  start: **switch** alpha := L1, L1, L12; **switch** beta := L9,
          L5, L6;
          switch gamma := L10, L11; **switch** delta := L8,
          L8, L13;
  comment   For interpolation, differentiation or integration set
          jt = 1, 2, or 3 respectively;
          go to alpha [jt];
  L1:   **if** xarg $\geq$ xa [nop] **then go to** L2;
          if xarg $\geq$ xa [nop−1] **then go to** L2;
          if xarg $\leq$ xa [1] **then go to** L3;
          if xarg $\leq$ xa [2] **then go to** L3; **go to** L4;

  L2:   jm := nop−1; js := 1; **go to** term;
  L3:   jm := 2; js := 1; **go to** term;
  comment   Locate argument;
  L4:   **for** ia := 2 **step** 1 **until** nop **do begin**
          if xa [ia] > xarg **then go to** L7; jm := ia **end**;
  comment   Before loop is complete xarg $\leq$ xa [ia];
  L5:   ca := a; cb := b; cc := c; js := 3; im :=
          jm+1; **go to** term;
  L6:   a := (ca+a)/2; b := (cb+b)/2; c := (cc+c)/2;
          go to L9;
  L7:   js := 2; **go to** term;
  L8:   **go to** beta [js];
  L9:   **go to** gamma [jt];
  comment   Interpolation, jt = 1;
  L10:   da := a $\times$ xarg $\uparrow$ 2 + b $\times$ xarg + c; **go to** exit1;
  comment   Differentiation, jt = 2;
  L11:   dif := 2 $\times$ xarg + b; **go to** exit2;
  comment   Integration, jt = 3;
  L12:   sum := 0; syl := xlo; jul := nop − 1;
          ib := 2;
  L16:   **for** jm := ib **step** 1 **until** jul **do begin**;
  comment   Lagrange formulae;
          term1 := ya [jm − 1]/((xa [jm − 1] − xa[jm]) $\times$
          (xa[jm − 1] − xa[jm + 1]));
          term2 := ya [jm]/((xa [jm] − xa [jm − 1]) $\times$
          (xa[jm] − xa [jm + 1]));
          term3 := ya [jm + 1]/((xa [jm + 1] − xa [jm − 1]) $\times$
          (xa [jm + 1] − xa [jm]));
          a := term1 + term2 + term3;
          b := −(xa [jm] + xa [jm + 1]) $\times$ term1 − (xa
          [jm − 1] + xa [jm + 1]) $\times$ term2 − (xa [jm − 1] +
          xa [jm]) $\times$ term3;
          c := xa [jm] $\times$ xa [jm + 1] $\times$ term1 + xa [jm − 1] $\times$
          xa [jm + 1] $\times$ term2 + xa [jm − 1] $\times$ xa [jm] $\times$
          term3; **go to** delta [jt];
  L13:   **if** jm $\neq$ 2 **then go to** L14;
          ca := a; cb := b; cc := c; **go to** L15;
  L14:   ca := (a + ca)/2; cb := (b + cb)/2; cc :=
          (c + cc)/2;
  L15:   syu := xa [jm];
          sum := sum + ca $\times$ (syu $\uparrow$ 3 − syl $\uparrow$ 3)/3 + cb $\times$
          (syu $\uparrow$ 2 − syl $\uparrow$ 2)/2 + cc $\times$ (syu − syl);
          ca := a; cb := b; cc := c; syl := syu **end**;
  comment   End of loop on [jm] index;
          sum := sum + ca $\times$ (xup $\uparrow$ 3-syl $\uparrow$ 3)/3 + cb $\times$
          (xup $\uparrow$ 2-syl $\uparrow$ 2)/2 + cc $\times$ (xup − syl); **go**
          to exit3;
  term: ib := jm; jul := ib; **go to** L16;
  comment   The results for interpolation, differentiation, and
          integration are da, dif, and sum respectively;
  exit1:   AVINT := da; **go to** exit;
  exit2:   AVINT := dif; **go to** exit;
  exit3:   AVINT := sum;
  exit: **end**

CERTIFICATION OF ALGORITHM 77
AVINT (Paul E. Hennion, *Comm. ACM 5*, Feb., 1962)
VICTOR E. WHITTIER
Computations Res. Lab., The Dow Chemical Co., Midland, Mich.

AVINT was transliterated into BAC-220 (a dialect of ALGOL-58) and was tested on the Burroughs 220 computer. The following minor errors were found:
1. The first statement following label L11 should read:
   dif := $2 \times$ a $\times$ xarg $+$ b;
2. The semicolon (;) at the end of the line beginning with the label L16 should be deleted.
3. There appears to be a confusion between "1" (numeric) and "l" (alphabetic) following label L12. This portion of the program should read:
   L12: sum := 0; syl := xlo; jul := nop $- 1$; ib := 2;

After making the above corrections the procedure was tested for interpolation, differentiation, and integration using e^x, log X, and sin X in the range $(1.0 \leq X \leq 5.0)$. Twenty-one values of each of these functions, evenly spaced with respect to X and accurate to at least 7 significant digits, were tabulated in the above range. Then the procedure was tested. The following table indicates approximately the accuracy obtained:

	Number of Significant Digits		
Function	*Interpolation*	*Differentiation*	*Integration*
e^x	≥ 4*	≥ 2	≥ 4
log X	≥ 4*	≥ 2	≥ 3
sin X	≥ 4*	≥ 2	≥ 4

* Except for interpolation between the first two points in the table.

The above results are quite reasonable in view of the relatively large increment in X. Tests using smaller increments in X and uneven spacing of X were also satisfactory.

It was also discovered that for integration the following restrictions must be observed:
1. xlo $\leq$ xa (1).
2. xup $\geq$ xa (nop).

REMARK ON ALGORITHM 77
INTERPOLATION, DIFFERENTIATION, AND INTEGRATION [P. E. Hennion, *Comm. ACM*, Feb., 1962]
P. E. HENNION
Giannini Controls Corp., Berwyn, Penn.

It was brought to my attention through the CERTIFICATION OF ALGORITHM 77 AVINT [V. E. Whittier, *Comm. ACM*, June, 1962] that restrictions on the upper and lower limits of integration existed, i.e., (1) $x10 \leq xa$ (1), (2) $xup \geq xa(nop)$. To remove these restrictions the following two changes should be made.
1. Before line $L16$: and after the statement $ib := 2$; place the following code:
   for $ia := 1$ **step** 1 **until** nop **do begin**
   if $xa(ia) \geq x10$ **then go to** $L17$; $ib := ib + 1$; **end**;
L17: $ju\,1 := nop + 1$; **for** $ia := 1$ **step** 1 **until** nop **do begin**
   $ju\,1 := ju\,1 - 1$; **if** $xa(ju\,1) > xup$ **end**; $ju\,1 := ju\,1 - 1$;
2. Change line $L13$: to read:
L13:   **if** $jm \neq ib$ **then go to** $L14$;

REMARK ON ALGORITHM 77
INTERPOLATION, DIFFERENTIATION, AND INTEGRATION [P. E. Hennion, *Comm. ACM 5*, Feb. 1962]
P. E. HENNION
Giannini Controls Corp., Berwyn, Penn.

It was brought to my attention through the CERTIFICATION OF ALGORITHM 77 AVINT (V. E. Whittier, *Comm. ACM*, June, 1962) that restrictions on the upper and lower limits of integration existed, i.e., (1) $xlo \leq xa(1)$, (2) $xup \geq xa(nop)$. To remove these restrictions the following two changes should be made.
1. Replace the two lines starting at line $L12$: and ending after the statement $ib := 2$; with the following code:
L12: $sum := 0$; $syl := xlo$; $ib := 2$, $ju1 := nop$;
   for $ia := 1$ **step** 1 **until** nop **do begin**
   if $xa [ia] \geq xlo$ **then go** *to* $L17$; $ib := ib + 1$; **end**;
L17:   **for** $ia := 1$ **step** 1 **until** nop **do begin**
   if $xup \geq xa [ju1]$ **then go to** $L18$; $ju1 := ju1 - 1$; **end**;
L18: $ju1 := ju1 - 1$;

2. Change line $L13$: to read
L13:   **if** $jm \neq ib$ **then go to** $L14$;

ALGORITHM 78
RATIONAL ROOTS OF POLYNOMIALS WITH INTEGER COEFFICIENTS
C. PERRY
University of California at San Diego, La Jolla, California

comment This ALGOL procedure, named ratfact, for finding rational roots of polynomials with integer coefficients is a pedagogical example illustrating the use of the **for** statement described in section 4.6.3. Also, an extension suggested by J. Peck of the well-known polynomial evaluation by nesting, i.e. Horner's method, is used. The polynomial $f(x) = a_0 + a_1x + \cdots + a_nx^n$ with integer coefficients and with $a_0 a_n \neq 0$ has a lowest term rational root p/q if and only if $a_0 q^n + a_1 q^{n-1} p + \cdots + a_{n-1} q\ p^{n-1} + a_n p^n = 0$, also q must be a factor of a_n and p a factor of a_0. Procedure RATFACT outputs the nonzero rational roots p/q by execution of the procedure whose formal name is print. The output procedure uses the string whose formal name is format for control of the output format;

```
procedure ratfact (a, n, print, format);
  integer array a[0:n]; integer n; procedure print; string
    format;
  begin integer i, p, q, r, t, f, g;
  p loop:  for p := 1 step 1 until abs (a[0]) do
    begin comment if p is not a factor of a [0] or q is not a factor
      of a[n] then skip to the end of the loop for advance in the
      respective for list;
      if a[0] ≠ (a[0]÷p)×p then go to 1
      else q loop: for q := 1 step 1 until abs (a[n]) do
        begin if a[n] ≠ (a[n]÷ q)×q then go to 2
          else
          begin comment root test and print;
            comment start polynomial evaluation;
              f := g := a[0];  t := p;
              for i := 1 step 1 until n do
              begin r := a[i]×t;
                    f := f×q+r;
                    g := −g×q+r;
                    t := t×p;
              end   polynomial evaluation;
              comment computing r saves  one subscript
                evaluation;
              if f=0 then print (format, p, q);
              if g=0 then print (format,−p, q);
              comment   print is the formal name of the procedure
                to be used to output the variables in the format
                specified by the string whose formal name is format;
          end root test and print;
        2:  end q loop;
1:  end p loop;
  end ratfact, without overflow test.
```

REMARK ON ALGORITHM 78
RATIONAL ROOTS OF POLYNOMIALS WITH INTEGER COEFFICIENTS [C. Perry, *Comm. ACM*, Feb. 1962]
D. M. COLLISON
Elliott Bros. (London) Ltd., Borehamwood, Herts., England

The algorithm was successfully run using the Elliott ALGOL translator on the National-Elliott 803. It was noticed that a multiple rational root will only be printed once by the procedure.

CERTIFICATION OF ALGORITHM 78.
RATFACT (C. Perry, *Comm. ACM 5*, Feb. 1962)
M. H. HALSTEAD
Navy Electronics Laboratory, San Diego, Calif.

RATFACT was copied in the Navy Electronics Laboratory International ALGOL Compiler, NELIAC, and tested on the UNIVAC M-490 Countess and the CDC 1604. Polynomials of order 2 through 6 were tested. No corrections were found necessary. It was noted that a polynomial whose coefficients included a common factor would produce superfluous values of p/q, in which this fraction was indeed a root, but one in which p and q contained a common factor.

ALGORITHM 79
DIFFERENCE EXPRESSION COEFFICIENTS

THOMAS P. GIAMMO

Space Technology Laboratories, Inc., Los Angeles, California

```
procedure  dicol (k, n, xp, xtab, coef);
value k, n;  integer k, n;  real xp;
array xtab, coef;
comment  dicol produces the coefficients for the n ordinates
    (corresponding to the abscissae, xtab) in the n-point finite
    difference expression for the k-th derivative evaluated at xp.
    The method used is to determine the analytic expression for
    the k-th derivative of each coefficient in the n-point Lagrangian
    interpolation formula and evaluate it at xp. Note that k=0
    will produce the Lagrangian interpolation coefficients them-
    selves;
begin integer array xuse [1 : n−1];  real factk, sum, denom,
    part;
integer i, terms, j, m, high;
factk := 1.0;  for i := 2 step 1 until k do factk := i×factk;
terms := n−k−1;  if terms <0 then go to Z;
for j := 1 step 1 until n do
loop:  begin sum := 0;  denom := 1.0;  part := 1.0;
        for i := 1 step 1 until n do
        if i ≠ j then denom := denom× (xtab [j] − xtab [i]);
        if terms = 0 then go to Y;
        m := 1;  high := 1;
    A:  if (high = j)∨(xtab [high] = xp) then
        A1:  begin high := high + 1;  go to A end A1;
        if high >.n then A2:  begin m := m−1;  if m>0
          then
        A3:  begin high := xuse [m]+1;  go to A end A3;
        go to X end A2;
        xuse [m] := high;  m := m+1;
        if m≦terms then begin high := high + 1;  go to
          A end;
        for i := 1 step 1 until terms do
          part := part× (xp − xtab [xuse [i]]);
        sum := sum + part;  m := terms;  part := 1.0;
        high := xuse [terms] + 1;  go to A;
    Y:  sum := 1.0;
    X:  coef [j] := sum × factk/denom end loop;
      go to EXIT;
    Z:  for i := 1 step 1 until n do coef [i] := 0;
    EXIT:  end dicol
```

n	Approximate Number of Machine Operations
4	1.3×10^3
6	6.9×10^3
8	3.8×10^4
10	1.8×10^5
12	8.6×10^5

The author indicated in a letter that the procedure was developed for use with small n and small k.

CERTIFICATION OF ALGORITHM 79
DIFFERENCE EXPRESSION COEFFICIENTS
[Thomas Giamo, *Comm. ACM*, Feb. 1962]

EVA S. CLARK

University of California at San Diego, La Jolla, California

The procedure was translated into FORTRAN and run on the CDC 1604. Reasonable accuracy was obtained for $k = 0, 4 \leq n \leq 12$. For increasing n and increasing k, the accuracy diminished. It was found that the execution time increased rapidly as n was increased. For $k = 0$, the following results were obtained:

COLLECTED ALGORITHMS FROM CACM

ALGORITHM 80
RECIPROCAL GAMMA FUNCTION OF REAL
   ARGUMENT

WILLIAM HOLSTEN
University of California at San Diego, La Jolla, California

real procedure RGR(x); **real** x; **real procedure** RGAM;
comment Procedure RGAM computes the real reciprocal
   Gamma function of real x for $-1 < x < 1$, utilizing Horner's
   method for polynomial evaluation of the approximation polynomial. RGR extends the range of RGAM by use of the formulae
   (1) $1/\text{Gamma}(x-1) = (x-1)/\text{Gamma}(x)$ for $x < -1$,
   (2) $1/\text{Gamma}(x+1) = 1/x \times \text{Gamma}(x)$ for $x < 1$.;

```
        begin  real y;
            if x = 0.then begin RGR := 0;  go to EXIT end
            if x = 1 then begin RGR := 1;  go to EXIT end
            if x < 1 then go to BB;
            y := 1;
AA:         x := x - 1;  y := y × x;  if x > 1 then go to AA;
            if x = 1 then begin RGR := 1/y;  go to EXIT end
            RGR := RGAM(x)/y;  go to EXIT;
BB:         if x = -1 then begin RGR := 0;  go to EXIT end
            if x > -1 then begin RGR := RGAM(x);
                go to EXIT end
            y := x;
CC:         x := x + 1;  if x < -1 then begin y := y × x;
                go to CC end
            RGR := RGAM(x) × y;
EXIT:   end RGR;

real procedure RGAM(x);  real x;  integer i;
    real array B[0:13];
```

comment The algorithm for this routine was adapted from
   "University of Illinois Digital Computer, Auxiliary Library
   Routine B-17-328", by John Ehrman. Reference may also be
   made to Algorithm 34, dated February, 1961. Approximation
   accuracy is $\pm 2^{-35}$.;

```
begin real z;
    B[ 0] :=   1.00000 00000 00;  B[ 1] := - .42278 43350 92;
    B[ 2] := - .23309 37363 65;  B[ 3] := + .19109 11011 62;
    B[ 4] := - .02455 24908 87;  B[ 5] := - .01764 52421 18;
    B[ 6] := + .00802 32781 13;  B[ 7] := - .00080 43413 35;
    B[ 8] := - .00036 08514 96;  B[ 9] := + .00014 56243 24;
    B[10] := - .00001 75279 17;  B[11] := - .00000 26257 .21;
    B[12] := + .00000 13285 54;  B[13] := - .00000 01812 20;
    z: = B[13];
    for i := 12 step -1 until 0 do z := z × x + B[i];
    RGAM := z × x × (x + 1)
end RGAM;
```

REMARKS ON:
ALGORITHM 34 [S14]
GAMMA FUNCTION
   [M. F. Lipp, *Comm. ACM 4* (Feb. 1961), 106]
ALGORITHM 54 [S14]
GAMMA FUNCTION FOR RANGE 1 TO 2
   [John R. Herndon, *Comm. ACM 4* (Apr. 1961), 180]

ALGORITHM 80 [S14]
RECIPROCAL GAMMA FUNCTION OF REAL
ARGUMENT
   [William Holsten, *Comm. ACM 5* (Mar. 1962), 166]
ALGORITHM 221 [S14]
GAMMA FUNCTION
   [Walter Gautschi, *Comm. ACM 7* (Mar. 1964), 143]
ALGORITHM 291 [S14]
LOGARITHM OF GAMMA FUNCTION
   [M. C. Pike and I. D. Hill, *Comm. ACM 9* (Sept. 1966),
   684]

M. C. PIKE AND I. D. HILL (Recd. 12 Jan. 1966)
Medical Research Council's Statistical Research Unit,
University College Hospital Medical School,
London, England

Algorithms 34 and 54 both use the same Hastings approximation, accurate to about 7 decimal places. Of these two, Algorithm
54 is to be preferred on grounds of speed.

Algorithm 80 has the following errors:
(1) *RGAM* should be in the parameter list of *RGR*.
(2) The lines
   if $x = 0$ **then begin** *RGR* := 0; **go to** *EXIT* **end**
and
   if $x = 1$ **then begin** *RGR* := 1; **go to** *EXIT* **end**
should each be followed either by a semicolon or preferably by an
else.
(3) The lines
   if $x = 1$ **then begin** *RGR* := 1/y; **go to** *EXIT* **end**
and
   if $x < -1$ **then begin** $y := y \times x$; **go to** *CC* **end**
should each be followed by a semicolon.
(4) The lines
   BB: **if** $x = -1$ **then begin** *RGR* := 0; **go to** *EXIT* **end**
and
   if $x > -1$ **then begin** *RGR* := *RGAM*(x); **go to** *EXIT* **end**
should be separated either by **else** or by a semicolon and this
second line needs terminating with a semicolon.
(5) The declarations of **integer** i and **real array** $B[0:13]$ in *RGAM*
are in the wrong place; they should come immediately after
   begin real z;

With these modifications (and the replacement of the array B
in *RGAM* by the obvious nested multiplication) Algorithm 80 ran
successfully on the ICT Atlas computer with the ICT Atlas
ALGOL compiler and gave answers correct to 10 significant digits.

Algorithms 80, 221 and 291 all work to an accuracy of about 10
decimal places and to evaluate the gamma function it is therefore
on grounds of speed that a choice should be made between them.
Algorithms 80 and 221 take virtually the same amount of computing time, being twice as fast as 291 at $x = 1$, but this advantage
decreases steadily with increasing x so that at $x = 7$ the speeds are
about equal and then from this point on 291 is faster—taking only
about a third of the time at $x = 25$ and about a tenth of the time
at $x = 78$. These timings include taking the exponential of *log-*

gamma.

For many applications a ratio of gamma functions is required (e.g. binomial coefficients, incomplete beta function ratio) and the use of algorithm 291 allows such a ratio to be calculated for much larger arguments without overflow difficulties.

ALGORITHM 81
ECONOMISING A SEQUENCE 1
Brian H. Mayoh
Digital Computer Laboratory, University of Illinois,
Urbana, Ill.

```
procedure ECONOMISER 1 (desired property, costs, n, C);
  array costs;  integer n;
    Boolean procedure desired property;
  Boolean array C;
begin comment Given a finite, monotonely increasing
  sequence of positive numbers, looked upon as prices, ECONO-
  MISER 1 selects the cheapest subsequence with a given prop-
  erty. The formal parameters are: Desired property, a function
  designator to answer the question: Does the subsequence held
  in array C possess the required property? n is (number of ele-
  ments in the sequence) + 1.  Costs is an array of size [1:n].
  Costs[1] to costs[n−1] hold the numbers of the sequence and
  costs[n] is any arbitrary number greater than the sum of all
  other elements of costs.  C is an array of the same size and indi-
  cates a subsequence by the rule: C[i] ≡ element i of the original
  sequence is in the subsequence. At exit from ECONOMISER 1,
  C indicates the cheapest subsequence. It is supposed that the
  original sequence has the desired property.;
  integer d, j, k, ℓ;  real i;
  for j := 1 step 1 until n do C[j] := j = 1;  ⌐d := 0;
  reenter:  d := d+1;
  INSIDE:  begin own real array prices [1:d];
              own Boolean array alternatives[1:d, 1:n];
              procedure ENTER SUCCESSORS;
              begin k := n−1;
                A:  if ⌐ C[k] then
                      begin k := k−1;  go to A end;  i := 0;
                      for j := 1 step 1 until n do
                      begin alternatives[ℓ,j]
                          := j ≠ k ∧ j ≠ k−1 ≡ C[j];
                          if alternatives[ℓ,j] then
                          i := i + costs[j]
                      end;
                B:  k := k−1;
                      go to if k = 0 then find cheapest
                        else if C[k] then (if k=1 then
                          find cheapest else B)
                        else if k=1 then E
                          else if C[k−1] then D
                        else find cheapest;
                D:  C[k−1] := false;
                E:  C[k] := true;  go to reenter
              end of ENTER SUCCESSORS;
              i := 0;  for j := 1 step 1 until n do
              begin alternatives[d,j] := C[j];  if C[j] then
                  i := i + costs[j]
              end;  prices[d] := i;
          find cheapest:  i := 0;  for j := 1 step 1 until d do
              begin if prices[j] < i then
                begin ℓ := j;  i := prices[ℓ] end
              end;
              for j := 1 step 1 until n do
              C[j] := alternatives[ℓ,j];
              if ⌐ desired property then
                ENTER SUCCESSORS
          end of INSIDE;
end of ECONOMISER 1.
```

ALGORITHM 82
ECONOMISING A SEQUENCE 2
Brian H. Mayoh
Digital Computer Laboratory, University of Illinois,
Urbana, Ill.

```
procedure  ECONOMISER 2 (desired property, costs, n, C, r,
  Reject list);  Boolean procedure desired property;
  integer n, r;  array costs;  Boolean array Reject list;
begin comment  In some applications of ECONOMISER 1, it
  is simple to establish that some subsequences are redundant in
  the sense that any sequence containing them is certainly not
  the cheapest subsequence with the desired property. For such
  applications ECONOMISER 2 avoids all unnecessary calls of
  desired property. The new formal parameters are:  r a variable
  whose value is initially 0 and is increased by 1 every time that
  desired property discovers a new redundant subsequence.
  Reject list an array of size [1:r,1:n]. Reject list [a,b] carries the
  answer to: Is element b of the original sequence in the a^th
  redundant subsequence found by desired property?;
  real i;  integer d, j, k, ℓ;  Boolean gapfilled, first time;
  procedure INSIDE (entrymaker);  Boolean entrymaker:
  begin own real array prices[1:d];
        own Boolean array alternatives[1:d,1:n];
        procedure ENTER SUCCESSORS;
        begin integer c;  Boolean array ssq[1:n];
            for j := 1 step 1 until n do ssq[j] := C[j];
            c := n−1;
    A:  if ¬ ssq[c] then begin c := c−1;  go to A end:
        C[c] := false;  C[c+1] := true;
        INSIDE (true);
        gapfilled := true;
    B:  c := c−1;
        go to if c=0 then F else if ssq[c] then
          (if c=1 then F else B) else if c=1 then
          E else if ssq[c−1] then D else F;
    D:  ssq[c−1] := false;
    E:  for j := 1 step 1 until n do C[j] := ssq[j] ≡ j≠c;
        INSIDE (true);
    F:  end of ENTER SUCCESSORS;
        if entrymaker then
        begin for j := 1 step 1 until r do
            begin for k := 1 step 1 until n do
              begin if ¬ C[k] ∧ Reject list[j,k] then
                go to G end;
              ENTER SUCCESSORS;  go to H;
    G:  end;
        i := 0;  if gapfilled then d := d+1;
        for j := 1 step 1 until n do
        begin alternatives[if gapfilled then
          d else ℓ, j] := C[j];
          if C[j] then i := i + costs[j]
        end;  prices[if gapfilled then d else ℓ] := i
      end;  if first time ∨ ¬ entrymaker then
      begin i := 0;  gapfilled := first time := false;
        for j := 1 step 1 until d do
        begin if prices[j] < i then
          begin ℓ := j;  i := prices[ℓ] end
        end;
            for j := 1 step 1 until n do
              C[j] := alternatives[ℓ,j];
          if desired property then go to found;
          ENTER SUCCESSORS;  go to reenter
        end;
  H:  end of INSIDE;
    for j := 1 step 1 until n do C[j] := j=1;
    d := 0;  first time := gapfilled := true;
reenter:  INSIDE (first time);
found:
end of ECONOMISER 2;
```

ALGORITHM 83
OPTIMAL CLASSIFICATION OF OBJECTS
BRIAN H. MAYOH
Digital Computer Laboratory, University of Illinois,
Urbana, Ill.

```
procedure OPTIMUM COVERING FINDER (Pattern, popu-
    lation, set number, set prices, chosen sets, bounds, overflow);
    Boolean array Pattern, chosen sets; integer population,
    set number, bounds; array set prices; label overflow;
begin comment The number of objects in some given set is
    given by population. The procedure is given a classification of
    these objects by a collection of overlapping subsets. A cost
    is assigned to each subset. Then OPTIMUM COVERING
    FINDER selects the cheapest subcollection such that every
    object is contained in at least one of the subsets of the sub-
    collection. set prices[i] carries the cost of subset i. Pattern
    is an array of size [1:set number,1:population] such that Pat-
    tern[a,b] ≡ does subset a include object b. chosen sets[i] finally
    carries the answer to the question: Is set i in the cheapest
    subcollection? The programmer must restrict the amount of
    space available to the procedure by setting bounds. From ex-
    perience bounds = set number ↑ 2 suffices to avoid most alarm
    exits to overflow.;
    Boolean array C[1:population], D[1:bounds, 1:population],
      R, S[1:bounds,1:set number];
    integer a, b, d, r, s;
    Boolean procedure HAVE WE A COVERING;
    begin procedure ADD to (Q,q,f); integer q;
            real f; Boolean array Q;
        begin if q=bounds then go to overflow else q := q+1;
          for a := 1 step 1 until set number do Q[q,a] := f
        end; for a := 1 step 1 until population do
                C[a] := false;
        for a := 1 step 1 until set number do
        begin if chosen sets[a] then
          for b := 1 step 1 until population do
          C[b] := C[b] ∨ Pattern[a,b]
        end; for a := 1 step 1 until population do
        begin if ¬ C[a] then go to E end;
        go to found;
    E:  for d := 1 step 1 until s do
        begin for b := 1 step 1 until population do
          begin if C[b] ∧ ¬ D[d,b] then go to try another end;
          ADD to (R, r, chosen sets[a]);
          for b := 1 step 1 until set number do
          begin if chosen sets[b] ∧ ¬ S[d,b] then
            ADD to (R, r, S[d,a] ∨ a=b)
          end; go to F;
        try another:
        end of for statement labelled E;
        ADD to (S, s, chosen sets[a]);
        for a := 1 step 1 until population do D[s,a] := C[a];
    F:  HAVE WE A COVERING := false
    end; r := s := 0;
    ECONOMISER 2 (HAVE WE A COVERING, set prices,
      set number, r, R, chosen sets);
found: end
```

ALGORITHM 84
SIMPSON'S INTEGRATION

Paul E. Hennion
Giannini Controls Corporation
Astromechanics Research Division, Berwyn, Penn.

real procedure SIM (n, a, b, y);
value n, a, b; **real** a, b; **integer** n; **array** y;
comment This is a method for obtaining the approximate value
of the definite integral of a continuous function when the integral cannot be evaluated in elementary functions. Given
$y = f(x)$ and the $\int_a^b y\,dx$ to be evaluated. Plot the curve $f(x)$,
and divide $[a, b]$ evenly into n equal parts, erecting the ordinates $y_0, y_1, \cdots, y_n$. Then the approximate value of the
definite integral by Simpson's rule states that:

$$\int_a^b f(x)\,dx = \frac{b-a}{3n}(y_0 + 4y_1 + 2y_2 + \cdots + 4y_{n-1} + y_n);$$

begin real s; **integer** i;
  s := (y[0] − y[n])/2;
  for i := 1 **step** 2 **until** n − 1 **do** s := s + 2 × y[i] + y[i+1];
  SIM := 2 × (b − a) × s/(3 × n)
end

CERTIFICATION OF ALGORITHM 84
SIMPSON'S INTEGRATION [P. E. Hennion, *Comm. ACM 5* (Apr. 1962)]

A. P. Relph
The English Electric Co., Whetstone, England

Simpson's Integration was translated using the Deuce Algol
compiler and, with no corrections, gave satisfactory results.

It is not stated in the comment that integer *n* needs to be even.

REMARK ON ALGORITHM 84
SIMPSON'S INTEGRATION [Paul E. Hennion. *Comm. ACM*, Apr. 1962]

Richard George*
Particle Accelerator Div., Argonne National Lab.,
Argonne, Ill.
* Work supported by the U. S. Atomic Energy Commission.

In performing integration by the use of Simpson's rule, it is well
known that the interval $[a, b]$ must be divided evenly into *n* equal
parts, and that *it is essential for n to be an even number*.

In the published algorithm, there is neither a comment on this
important restriction, nor a programmed test for the parity of *n*.
It is therefore a potential trap for the unwary programmer.

CERTIFICATION OF ALGORITHM 84
SIMPSON'S INTEGRATION [P. E. Hennion, *Comm. ACM*, Apr. 62]

Peter G. Behrenz
Matematikmaskinnämnden, Stockholm, Sweden

SIM was successfully run on FACIT EDB using Facit-Algol
1, which is a realization of Algol 60 for FACIT EDB. No changes
in the program were necessary. To test SIM some polynomials
were integrated.

ALGORITHM 85
JACOBI
Thomas G. Evans
Bolt, Beranek, and Newman*, Cambridge, Mass.

---

* This work has been sponsored by the Air Force Cambridge Research Laboratories, OAR (USAF), Detection Physics Laboratory, under contract AF 19(628)-227.

procedure JACOBI (A, S, n, rho);
value n, rho; **integer** n; **real** rho; **real array** A, S;
comment This procedure finds all eigenvalues and eigenvectors of a given square symmetric matrix by a modified Jacobi (iterative) method (cf. J. Greenstadt, *"The determination of the characteristic roots of a matrix by the Jacobi method,"* in *Mathematical Methods for Digital Computers*, A. Ralston and H. S. Wilf, eds.). JACOBI is given a square symmetric matrix of order n stored in the array A. The initial contents of the array S are immaterial, as S is initialized by the procedure. At exit the kth column of the array S contains the kth of the n eigenvectors of the given matrix, and the diagonal element A[k, k] of the array A is the corresponing k^{th} eigenvalue. The parameter rho is the "accuracy requirement" introduced in the above reference, where a detailed flow chart of the method is given. The significance of rho is that the iteration terminates when, for every off-diagonal element A[i, j], abs (A[i, j]) < (rho/n) × norm1, where norm1 is a function only of the off-diagonal elements of the original matrix;
begin real norm1, norm2, thr, mu, omega, sint, cost, int1, v1, v2, v3;
  integer i, j, p, q, ind;
  comment Set array S = n × n identity matrix;
  for i := 1 **step** 1 **until** n **do**
  for j := 1 **step** 1 **until** i **do**
    if i = j **then** S[i, j] := 1.0
    else S[i, j] := S[j, i] := 0.0;
  comment Calculate initial norm (norm1), final norm (norm2), and threshold (thr);
  int1 := 0.0;
  for i := 2 **step** 1 **until** n **do**
    for j := 1 **step** 1 **until** i−1 **do**
      int1 := int1 + 2.0 × A[i, j] ↑ 2;
  norm1 := sqrt (int1); norm2 := (rho/n) × norm1;
  thr := norm1; ind := 0;
main: thr := thr/n;
  comment The sweep through the off-diagonal elements begins here;
main1: **for** q := 2 **step** 1 **until** n **do**
    for p := 1 **step** 1 **until** q−1 **do**
      if abs (A[p, q]) ≧ thr **then**
        begin ind := 1; v1 := A[p, p]; v2 := A[p, q];
        v3 := A[q, q]; mu := 0.5 × (v1−v3);
        omega := (**if** mu = 0.0 **then** 1 **else** sign (mu)) ×
          (−v2)/sqrt(v2↑2 + mu↑2);
        sint := omega/sqrt(2.0 × (1.0 + sqrt(1.0 − omega↑2)));
        cost := sqrt (1.0 − sint↑2);
        for i := 1 **step** 1 **until** n **do**
        begin int1 := A[i, p] × cost − A[i, q] × sint;
          A[i, q] := A[i, p] × sint + A[i, q] × cost;
          A[i, p] := int1;
          int1 := S[i, p] × cost − S[i, q] × sint;

          S[i, q] := S[i, p] × sint + S[i, q] × cost;
          S[i, p] := int1
        end;
        for i := **step** 1 **until** n **do**
          begin A[p, i] := A[i, p]; A[q, i] := A[i q] **end;**
        A[p, p] := v1 × cost↑2 + v3 × sint↑2 − 2.0 × v2 × sint × cost;
        A[q, q] := v1 × sint↑2 + v3 × cost↑2 + 2.0 × v2 × sint × cost;
        A[p, q] := A[q, p] := (v1 − v3) × sint × cost + v2 × (cost↑2 − sint↑2)
      end;
  comment Now test to see if current tolerance exceeded and, if not, whether final tolerance reached;
  if ind = 1 **then begin** ind := 0; **go to** main1 **end**
  else if thr > norm2 **then go to** main
end JACOBI

CERTIFICATION OF ALGORITHM 85
JACOBI [T. G. Evans, *Comm. ACM*, Apr. 1962]
J. S. Hillmore
Elliott Bros. (London) Ltd., Borehamwood, Herts., England

The statement
$$omega := (\textbf{if } mu = 0.0 \textbf{ then } 1 \textbf{ else } sign (mu)) \times (−V2)/sqrt(V2 \uparrow 2 + mu \uparrow 2);$$
was changed to
$$omega := \textbf{if } mu = 0.0 \textbf{ then } −1.0 \textbf{ else } − sign (mu) \times V2/sqrt (V2 \uparrow 2 + mu \uparrow 2);$$
When $mu = 0$, the original statement reduces to
$$omega := − V2/sqrt (V2 \uparrow 2);$$
and a truncation error in the evaluation of the square root can make the magnitude of omega slightly greater than unity. As a result, an error stop occurs during execution of the next statement when an attempt is made to evaluate $sqrt (1 − omega \uparrow 2)$.

In its modified form the algorithm has been successfully run using the Elliott ALGOL translator on the National-Elliott 803. Matrices of order up to fifteen have been solved, yielding eigenvalues and eigenvectors with an overall accuracy of seven decimal places.

CERTIFICATION OF ALGORITHM 85
JACOBI [Thomas G. Evans, *Comm. ACM* (Apr. 1962), 208]
P. Naur
Regnecentralen, Copenhagen, Denmark

We have first run this algorithm in the GIER ALGOL system with the following corrections included:

1. The change given by J. S. Hillmore [*Comm. ACM 5* (Aug. 1962), 440] with capital *V* changed to *v*.

2. The 4th **for** clause corrected to read:

$$\textbf{for } j := 1 \textbf{ step } 1 \textbf{ until } i - 1 \textbf{ do}$$

3. The last **for** clause corrected to read:

$$\textbf{for } i := 1 \textbf{ step } 1 \textbf{ until } n \textbf{ do}$$

On closer examination we have found, however, that a significant number of superfluous operations could be eliminated in the innermost loop by rewriting the two **for** statements at the center of the algorithm as a single **for** statement, to read as follows:

```
.....
cost := sqrt (1−sint ↑ 2);
for i := 1 step 1 unti! n do
  begin if i ≠ p ∧ i ≠ q then
    begin int1 := A[i,p];  mu := A[i,q];
    A[q,i] := A[i,q] := int1 × sint + mu × cost;
    A[p,i] := A[i,p] := int1 × cost − mu + sint
    end;
  intl := S[i,p];  mu := S[i,q];
  S[i,q] := int1 × sint + mu × cost;
  S[i,p] := int1 × cost − mu × sint
  end;
A[p,p] := v1 × cost ↑ 2 + v3 × sint ↑ 2 − 2 × v2 × sint × cost;
.....
```

This revision is particularly advantageous in systems having a comparatively slow subscript mechanism, such as GIER ALGOL, because it eliminates more than 3 out of 8 references to subscripted variables.

JACOBI has been tried with two different sets of matrices having known eigenvalues. In both cases a test program was set up to find the range of errors of the eigenvalues computed by JACOBI. In addition, the relations $Av - \lambda v = 0$ (A is the given matrix, v an eigenvector, and λ the corresponding eigenvalue) and $A - (ST)$ LAMBDA $S = 0$ (S is the matrix having the eigenvectors as columns and ST its transpose, and LAMBDA is the diagonal matrix of the eigenvalues) were used as checks. The test matrices were TESTMATRIX calculated by the revised algorithm 52 given in *Comm. ACM 6* (Jan. 1963), 39, and the following matrix suggested by Mr. H. B. Hansen:

$$\text{HBH TESTMATRIX } [j,i] = \text{HBH TESTMATRIX } [i,j]$$
$$= n + 1 - j \qquad j \geq i$$

having the eigenvalues $0.5/(1 - \cos((2 \times i - 1) \times pi/(2 \times n + 1)))$.

The results were as shown in Table 1 (GIER ALGOL works with floating numbers of 29 significant bits).

The compile time for the program which produced one of these tables was about 40 seconds. Run times were as follows:

Rho	n	*Original algorithm* TESTMATRIX ALG. 52 (seconds)	HBH TESTMATRIX	*Revised algorithm* HBH TESTMATRIX (seconds)
10^{-3}	5			3
	10			22
	15			70
10^{-5}	5	3		5
	10	5	41	29
	15	13	148	99
10^{-8}	5	4	7	6
	6	5	12	
	7	5	18	
	8	5	25	
	10	13		38
	15	22		116

From these figures it looks as if TESTMATRIX, Algorithm 52, is atypical as far as solution by means of JACOBI is concerned. The much higher accuracy obtained for this matrix as compared with the HBH matrix points in the same direction.

For further comparison it may be mentioned that the algorithms published by J. H. Wilkinson [*Num. Math. 4* (1962), 354-376] also have been tested successfully with GIER ALGOL. Wilkinson's algorithms reduce the matrix to tridiagonal form by means of Householder's method and use Sturm sequences to find the eigenvalues and inverse iteration to find the eigenvectors. In GIER ALGOL this method is about 1.3 times as fast as JACOBI for the range of matrices considered here. JACOBI has the advantage that the eigenvectors are properly orthogonal, even in the case of multiple eigenvalues, and also has a much simpler logic. On the other hand if only some of the eigenvalues and/or eigenvectors are sought Wilkinson's algorithms will often offer much higher speed than JACOBI, which always finds them all.

TABLE 1
HBH TESTMATRIX

	Range of true errors of eigenvalues				Range of deviations from relation $Av - \lambda v = 0$						Range of deviations from relation $A - (ST)\,\Lambda\,S = 0$					
Order	j	error[j]	j	error[j]	Element	Vector	Error	Element	Vector	Error	Element	Vector	Error	Element	Vector	Error
							$rho = 1.0_{10}-3$									
5	1	$-1.1_{10}-6$	3	$5.2_{10}-8$	1	1	$-1.7_{10}-4$	1	3	$2.0_{10}-4$	1	1	$-2.5_{10}-4$	5	5	$1.0_{10}-4$
10	9	$-7.9_{10}-5$	8	$3.5_{10}-5$	7	2	$-3.3_{10}-3$	6	6	$3.0_{10}-3$	1	1	$-4.2_{10}-3$	6	7	$3.2_{10}-3$
15	15	$-9.2_{10}-5$	12	$3.7_{10}-5$	6	3	$-1.7_{10}-3$	11	13	$1.7_{10}-3$	9	15	$-1.5_{10}-3$	8	9	$1.8_{10}-3$
							$rho = 1.0_{10}-5$									
5	1	$-1.1_{10}-6$	3	$6.0_{10}-8$	2	5	$-1.3_{10}-7$	5	2	$4.1_{10}-8$	1	2	$-1.6_{10}-7$	4	5	$4.5_{10}-8$
10	1	$-1.2_{10}-5$	2	$2.2_{10}-7$	7	3	$-2.7_{10}-5$	2	8	$2.2_{10}-5$	7	7	$-2.4_{10}-5$	2	8	$2.3_{10}-5$
15	1	$-3.5_{10}-5$	4	$3.9_{10}-7$	11	9	$-6.4_{10}-6$	7	2	$4.8_{10}-6$	11	12	$-5.3_{10}-6$	12	12	$4.7_{10}-6$
							$rho = 1.0_{10}-8$									
5	1	$-1.1_{10}-6$	3	$6.0_{10}-8$	2	5	$-1.3_{10}-7$	4	2	$6.5_{10}-9$	2	2	$-1.3_{10}-7$	4	4	$3.0_{10}-8$
10	1	$-1.2_{10}-5$	2	$2.2_{10}-7$	1	10	$-1.1_{10}-6$	4	2	$6.4_{10}-8$	1	2	$-5.7_{10}-7$	9	9	$8.2_{10}-8$
15	1	$-3.5_{10}-5$	4	$3.9_{10}-7$	1	14	$-3.4_{10}-6$	4	2	$3.9_{10}-7$	2	2	$-1.3_{10}-6$	15	15	$8.9_{10}-8$

TESTMATRIX, Algorithm 52

	Range of true errors of eigenvalues				Range of deviations from relation $Av - \lambda v = 0$						Range of deviations from relation $A - (ST)\,\Lambda\,S = 0$					
Order	j	error[j]	j	error[j]	Element	Vector	Error	Element	Vector	Error	Element	Vector	Error	Element	Vector	Error
							$rho = 1.0_{10}-5$									
5	4	$-1.0_{10}-8$	1	$.0$	5	5	$-3.3_{10}-8$	5	4	$4.3_{10}-8$	5	5	$-5.1_{10}-8$	4	4	$3.9_{10}-8$
10	8	$-1.1_{10}-8$	4	$.0$	7	7	$-1.2_{10}-8$	9	6	$1.3_{10}-8$	7	8	$-5.1_{10}-9$	6	6	$2.0_{10}-8$
15	13	$-1.1_{10}-8$	6	$.0$	14	14	$-9.3_{10}-9$	10	10	$9.4_{10}-9$	8	9	$-1.9_{10}-9$	10	10	$1.3_{10}-8$
							$rho = 1.0_{10}-8$									
3	3	$-7.5_{10}-9$	1	$3.7_{10}-9$	3	1	$-2.8_{10}-9$	2	2	$9.3_{10}-9$	1	3	$.0$	1	2	$1.9_{10}-8$
4	4	$-5.6_{10}-9$	3	$.0$	2	2	$-4.5_{10}-9$	3	4	$3.3_{10}-9$	2	2	$.0$	2	3	$9.3_{10}-9$
5	4	$-1.0_{10}-8$	1	$.0$	5	4	$-4.9_{10}-9$	4	4	$5.8_{10}-9$	1	1	$-7.5_{10}-9$	3	4	$7.5_{10}-9$
6	4	$-4.7_{10}-9$	4	$.0$	4	3	$-2.8_{10}-9$	5	4	$3.6_{10}-9$	1	6	$-2.3_{10}-10$	4	5	$9.3_{10}-9$
7	4	$-5.1_{10}-9$	5	$.0$	6	6	$-2.8_{10}-9$	4	4	$3.4_{10}-9$	5	7	$-1.2_{10}-10$	5	6	$7.5_{10}-9$
8	7	$-7.5_{10}-9$	5	$.0$	5	5	$-6.0_{10}-9$	5	6	$3.2_{10}-9$	8	8	$-1.2_{10}-10$	7	7	$9.3_{10}-9$
9	6	$-4.4_{10}-9$	7	$.0$	6	5	$-5.1_{10}-9$	7	6	$3.2_{10}-9$	5	5	$-7.5_{10}-9$	8	8	$1.5_{10}-8$
10	8	$-1.5_{10}-8$	8	$.0$	8	9	$-9.3_{10}-9$	9	7	$7.2_{10}-9$	6	7	$-2.3_{10}-9$	9	9	$2.0_{10}-8$
11	10	$-7.5_{10}-9$	1	$.0$	9	10	$-6.5_{10}-9$	8	11	$3.0_{10}-9$	1	1	$-3.1_{10}-9$	8	8	$7.5_{10}-9$
12	8	$-5.0_{10}-9$	11	$.0$	10	6	$-7.6_{10}-9$	10	8	$2.4_{10}-9$	6	6	$-1.7_{10}-8$	4	4	$1.3_{10}-8$
13	12	$-1.1_{10}-8$	10	$.0$	10	11	$-6.9_{10}-9$	12	10	$9.1_{10}-9$	7	7	$-3.0_{10}-8$	12	12	$3.2_{10}-8$
14	10	$-1.5_{10}-8$	4	$.0$	13	13	$-1.1_{10}-8$	10	10	$6.7_{10}-9$	9	10	$-3.5_{10}-9$	6	6	$1.7_{10}-8$
15	13	$-1.1_{10}-8$	6	$.0$	14	14	$-1.1_{10}-8$	11	10	$3.5_{10}-9$	8	9	$-3.0_{10}-9$	6	11	$7.5_{10}-9$

ALGORITHM 86
PERMUTE
J. E. L. Peck and G. F. Schrack
University of Alberta, Calgary, Alberta, Canada

```
procedure PERMUTE (x, n);
array x;  integer n;
comment  Each call of PERMUTE executes a permutation of
  the first n components of x. It assumes a nonlocal Boolean
  variable 'first', which when true causes the procedure to initial-
  ise the signature vector p. Thereafter 'first' remains false until
  after n! calls;
begin  own  integer  array  p[2:n];  integer i, k;
  if first then
  begin for i := 2 step 1 until n do
            p[i] := i;  first := false
  end initialise;
  for k := 2 step 1 until n do
    begin integer km;  real t;
      t := x[1];  km := k − 1;
      for i := 1 step 1 until km do
          x[i] := x[i+1];
      x[k] := t;  p[k] := p[k] − 1;
      if p[k] ≠ 0 then go to EXIT;
      p[k] := k
    end k;
  first := true;
EXIT: end PERMUTE
```

CERTIFICATION OF ALGORITHM 86
PERMUTE [J. E. L. Peck and G. F. Schrock, *Comm.
  ACM*, Apr. 1962]
D. M. Collison
Elliott Bros. (London) Ltd., Borehamwood, Herts.,
England

The algorithm was successfully run using the Elliott Algol
translator on the National-Elliott 803. Values of n used were 0, 1,
2, 3, 4.

ALGORITHM 87
PERMUTATION GENERATOR
JOHN R. HOWELL
Orlando Aerospace Division, Martin Marietta Corp.,
Orlando, Florida

```
procedure PERMUTATION (N, K);
value K, N;  integer K;  integer array N;
comment This procedure generates the next permutation in
  lexicographic order from a given permutation of the K marks
  0, 1, ··· , (K−1) by the repeated addition of (K−1) radix K.
  The radix K arithmetic is simulated by the addition of 9 radix
  10 and a test to determine if the sum consists of only the original
  K digits. Before each entry into the procedure the K marks
  are assumed to have been previously specified either by input
  data or as the result of a previous entry. Upon each such entry a
  new permutation is stored in N[1] through N[K]. In case the
  given permutation is (K−1), (K−2), ··· , 1, 0, then the next
  permutation is taken to be 0, 1, ··· , (K − 1). A FORTRAN
  subroutine for the IBM 7090 has been written and tested for
  several examples;
begin integer i, j, carry;
  for i := 1 step 1 until K do
    if N[i] − K + i ≠ 0 then go to add;
  for i := 1 step 1 until K do N[i] := i − 1;
  go to exit;
add:  N[K] := N[K] + 9;
    for i := 1 step 1 until K−1 do
      begin if K > 10 then go to B;
          carry := N[K−i+1]÷10;  go to C;
      B:    carry := N[K−i+1]÷K;
      C:    if carry = 0 then go to test;
          N[K−i] := N[K−i] + carry;
          N[K−i+1] := N[K−i+1] −10 × carry
      end i;
test:  for i := 1 step 1 until K do if N[i] − (K − 1) > 0
    then go to add;
    for i := 1 step 1 until K−1 do
      for j := i+1 step 1 until K do
        if N[i]−N[j] = 0 then go to add;
exit:  end PERMUTATION GENERATOR
```

CERTIFICATION OF ALGORITHM 87
PERMUTATION GENERATOR [John R. Howell,
  Comm. ACM, Apr. 1962]
D. M. COLLISON
Elliott Bros. (London) Ltd., Borehamwood, Herts.,
  England

The array N was removed from the value list in order that the
permutations might be available outside the procedure. The
algorithm was then run successfully with the Elliott ALGOL translator on the National-Elliott 803. It was rather slower than
Algorithm 86.

CERTIFICATION OF ALGORITHM 87
PERMUTATION GENERATOR [John R. Howell,
  Comm. ACM (Apr. 1962)]
G. F. SCHRACK and M. SHIMRAT
University of Alberta, Calgary, Alb., Canada

PERMUTATION GENERATOR was translated into FORTRAN
for the IBM 1620 and it performed satisfactorily. The algorithm
was timed for several small values of n. For purposes of comparison
we include the times (in seconds) for PERMULEX (Algorithm
102).

	n	3	4	5	6	7
PERMUTATION GENERATOR		3	41	558	—	—
PERMULEX		—	3	6	37	278

As can be seen from this table, PERMUTATION GENERATOR is
considerably slower. It is probable that one could speed up
PERMUTATION GENERATOR to a great extent by rearranging
the algorithm in such a manner that the digits of a number to a
certain base are permuted rather than the elements of a sequence.

REMARKS ON:
ALGORITHM 87 [G6]
PERMUTATION GENERATOR
  [John R. Howell, *Comm. ACM* 5 (Apr. 1962), 209]
ALGORITHM 102 [G6]
PERMUTATION IN LEXICOGRAPHICAL ORDER
  [G. F. Schrak and M. Shimrat, *Comm. ACM* 5 (June
  (1962), 346]
ALGORITHM 130 [G6]
PERMUTE
  [Lt. B. C. Eaves, *Comm. ACM* 5 (Nov. 1962), 551]
ALGORITHM 202 [G6]
GENERATION OF PERMUTATIONS IN
LEXICOGRAPHICAL ORDER
  [Mok-Kong Shen, *Comm. ACM* 6 (Sept. 1963), 517]

R. J. ORD-SMITH (Recd. 11 Nov. 1966, 28 Dec. 1966 and
  17 Mar. 1967)
Computing Laboratory, University of Bradford, England

A comparison of the published algorithms which seek to generate
successive permutations in lexicographic order shows that Algorithm 202 is the most efficient. Since, however, it is more than twice
as slow as transposition Algorithm 115 [H. F. Trotter, Perm,
Comm. ACM 5 (Aug. 1962), 434], there appears to be room for improvement. Theoretically a "best" lexicographic algorithm
should be about one and a half times slower than Algorithm 115.
See Algorithm 308 [R. J. Ord-Smith, Generation of Permutations
in Pseudo-Lexicographic Order, *Comm. ACM* 10 (July 1967), 452]
which is twice as fast as Algorithm 202.

ALGORITHM 87 is very slow.

ALGORITHM 102 shows a marked improvement.

ALGORITHM 130 does not appear to have been certified before. We find that, certainly for some forms of vector to be permuted, the algorithm can fail. The reason is as follows.

At execution of $A[f] := r$; on line prior to that labeled *schell*, f has not necessarily been assigned a value. f has a value if, and only if, the Boolean expression $B[k] > 0 \wedge B[k] < B[m]$ is **true** for at least one of the relevant values of k. In particular when matrix A is set up by $A[i] := i$; for each i the Boolean expression above is **false** on the first call.

ALGORITHM 202 is the best and fastest algorithm of the exicographic set so far published.

A collected comparison of these algorithms is given in Table I. t_n is the time for complete generation of $n!$ permutations. Times are scaled relative to t_8 for Algorithm 202, which is set at 100. Tests were made on an ICT 1905 computer. The actual time t_8 for Algorithm 202 on this machine *was* 100 seconds. r_n has the usual definition $r_n = t_n/(n \cdot t_{n-1})$.

TABLE I

Algorithm	t_6	t_7	t_8	r_6	r_7	r_8
87	118	—	—	—	—	—
102	2.1	15.5	135	1.03	1.08	1.1
130	—	—	—	—	—	—
202	1.7	12.4	100	1.00	1.00	1.00

ALGORITHM 88
EVALUATION OF ASYMPTOTIC EXPRESSION FOR THE FRESNEL SINE AND COSINE INTEGRALS
JOHN L. CUNDIFF
Engineering Experiment Station, Georgia Institute of Technology, Atlanta, Ga.

real procedure FRESNEL (u) Result: (frcos, frsin); **value** (u);

comment This procedure evaluates the Fresnel sine and cosine integrals for large u by expanding the anymptotic series given by

$$S(u) = \frac{1}{2} - \frac{\cos(x)}{\sqrt{2\pi x}}\left[1 - \frac{1\cdot3}{(2x)^2} + \frac{1\cdot3\cdot5\cdot7}{(2x)^4} - \cdots\right]$$
$$- \frac{\sin(x)}{\sqrt{2\pi x}}\left[\frac{1}{2x} - \frac{1\cdot3\cdot5}{(2x)^3} + \frac{1\cdot3\cdot5\cdot7\cdot9}{(2x)^5} - \cdots\right]$$

and

$$C(u) = \frac{1}{2} - \frac{\sin(x)}{\sqrt{2\pi x}}\left[1 - \frac{1\cdot3}{(2x)^2} + \frac{1\cdot3\cdot5\cdot7}{(2x)^4} - \cdots\right]$$
$$- \frac{\cos(x)}{\sqrt{2\pi x}}\left[\frac{1}{2x} - \frac{1\cdot3\cdot5}{(2x)^3} + \frac{1\cdot3\cdot5\cdot7\cdot9}{(2x)^5} - \cdots\right]$$

in which $x = \pi u^2/2$. Reference: PEARCEY, T. *Table of the Fresnel Integral to Six Decimal Places*. The Syndics of the Cambridge University Press, Melbourne, Australia (1956).;

```
begin    pi := 3.14159265;  arg := pi × (u↑2)/2;  temp := 1;
         argsq := 1/(4 × (arg↑2));  term := −3 × argsq;
         series := 1 + term;  N := 3;
first:   if temp = series then go to second;  temp := series;
         termi := term;
         term := −termi × (4 × N − 7) × (4×N − 5) × (argsq);
         if abs(term) > abs(termi) then go to second;
         series := temp + term;  N := N + 1;  go to first;
second:  series2 := ½ × arg;  temp := 0; term := series2;
         N := 2;
loop:    if series2 = temp then go to exit;  termi := term;
         term := −termi × argsq × (4×N−5) × (4×N−3);
         if abs(term) > abs(termi) then go to exit;
         temp := series2; series2 := temp + term;
         N := N + 1;  go to loop;
exit:    if u < 0 then half := −½ else half := ½;
         frcos := half + (sin(arg) × series − cos(arg) + series2)/
         (pi × u);
         frsin := half − (cos(arg) × series2 + sin(arg) × series)/
         (pi × u)
end FRESNEL;
```

REMARK ON ALGORITHMS 88, 89 AND 90
EVALUATION OF THE FRESNEL INTEGRALS
[J. L. Cundiff, *Comm. ACM*, May 1962]
MALCOLM D. GRAY
The Boeing Co., Seattle, Wash.

While coding these algorithms in FORTRAN for the IBM 7094, modifications were required (both in the formulation and in the language) before execution with any degree of speed and accuracy could be obtained. In the process it was found that the reference, *Pearcy*, contains an error in the formula for $C(u)$. This error is contained in Algorithm 88 in the formula

$$C(u) = \frac{1}{2} - \frac{\sin(x)}{\sqrt{2\pi x}}[\quad] - \cdots$$

The first minus sign above should be a plus sign.

After the necessary modifications were made, the three algorithms were found to be too large and uneconomical for our usage. A single algorithm, incorporating these three procedures, was written and is in current usage in a computer program which requires several thousand evaluations of each Fresnel integral.

ALGORITHM 89
EVALUATION OF THE FRESNEL SINE INTEGRAL
JOHN L. CUNDIFF
Engineering Experiment Station, Georgia Institute of
  Technology, Atlanta, Ga.

real procedure FRESNELSIN (u) Result: (frsin); **value** u;
comment This algorithm computes the Fresnel sine integral
  defined by,

$$S(u) = \int_0^u \sin \pi t^2/2 \; dt,$$

by evaluating the series expansion

$$S(x) = \sqrt{\frac{2x}{\pi}} \left[\frac{x}{3} - \frac{x^3}{7 \cdot 3!} + \frac{x^5}{11 \cdot 5!} - \frac{x^7}{15 \cdot 7!} + \cdots \right]$$

where $x = \pi u^2/2$. Reference: PEARCEY, T. *Table of the
Fresnel Integral to Six Decimal Places.* The Syndics of the
Cambridge University Press, Melbourne, Australia (1956).;

```
begin  Pi2 := 1.5707963;  x := Pi2 x (u ↑ 2);  frsin := x/3;
       frsqr := x ↑ 2;  N := 3;  term := (−x × frsqr)/6;
       frsini := frsin + term/7;
Loop:  if frsin = frsini then go to exit;  frsin := frsini;
       term := −term × frsqr/((2×N−1) × (2×N−2));
       frsini := frsin + term/(4×N−1);  N := N + 1;
       go to Loop;
exit:    frsin := frsini × u
end FRESNELSIN;
```


REMARK ON ALGORITHMS 88, 89 AND 90
EVALUATION OF THE FRESNEL INTEGRALS
  [J. L. Cundiff, *Comm. ACM*, May 1962]
MALCOLM D. GRAY
The Boeing Co., Seattle, Wash.

While coding these algorithms in FORTRAN for the IBM 7094,
modifications were required (both in the formulation and in the
language) before execution with any degree of speed and accuracy
could be obtained. In the process it was found that the reference,
Pearcy, contains an error in the formula for $C(u)$. This error is
contained in Algorithm 88 in the formula

$$C(u) = \frac{1}{2} - \frac{\sin (x)}{\sqrt{2\pi x}} [\quad] - \cdots .$$

The first minus sign above should be a plus sign.

After the necessary modifications were made, the three algorithms were found to be too large and uneconomical for our
usage. A single algorithm, incorporating these three procedures,
was written and is in current usage in a computer program which
requires several thousand evaluations of each Fresnel integral.

ALGORITHM 90
EVALUATION OF THE FRESNEL COSINE INTE-
GRAL
JOHN L. CUNDIFF
Engineering Experiment Station, Georgia Institute of
Technology, Atlanta, Ga.

real procedure FRESNELCOS (u) result: (frcos); **value** (u);
comment This algorithm computes the Fresnel cosine integral
defined by

$$C(u) = \int_0^u \cos \frac{\pi t^2}{2}\, dt,$$

by evaluating the series expansion

$$C(u) = \sqrt{\frac{2x}{\pi}}\left[1 - \frac{x^2}{5 \cdot 2!} + \frac{x^4}{9 \cdot 4!} - \frac{x^6}{13 \cdot 6!} + \cdots\right],$$

where $x = \pi u^2/2$. Reference: PEARCEY, T. *Table of the Fresnel
Integral to Six Decimal Places.* The Syndics of the Cambridge
University Press, Melbourne, Australia (1956).;

```
begin  pi2 := 1.5707963;  x := pi2 × (u↑2);  frcos := 1;
       xsqr := x↑2;  N := 3;  term := −xsqr/2;
       frcoi := 1 + (term/5);
loop:  if frcoi =·frcos then go to exit;  term := −term ×
       xsqr/((2×N−2 × (2×N−3));  frcos := frcoi;  frcoi :=
       frcos + term/(4×N−3);  N := N + 1;  go to loop:
exits:  frcos := u × frcos
end FRESNELCOS;
```


REMARK ON ALGORITHMS 88, 89 AND 90
EVALUATION OF THE FRESNEL INTEGRALS
 [J. L. Cundiff, *Comm. ACM*, May 1962]
MALCOLM D. GRAY
The Boeing Co., Seattle, Wash.

While coding these algorithms in FORTRAN for the IBM 7094,
modifications were required (both in the formulation and in the
language) before execution with any degree of speed and accuracy
could be obtained. In the process it was found that the reference,
Pearcy, contains an error in the formula for $C(u)$. This error is
contained in Algorithm 88 in the formula

$$C(u) = \frac{1}{2} - \frac{\sin(x)}{\sqrt{2\pi x}}[\quad] - \cdots.$$

The first minus sign above should be a plus sign.

After the necessary modifications were made, the three algorithms were found to be too large and uneconomical for our
usage. A single algorithm, incorporating these three procedures,
was written and is in current usage in a computer program which
requires several thousand evaluations of each Fresnel integral.

ALGORITHM 91
CHEBYSHEV CURVE-FIT
ALBERT NEWHOUSE
University of Houston, Houston, Texas

```
procedure CHEBFIT(m, n, X, Y);  integer m, n;  array X, Y;
comment  This procedure fits the tabular function Y(X) (given
```

as m points (X, Y) by a polynomial $P = \sum_{i=0}^{n} A_i X^i$. This

polynomial is the best polynomial approximation of $Y(X)$ in
the Chebyshev sense. Reference: STIEFEL, E. *Numerical
Methods of Tchebycheff Approximation*, U. of Wisc. Press (1959),
217–232;

```
begin array X[1:m], Y[1:m], T[1:m], A[0:n], AX[1:n+2],
        AY[1:n+2], AH[1:n+2], BY[1:n+2], BH[1:n+2];
    integer array IN [1:n+2];  real TMAX, H;  integer i,
    j, k, imax;
    comment Initialize;
    k := (m−1)/(n+1);
    for 1 := 1 step 1 until n+1 do IN [i] := (i−1)×k + 1;
    IN[n+2] := m;
START:  comment Iteration begins;
    for i := 1 step 1 until n+2 do
        begin AX[i] := X[IN[i]];
              AY[i] := Y[IN[i]];
              AH[i] := (−1) ↑ (i−1)
        end i;
DIFFERENCE:  comment divided differences;
    for i := 2 step 1 until n+2 do
        begin
        for.j := i−1 step 1 until n+2 do
        begin BY[j] := AY[j];
              BH[j] := AH[j]
        end j;
        for j := i step 1 until n+2 do
        begin AY[j] := (BY[j] −BY[j−1])/
              (AX[j] −AX[j−i+1]);
              AH[j] := (BH[j] −BH[j−1])/
              (AX[j] −AX[j−i+1])
        end j;
    end i;
    H := −AY[n+2]/AH[n+2];
POLY:  comment polynomial coefficients;
    for i := 0 step 1 until n do
        begin A[i] := AY[i] +AH[i] ×H;
              BY[i] := 0
        end i;
    BY[1] := 1;  TMAX := abs(H);  imax := IN[1];
    for i := 1 step 1 until n do
        begin
        for j := 0 step 1 until i−1 do
            begin
            BY[i+1−j] := BY[i+1−j] −BY[i−j] ×X[IN[i]];
            A[j] := A[j] +A[i] ×BY[i+1−j]
            end j;
        end i;
ERROR:  comment compute deviations;
    for i := 1 step 1 until m do
        begin T[i] := A[n];
```

3. It should be noted that the $X[i]$ must form a monotonic sequence.

4. **comment** cannot follow the colon following a label. This occurs in four places.

5. The **end** following **go to** *FIT* must be removed.

In addition, a large number of details can be made more concise and unnecessary operations can be eliminated. Also, it seems desirable to produce the maximum deviation as a result.

CERTIFICATION OF ALGORITHM 91 [E2]
CHEBYSHEV CURVE-FIT [Albert Newhouse *Comm. ACM 5* (May 1962), 281; *6* (April 1963), 167; *7* (May 1964), 296]

J. BOOTHROYD (Recd. 15 May 1967 and 5 Sept. 1967)
University of Tasmania, Hobart, Tasmania, Australia.

In addition to the corrections noted by R. P. Hale [OP. CIT., April 1963] and P. Naur [OP. CIT., May 1964], the following changes are necessary:

1. The first statement should be $k := entier((m-1)/(n+1))$

2. A semi-colon should precede label $L1$.

With these changes the procedure ran successfully using Elliott 503 ALGOL.

Although this procedure is an implementation of a finite algorithm, roundoff errors may give rise to cyclic changes of the reference set causing the procedure to fail to terminate.

Algorithm 318 [J. Boothroyd, Chebyshev Curve-Fit(Revised), *Comm.ACM 10* (Dec. 1967), 801] avoids this cycling difficulty, uses less than half the auxiliary array space of Algorithm 91 and, on test, appears to be at least four times as fast.

ALGORITHM 92

SIMULTANEOUS SYSTEM OF EQUATIONS AND
MATRIX INVERSION ROUTINE

DEREK JOHANN ROEK
Applied Physics Laboratory of Johns Hopkins University,
Silver Spring, Maryland

```
procedure SIMULTANEOUS (U, W, C, X, B, n, kount, eps,
  absf) ;
array U, W, C, X, B ; integer n, kount ;
  real eps; real procedure absf;
comment This procedure solves the problem Ux := b for the
  vector x. It assumes the problem written in the form x'U' := b',
  where ' denotes transpose. The procedure is completed in n
  cycles and may be iterated kount times (kount ≦ 6). The trans-
  pose of U is in U[,] and the row vector b' is in B. The integer n
  is the dimension of U, and the solution row vector x' is in X.
  The matrix C is a check of accuracy. It should have b' in its
  first row, the first element b₁ of b' along its main diagonal,
  and zeros elsewhere. The real number eps checks to see how close
  the actual result is to this theoretical one. Also if we let b' :=
  (1, 0, ⋯ , 0), then this procedure finds the inverse W[,] of U.
  The function absf finds the absolute value of its argument. The
  procedure chooses the column vectors of U as the row vectors of
  W in the 0ᵗʰ cycle of the first iteration. For all subsequent itera-
  tions, the row vectors of W, computed at the nᵗʰ cycle of the
  last iteration, are the row vectors of W in the 0ᵗʰ cycle  ;
begin integer i, j, k, p ;  real bh, b1, Z ;
  for j := 1 step 1 until n do
      for i := 1 step 1 until n do W[j, i] := U[i, j];
S1:  for j := 1 step 1 until n do
         for i := 1 step 1 until n do C[i, j] := 0 ;
  for j := 1 step 1 until n do
     begin for k := 1 step 1 until n do
       begin C[j, j] := C[j, j] + W[j, k] × U[k, j] end;
       if j = 1 then Z := B[j]/C[j, j] else Z := 1/C[j, j];
       for k := 1 step 1 until n do
          begin X[k] := Z × W[j, k];
            W[j, k] := X[k]
          end k;
       for k := 1 step 1 until n do
          begin if k = j then go to S2 else
            for p := 1 step 1 until n do
               C[k, j] := C[k, j] + U[p, j] × W[k, p];
          if j = 1 then bh := B[j] else bh := 1;
          if k = 1 then b1 := B[j] else b1 := 0;
          for p := 1 step 1 until n do
          begin  X[p] := bh × W[k,  p] + (b1 − C[k,  j]) ×
          W[j, p];
            W[k, p] := X[p]
          end p;
S2:        if k = j ∧ j = n then go to S3
          end k;
     end j;
S3:  for j := step 1 until n do
         if absf(absf(C[j, j]) − absf(B[1])) > eps then go to S4;
      go to S6;
S4:  if kount > 0 then go to S5 else go to S6;
S5:  kount := kount − 1;
     go to S1;
S6:  for j := step 1 until n do
        X[j] := W[1, j];
S7:  end SIMULTANEOUS
```

ALGORITHM 93
GENERAL ORDER ARITHMETIC
MILLARD H. PERSTEIN
Control Data Corp., Palo Alto, Calif.

```
procedure  arithmetic (a, b, c, op);
integer  a, b, c, op;
comment  This procedure will perform different order arithmetic
    operations with b and c, putting the result in a. The order of the
    operation is given by op. For op = 1 addition is performed. For
    op = 2 multiplication, repeated addition, is done. Beyond these
    the operations are non-commutative. For op = 3 exponentiation,
    repeated multiplication, is done, raising b to the power c. Beyond
    these the question of grouping is important. The innermost
    implied parentheses are at the right. The hyper-exponent is
    always c. For op = 4 tetration, repeated exponentiation, is
    done. For op = 5, 6, 7, etc., the procedure performs pentation,
    hexation, heptation, etc., respectively.
        The routine was originally programmed in FORTRAN for the
    Control Data 160 desk-size computer. The original program
    was limited to tetration because subroutine recursiveness in
    Control Data 160 FORTRAN has been held down to four levels in
    the interests of economy.
        The input parameter, b, c, and op, must be positive integers,
    not zero;
begin own integer d, e, f, drop;
        if op = 1 then
        begin a := b + c;  go to 1
        end if op = 2 then d := 0;
        else d := 1;  e := c; drop := op − 1;
        for f := 1 step 1 until e do
        begin arithmetic (a, b, d, drop);
                d := a
        end;
1:      end arithmetic
```

CERTIFICATION OF ALGORITHM 93
GENERAL ORDER ARITHMETIC [Millard H. Per-
    stein, *Comm. ACM* (June 1962)]
RICHARD GEORGE
Particle Accelerator Div. Argonne National Laboratory,
    Argonne, Ill.

Algorithm 93 was programmed for the IBM 1620, using
"FORTRAN-recursion" (i.e., generous use of the copy rule). The
program ran without any modifications and was tested through
tetration. Further levels were available, but were too time-
consuming to reach.

ALGORITHM 94
COMBINATION

JEROME KURTZBERG
Burroughs Corp., Burroughs Laboratories, Paoli, Pa.

```
procedure  COMBINATION (J, N, K);  value N, K;  integer
    array J;  integer N, K;
comment  This procedure generates the next combination of N
```

comment This procedure generates the next combination of N integers taken K at a time upon being given N, K and the previous combination. The K integers in the vector $J(1) \cdots J(K)$ range in value from 0 to $N - 1$, and are always monotonically strictly increasing with respect to themselves in input and output format. If the vector J is set equal to zero, the first· combination produced is $N-K, \cdots, N-1$. That initial combination is also produced after $0, 1, \cdots, N-1$, the last value in that cycle;

```
begin  integer  B, L;
        B := 1:
mainbody:  if J(B)≧B then begin A := J(B) − B−1;
            for L := 1 step 1 until B do J(L) := L + A;
            go to exit end;
           if B = K then go to initiate;
           B := B + 1; go to mainbody;
initiate:   for B := 1 step 1 until K do J(B) := N − K − 1 + B
exit:       end COMBINATION
```

CERTIFICATION OF ALGORITHM 94
COMBINATION [J. Kurtzberg, *Comm. ACM*, June 1962]

RONALD W. MAY
University of Alberta, Calgary, Alberta, Canada

Algorithm 94 was translated into FORTRAN for the IBM 1620 and run successfully with no corrections. The variable A, however, has not been declared.

CERTIFICATION OF ALGORITHM 94
COMBINATION [J. Kurtzburg, *Comm. ACM*, June, 1962]

R. E. GRENCH*
Reactor Eng. Div., Argonne National Laboratory, Argonne, Ill.
* Work supported by U. S. Atomic Energy Commission

Four changes were required in the algorithm.

1. The last sentence in the comment should read: That initial combination is also produced after $0, 1, \cdots, K-1$, the last value in that cycle;
2. The integer A was declared;
3. Parentheses were replaced by brackets in the subscript expressions;
4. A semicolon was inserted at the end of the initiate statement.

After the above changes were made the body of Algorithm 94 was tested on an LGP-30 computer using the Dartmouth College ALGOL-30 translator. The body tested satisfactorily and the time required to generate one J when $K = 5$ and $N = 15$ was 30 seconds.

Various tests should be included if this algorithm is to be used as a procedure. These tests might include a statement to check if $K > N$ and if the initial value of J is correct These two possibilities were investigated and it was found that improper J's are generated.

ALGORITHM 95
GENERATION OF PARTITIONS IN PART-COUNT
  FORM
Frank Stockmal
System Development Corp., Santa Monica, Calif.

```
procedure partgen(c,N,K,G);   integer N,K;   integer array c;
  Boolean G;
comment  This procedure operates on a given partition of the
  positive integer N into parts ≤ K, to produce a consequent
  partition if one exists. Each partition is represented by the
  integers c[1] thru c[K], where c[j] is the number of parts of the
  partition equal to the integer j. If entry is made with G = false,
  procedure ignores the input array c, sets G = true, and pro-
  duces the first partition of N ones. Upon each successive entry
  with G = true, a consequent partition is stored in c[1] thru c[K].
  For N = KX, the final partition is c[K] = X. For N = KX+r,
  1 ≤ r ≤ K−1, final partition is c[K] = X, c[r] = 1. When entry
  is made with array c = final partition, c is left unchanged and G
  is reset to false;
begin integer a,i,j;
        if ¬ G then go to first;
        j := 2;
        a := C[1];
test:   if a < j then go to B;
        c[j] := 1 + c[j];
        c[1] := a − j;
zero:   for i := 2 step 1 until j − 1
        do c[i] := 0;
        go to EXIT;
B:      if j = K then go to last;
        a := a + j × c[j];
        j := j + 1;
        go to test;
first:  G := true;
        c[1] := N;
        j := K + 1;
        go to zero;
last:   G := false;
EXIT: end partgen
```

ALGORITHM 96
ANCESTOR
ROBERT W. FLOYD
Armour Research Foundation, Chicago, Ill

procedure ancestor (m, n); **value** n; **integer** n; **Boolean array** m;

comment Initially $m[i, j]$ is **true** if individual i is a parent of individual j. At completion, $m[i, j]$ is **true** if individual i is an ancestor of individual j. That is, at completion $m[i, j]$ is **true** if there are k, l, etc. such that initially $m[i, k], m[k, l], \cdots, m[p, j]$ are all **true**. Reference: WARSHALL, S. A theorem on Boolean matrices, $J.ACM$ 9(1962), 11–12;

```
begin
integer  i, j, k;
for i := 1 step 1 until n do
for j := 1 step 1 until n do
if m [j, i] then
for k := 1 step 1 until n do
if m [i, k] then
m [j, k] := true
end ancestor
```

CERTIFICATION OF ALGORITHM 96
ANCESTOR [Robert W. Floyd, *Comm. ACM*, June, 1962]
HENRY C. THACHER, JR.*
Argonne National Laboratory, Argonne, Ill.

* Work supported by the U.S. Atomic Energy Commission

The body of this procedure was tested on the LGP-30 using the Dartmouth translator. After inclosing conditional statements in **begin end** brackets (apparently necessary for this translator), the procedure operated satisfactorily for the following matrices:

The correctness of these results was confirmed by inspection of the network diagrams.

$n=5$, Time: 8'15"

```
FTTFF      FTTTT
FFFFT      FFFFT
FFFTF  →   FFFTT
FFFFT      FFFFT
FFFFF      FFFFF
```

$n = 6$, Time: 13'15"

```
FTTFFF      FTTTTT
FFFTFF      FFFTFT
FFFFTF  →   FFFTFT
FFFFFT      FFFFFT
FFFFFF      FFFFFF
```

$n = 9$, Time 31'2"

```
FTTFFFFFF      FTTTTTTTT
FFFFTFFFF      FFFFTTTTF
FFFTTFFFF      FFFTTTTTT
FFFFFFFFT      FFFFFTTTT
FFFFFTTFF  →   FFFFFTTTF
FFFFFFFTF      FFFFFFFTF
FFFFFFFTF      FFFFFFFTF
FFFFFFFFF      FFFFFFFFF
FFFFFTTFF      FFFFFTTTF
```

ALGORITHM 97
SHORTEST PATH
Robert W. Floyd
Armour Research Foundation, Chicago, Ill.

```
procedure  shortest path (m, n);  value n;  integer n;  array m;
comment   Initially m[i, j] is the length of a direct link from
    point i of a network to point j. If no direct link exists, m [i, j] is
    initially 10↑10. At completion, m [i, j] is the length of the shortest
    path from i to j. If none exists, m [i, j] is 10↑10. Reference: WAR-
    SHALL, S. A theorem on Boolean matrices. J, ACM 9 (1962), 11–12;
begin
integer i, j, k;  real inf, s;  inf := 10↑10;
for i := 1 step 1 until n do
for j := 1 step 1 until n do
if m [j, i] < inf then
for k := 1 step 1 until n do
if m [i, k] < inf then
begin s := m [j, i] + m [i, k];
if s < m [j, k] then m [j, k] := s
end
end shortest path
```

ALGORITHM 98
EVALUATION OF DEFINITE COMPLEX LINE
  INTEGRALS
JOHN L. PFALTZ
Syracuse University Computing Center, Syracuse, N. Y.

```
procedure  COMPLINEINTGRL(A, B, N, RSSUM);
    value A, B, N;  real A, B, N;  array RSSUM;
comment  COMPLINEINTGRL approximates the complex line
    integral by evaluating the partial Riemann-Stieltjes sum
```

$\sum_{t-1}^{n} f(z_k)[z_t - z_{t-1}]$ where $a \leq t \leq b$ and $z_k \in (z_{t-1}, z_t)$. The
programmer must provide 1) the procedures GAMMA(T, Z) to
calculate $z(t)$ on Γ, and FUNCT(Z, F) to calculate function
values, and 2) the end points A and B of the parametric interval
and N the number of subintervals into which $[a, b]$ is to be
partitioned;

```
begin integer I;  real T, DELT;  real array ZT, ZTL, DELZ,
        ZK, PART[1:2];  RSSUM[1] := 0.0;  RSSUM[2] := 0.0;
        DELT := (B − A)/N; T := A;
line:   GAMMA(T, ZT);
        if T = A then go to next;
        for I := 1 step 1 until 2 do
        begin
            DELZ[I] := ZT[I] − ZTL[I]; end;
        for I := 1 step 1 until 2 do
        begin
            ZK[I] := ZTL[I] + DELZ[I]/2.0;  end;
        FUNCT(ZK, FZ);
        PART[1] := FZ[1] × DELZ[1] − FZ[2] × DELZ[2];
        PART[2] := FZ[1] × DELZ[2] + FZ[2] × DELZ[1];
        for I := 1 step 1 until 2 do
        begin
            RSSUM[I] := RSSUM[I] + PART[I];  end;
        if T < B − (0.25 × DELT) then go to next else go to
            exit;
next:   for I := 1 step 1 until 2 do
        begin
            ZTL[I] := ZT[I];  end;
        T := T + DELT;
        go to line;
exit:   end COMPLINEINTGRL.
```

ALGORITHM 99
EVALUATION OF JACOBI SYMBOL
STEPHEN J. GARLAND AND ANTHONY W. KNAPP
Dartmouth College, Hanover, N. H.

```
procedure  Jacobi (n,m,r);  value n,m;
integer  n, m, r;
comment  Jacobi computes the value of the Jacobi symbol (n/m),
    where m is odd, by the law of quadratic reciprocity. The param-
    eter r is assigned one of the values −1, 0, or 1 if m is odd. If m
    is even, the symbol is undefined and r is assigned the value 2.
    For odd m the routine provides a test of whether m and n are
    relatively prime. The value of r is 0 if and only if m and n have
    a nontrivial common factor. In the special case where m is prime,
    r = −1 if and only if n is a quadratic nonresidue of m;
begin
  integer s;
  Boolean p, q;
  Boolean procedure parity (x);  value x;  integer x;
    comment  The value of the function parity is true if x is
      odd, false if x is even;
    begin
      parity := x ÷ 2 × 2 ≠ x
    end parity;
  if ¬ parity (m) then begin r := 2;  go to exit end;
  p := true;
  loop:  n := n − n ÷ m × m;
         q := false;
         if n ≦ 1 then go to done;
  even:  if ¬ parity (n) then
         begin
           q := ¬ q;
           n := n ÷ 2;
           go to even
         end n now odd;
         if q then if parity ((m↑2 − 1)÷8) then p := ¬ p;
         if n = 1 then go to done;
         if parity ((m−1) × (n−1) ÷ 4) then p := ¬ p;
         s := m;  m := n;  n := s;  go to loop;
  done:  r := if n = 0 then 0 else if p then 1 else −1;
exit:  end Jacobi
```

REMARK ON ALGORITHM 99
EVALUATION OF JACOBI SYMBOL [S. J. Gar-
    land and A. W. Knapp, *Comm. ACM 6*, June 1962]
RONALD W. MAY
University of Alberta, Calgary, Alberta, Canada

One syntactical error was found in this procedure. It occurs
in the second **if** statement following the **label** even. The statement

```
        if q then if parity ((m↑2−1)÷8) then
            p := ¬ p;
```

might be changed as follows.

```
        if q then go to CHECK;
next 1:   if n = 1 then go to done;
CHECK:  if parity ((m ↑ 2 − 1) ÷ 8) then
            p := ¬ p;
        go to next 1;
```

The two statements beginning with *CHECK* could be inserted
before the **label** done and after the statement **go to** loop;.

ALGORITHM 100
ADD ITEM TO CHAIN-LINKED LIST
Philip J. Kiviat
United States Steel Corp., Appl. Research Lab., Monroe-
ville, Penn.

```
procedure  inlist (t,info,m,list,n,first,flag,addr,listfull);
integer  n,m,first,flag,t;  integer array info,list,addr;
comment  inlist adds the information pair {t,info} to the chain-
   link structured matrix list (i,j), where t is an order key ≥ 0, and
   info(k) an information vector associated with t. info(k) has di-
   mension m, list(i,j) has dimensions (n × (m+3)). flag denotes
   the head and tail of list(i,j), and first contains the address of the
   first (lowest order) entry in list(i,j).  addr(k) is a vector con-
   taining the addresses of available (empty) rows in list(i,j).
   Initialization: list(i,m+2) = flag, for some i ≤ n. If list(i,j) is
   filled exit is to listfull;
begin integer i, j, link1, link2;
0:  if addr [1] = 0;  then go to listfull;  i := 1;
1:  if list [i,1] ≤ t
       then begin if list [i,2] ≠ 0 then begin link1 := m+2;
         link2 := m+3;  go to 2 end;  else begin if
         list [i,m+2] = flag then begin i := flag;
         link1 := m+3;  link2 := m+2;  go to 3 end;
         else begin i := i+1;  go to 1 end end end;
       else begin link1 := m+3;  link2 := m+2 end;
2:  if list [i,link2] ≠ flag
       then begin k := i;  i := list [i,link2];
         if (link2 = m+2 ∧ list [i,1] ≤ t) ∨
           (link2 ≠ m+2 ∧ list [i,1] > t) then go to 4;
         else go to 1 end;
       else begin list [i,link2] := addr [1] end;
3:  j := addr [1];  list [j,link1] := i;
     list [j,link2] := flag;  if link2 = m+2 then
     first := addr [1];  go to 5;
4:  j := addr [1];  list [j,link1] := list [i,link1];
     list [i,link1] := list [k,link2] := addr [1];
     list [j,link2] := i;
5:  list [j,1] := t;  for i := 1 step 1 until m do
     list [j,i+1] := info [i];  for i := 1 step 1 until n−1 do
     addr [i] := addr [i+1];  addr [n] := 0
end  inlist
```

ALGORITHM 101

REMOVE ITEM FROM CHAIN-LINKED LIST

Philip J. Kiviat

United States Steel Corp., Appl. Res. Lab., Monroeville,
  Penn.

```
procedure  outlist (vector,m,list,n,first,flag,addr);
integer  n,m,first,flag;  integer array vector,list,addr;
comment  outlist removes the first entry (information pair with
  lowest order key) from list(i,j) and puts it in vector(k);
begin integer i;
for i := 1 step 1 until m+1 do vector[i] := list [first,i];
for i := n−1 step −1 until 1 do addr [i+1] := addr [i];
addr [1] := first;
if list [first,m+3] = flag then
  begin list [1,m+2] := flag;  first := 1;
  for i := 1 step 1 until n do addr [i] := i end;
else begin first := list [first,m+3];
list [first,m+2] := flag end;
for i := 1 step 1 until m+3 do list [addr [1], i] := 0
end outlist
```

COLLECTED ALGORITHMS FROM CACM

ALGORITHM 102
PERMUTATION IN LEXICOGRAPHICAL ORDER
G. F. Schrack and M. Shimrat
University of Alberta, Calgary, Alberta, Canada

```
procedure  PERMULEX(n,p);
integer n;  integer array p;
comment Successive calls of the procedure will generate all
    permutations p of 1,2,3,···,n in lexicographical order. Before the
    first call, the non-local Boolean variable 'flag' must be set to
    true. If after an execution of PERMULEX 'flag' is false,
    additional calls will generate further permutations—if true, all
    permutations have been obtained;
begin integer array q[1:n];  integer i, k, t;  Boolean flag2;
if flag then
    begin for i := 1 step 1 until n do
    p[i] := i;  flag2 := true;  flag := false;
    go to EXIT
    end initialize;
if flag2 then
    begin t := p[n];  p[n] := p[n−1];  p[n−1] := t;
    flag2 := false;  go to EXIT
    end bypass;
flag2 := true; for i := n−2 step −1 until 1 do
    if p[i] < p[i+1] then go to A;
    flag := true;  go to EXIT;
A:  for k := 1 step 1 until n do q[k] := 0;
    for k := i step 1 until n do q[p[k]] := p[k];
    for k := p[i] + 1 step 1 until n do
    if q[k] ≠ 0 then go to B;
B:  p[i] := k; q[k] := 0;
    for k := 1 step 1 until n do
    if q[k] ≠ 0 then begin i := i + 1; p[i] := q[k] end
    else if i ≧ n then go to EXIT;
EXIT:
end PERMULEX
```

REMARKS ON:

ALGORITHM 87 [G6]
PERMUTATION GENERATOR
[John R. Howell, *Comm. ACM 5* (Apr. 1962), 209]
ALGORITHM 102 [G6]
PERMUTATION IN LEXICOGRAPHICAL ORDER
[G. F. Schrak and M. Shimrat, *Comm. ACM 5* (June
(1962), 346]
ALGORITHM 130 [G6]
PERMUTE
[Lt. B. C. Eaves, *Comm. ACM 5* (Nov. 1962), 551]
ALGORITHM 202 [G6]
GENERATION OF PERMUTATIONS IN
LEXICOGRAPHICAL ORDER
[Mok-Kong Shen, *Comm. ACM 6* (Sept. 1963), 517]

R. J. Ord-Smith (Recd. 11 Nov. 1966, 28 Dec. 1966 and
    17 Mar. 1967)
Computing Laboratory, University of Bradford, England

A comparison of the published algorithms which seek to generate
successive permutations in lexicographic order shows that Algorithm 202 is the most efficient. Since, however, it is more than twice
as slow as transposition Algorithm 115 [H. F. Trotter, Perm,
Comm. ACM 5 (Aug. 1962), 434], there appears to be room for improvement. Theoretically a "best" lexicographic algorithm
should be about one and a half times slower than Algorithm 115.
See Algorithm 308 [R. J. Ord-Smith, Generation of Permutations
in Pseudo-Lexicographic Order, *Comm. ACM 10* (July 1967), 452]
which is twice as fast as Algorithm 202.

ALGORITHM 87 is very slow.

ALGORITHM 102 shows a marked improvement.

ALGORITHM 130 does not appear to have been certified before.
We find that, certainly for some forms of vector to be permuted,
the algorithm can fail. The reason is as follows.

At execution of $A[f] := r$; on line prior to that labeled *schell*, f
has not necessarily been assigned a value. f has a value if, and
only if, the Boolean expression $B[k] > 0 \wedge B[k] < B[m]$ is **true** for
at least one of the relevant values of k. In particular when matrix
A is set up by $A[i] := i$; for each i the Boolean expression above is
false on the first call.

ALGORITHM 202 is the best and fastest algorithm of the
exicographic set so far published.

A collected comparison of these algorithms is given in Table I.
t_n is the time for complete generation of $n!$ permutations. Times
are scaled relative to t_8 for Algorithm 202, which is set at 100.
Tests were made on an ICT 1905 computer. The actual time t_8
for Algorithm 202 on this machine *was* 100 seconds. r_n has the
usual definition $r_n = t_n/(n \cdot t_{n-1})$.

TABLE I

Algorithm	t_6	t_7	t_8	r_6	r_7	r_8
87	118	—	—	—	—	—
102	2.1	15.5	135	1.03	1.08	1.1
130	—	—	—	—	—	—
202	1.7	12.4	100	1.00	1.00	1.00

ALGORITHM 103
SIMPSON'S RULE INTEGRATOR
GUY F. KUNCIR
UNIVAC Division, Sperry Rand Corp., San Diego, Calif.

procedure SIMPSON (a, b, f, I, i eps, N);
  value a, b, eps, N; **integer** N;
  real a, b, I, i, eps; **real procedure** f;
comment This procedure integrates the function $f(x)$ using a
  modified Simpson's Rule quadrature formula. The quadrature is
  performed over j subintervals of $[a,b]$ forming the total area I.
  Convergence in each subinterval of length $(b-a)/2^n$ is indicated
  when the relative difference between successive three-point and
  five-point area approximations

$$A_{3,j} = (b-a)(g_0 + 4g_2 + g_4)/(3 \cdot 2^{n+1})$$
$$A_{5,j} = (b-a)(g_0 + 4g_1 + 2g_2 + 4g_3 + g_4)/(3 \cdot 2^{n+2})$$

  is less than or equal to an appropriate portion of the over-all
  tolerance eps (i.e., $|(A_{5,j} - A_{3,j})/A_{5,j}| \leq eps/2^n$ with $n \leq N$).
  SIMPSON will reduce the size of each interval until this condition is satisfied.

    Complete integration over $[a,b]$ is indicated by $i = b$. A value
  $a \leq i < b$ is indicates that the integration was terminated, leaving I the true area under f in $[a,i]$. Further integration over $\lfloor i,b \rfloor$
  will necessitate either the assignment of a larger N, a larger eps,
  or an integral substitution reducing the slope of the integrand in
  that interval. It is recommended that this procedure be used
  between known integrand maxima and minima.;
begin integer m, n; **real** d, h; **array** g[0:4], A[0:2], S[1:N, 1:3];
  I := i := m := n := 0;
  g[0] := f(a);
  g[2] := f((a + b)/2);
  g[4] := f(b);
  A[0] := (b − a) × (g[0] + 4 × g[2] + g[4])/2;
AA: d := 2↑n; h := (b − a)/4/d;
    g[1] := f(a + h × (4 × m + 1));
    g[3] := f(a + h × (4 × m + 3));
    A[1] := h × (g[0] + 4 × g[1] + g[2]);
    A[2] := h × (g[2] + 4 × g[3] + g[4]);
    if abs (((A[1] + A[2]) − A[0])/(A[1] + A[2])) > eps/d
      then begin m := 2 × m; n := n + 1;
          if n > N **then go to** CC;
          A[0] := A[1]; S[n,1] := A[2];
          S[n,2] := g[3]; S[n,3] := g[4];
          g[4] := g[2]; g[2] := g[1]; **go to** AA
        end
    else begin I := I + (A[1] + A[2])/3;
        m := m + 1; i := a + m × (b − a)/d;
BB: **if** m = 2 × (m ÷ 2) **then**
         begin m := m ÷ 2; n := n − 1; **go to** BB **end**
       if (m ≠ 1) ∨ (n ≠ 0) **then**
         begin A[0] := S[n,1]; g[0] := g[4];
            g[2] := S[n,2]; g[4] := S[n,3]; **go to** AA **end**
      end
CC: **end** SIMPSON

ALGORITHM 104
REDUCTION TO JACOBI
H. Rutishauser
Eidg. Technische Hochschule, Zurich, Switzerland

procedure m21 (n, a, b, c, *inform*); **value** n;
  integer n; **array** a, b, c; **procedure** *inform*;
  comment: m21 transforms symmetric bandmatrix

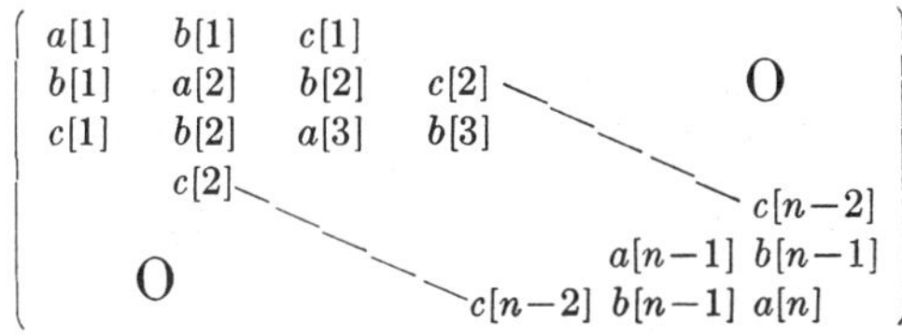

represented by the arrays a, b, c by orthogonal transformation
to Jacobi form which is represented by the arrays a, b. The
method is described in H. Rutishauser, "On Jacobi rotation
patterns," to appear in Proc. Symposium in Experimental
Arithmetic, Chicago, Apr. 12–14, 1962, Sect. 5. Note that declarations must be given for the arrays a, b, c with subscripts
ranging from 1 to n. Also procedure *inform* must be declared.
It may serve to use the Jacobi rotations occurring inside m21
also for other purposes;
begin
  real p, g, d, s;
  integer k, j;
  b[n] := c[n] := c[n−1]: = 0;
  for k := 2 **step** 1 **until** n−1 **do**
  begin
    for j := k **step** 2 **until** n−1 **do**
    begin
      if k=j **then**
      begin
        p := sqrt(b[k−1]↑2 + c[k−1]↑2);
        if p=0 **then go to** ex;
        d := b[k−1]/p;
        s := −c[k−1]/p;
        b[k−1] := p;
        c[k−1] := 0
      end k=j
    else
      begin
        p := sqrt(c[j−2]↑2+g↑2);
        if p = 0 **then go to** ex;
        d := c[j−2]/p;
        s := −g/p;
        c[j−2] := p;
        p := d×b[j−1]−s×c[j−1];
        c[j−1] := s×b[j−1]+d×c[j−1];
        b[j−1] := p
      end j≠k;
  common: g := 2×b[j]×d×s;
      p := a[j]×d×d−g+a[j+1]×s×s;
      b[j] := (a[j]−a[j+1])×d×s+b[j]×(d×d−s×s);
      a[j+1] := a[j]×s×s+g+a[j+1]×d×d;
      a[j] := p;
      p := d×c[j]−s×b[j+1];
      b[j+1] := s×c[j]+d×b[j+1];

c[j] := p;
g := −s×c[j+1];
c[j+1] := d×c[j+1];
inform (n, j, d, s);
comment: The Jacobi rotation which has been performed
  in this turn of the j-loop is $A := U^T A U$ with

$$
U = \begin{pmatrix}
1 & & & & & \\
 & 1 & & & O & \\
 & & d & s & & \\
 & & -s & d & & \\
 & O & & & 1 & \\
 & & & & & 1
\end{pmatrix},
$$

  where the d's and s's are located at the crosspoints of
  rows and columns j and $j+1$;

    end j;
ex: **end** k
end m21

ALGORITHM 105
NEWTON MAEHLY
F. L. BAUER AND J. STOER
Johannes Gutenberg-Universität, Mainz, Germany

```
procedure Newton Maehly (a, n, z, eps);
  value n, eps;
  array a, z;
  integer n;
  real eps;
  comment The procedure determines all zeros z[1:n] of the
    polynomial p(x) := a[0] × x↑n + ⋯ + a[n] of order n, if p(x)
    has only real zeros which have to be all different. The zeros
    z[i] are ordered according to their magnitude: z[1]>z[2]>
    ⋯>z[n]. The approximations for each zero will be improved
    by iteration as long as abs(x1−x0)>eps × abs(x1) holds
    for two successive approximations x0 and x1;
begin  real aa, pp, qq, x0, x1;
  integer i, m, s;
  array b, p, q[0:n−1];
  procedure Horner(p, q, n, x, pp, qq);
    value n, x;
    array p, q;
    real pp, x, qq;
    integer n;
  begin real s, s1;
    integer i;
    s := s1 := 0;
    for i := 0 step 1 until n−1 do
    begin s := s×x+p[i];  s1 := s1×x+q[i];  end ;
    pp := s×x+p[n];  qq := s1;
  end;
  p[0] := aa := a[0];  x0 := pp := 0;  s := sign(a[0]);
  for i := 1 step 1 until n do
    if s × a[i]<0 then
      begin if pp=0 then pp := i;
            if x0<abs(a[i]) then x0 := abs(a[i]);
      end;
  x0 := if pp=0 then 0 else 1+exp(ln(abs(xo/aa))/pp);
  comment  x0 is a first approximation for the largest zero which
    may be printed out at this point of the program;
  for i := 0 step 1 until n−1 do b[i] := (n−1)×a[i];
  for m := 1 step 1 uhtil n do
    begin
      iteration:
      Horner (a, b, n, x0, pp, qq);  x1 := x0−pp/qq;
      if abs(x1−x0)>eps×abs(x1) then
        begin x0 := x1;
          comment  x0 is the last approximation for the zero
            being improved, which may be printed out at this
            point;
          go to iteration;
        end;
      z[m] := x1;
      comment  z[m] := x1 is the mth zero of the polynomial;
      pp := b[0] := b[0] − aa;  q[0] := pp;
      if m<n then
        begin for i := 1 step 1 until n−1 do
                begin pp := p[i] := x1×p[i−1]+a(i);
                      pp := b[i] := b(i)−pp;
                      q[i] := x1×q[i−1]+pp;
```

```
        end;
      Horner (p, q, n−1, x1, pp, qq);
        x0 := x1−pp/qq;
      comment x0 is a first approximation for the
        next zero;
    end
  end
end Newton Maehly;
```

CERTIFICATION OF ALGORITHM 105
NEWTON MAEHLY [F. L. Bauer and J. Stoer, Comm.
  ACM, July 1962]
JOANNE KONDO
Burroughs Corp., Pasadena, Calif.

Algorithm 105 was successfully run on Burroughs 220 computer
after the following correction had been made:

```
    for i := 0 step 1 until n − 1 do b[i] := (n−1) × a[i]
```

changed to

```
    for i := 0 step 1 until n−1 do b[i] := (n−i) × a[i].
```

The following polynomials were tested for real roots using this
algorithm:

polynomial	epsilon	accuracy
(1) $x^3 - 2x^2 - 5x + 6$	0.0000001	10^{-8}
(2) $x^5 - 15x^4 + 85x^3 - 225x^2 + 274x - 120$	0.000001	10^{-5}

ALGORITHM 106
COMPLEX NUMBER TO A REAL POWER
MARGARET L. JOHNSON AND WARD SANGREN
Computer Applications, Inc., San Diego, California

```
procedure  POWC (x, y, w, A, B);  value x, y, w;
  real x, y, w, A, B;
comment  This procedure takes a complex number (x+iy) to
  a real power w. The result is A+iB=(x+iy)ʷ. This procedure
  must be used with caution because although it is formally cor-
  rect, it may not give the desired results. For example, if w is a
  reciprocal integer it does not follow that the desired power
  (a root) will be calculated;
begin real    THETA, PHI, R;
  if x>0 then begin THETA := 0.0;  go to SOL 1 end;
  if x<0∧y≧0 then begin THETA := 3.1415927;
    go to SOL 1 end;
  if x<0∧y<0 then begin THETA := 3.1415927;
    go to SOL 1 end;
  if x=0∧y=0 then begin A := B := 0.0;  go to RETURN end:
  if x=0∧y<0 then begin PHI := 1.5707963;  go to SOL 2 end;
  if x=0∧y>0 then begin PHI := -1.5707963;
    go to SOL 2 end;
  SOL 1:  PHI := arctan (y/x)+THETA;
  SOL 2:  R := sqrt (x×x+y×y);
          R := exp (w×ln(R));
          A := R×cos (w×PHI);
          B := R×sin (w×PHI);
  RETURN:  end POWC
```

REMARK ON ALGORITHM 106
COMPLEX NUMBER TO A REAL POWER [Mar-
  garet L. Johnson and Ward Sangren, *Comm. ACM*
  5, Jul. 1962]
GRANT W. ERWIN, JR.
The Boeing Co., Renton, Wash.

The comment "if W is a reciprocal integer it does not follow
that the desired power (a root) will be calculated" might better
read "if W is the reciprocal of an integer N, the procedure will
calculate an nth root, but possibly not the particular nth root
desired. E.g. $w = \frac{1}{3}, x = -1, y = 0$ uields $A = \frac{1}{2}, B = \frac{1}{2}\sqrt{3}$ rather
than the simpler $A = -1, B = 0$."

The comment should be made that it is assumed that the arctan
function yields a result between $-\pi/2$ and $\pi/2$.

The following four corrections should be made:

(1)   **if** $x<0 \wedge y < 0$ **then begin** THETA: = 3.1415927;
        should read:
                        ··· THETA: = -3.1415927;

(2)   **go to** RETURN **end**:
        should read:
                        go to RETURN **end**;

(3)   **if** $x = 0 \wedge y<0$ ···
        should read:
                        if x = 0 ∧ y>0 ···

(4)   **if** $x = 0 \wedge y > 0$
        should read:
                        if $x = 0 \wedge y < 0$ ···

ALGORITHM 107
GAUSS'S METHOD
JAY W. COUNTS
University of Missouri, Columbia, Mo.

```
    procedure gauss (u, a, y);
    real array a, y;  real temp;  integer u;
    comment This procedure is for solving a system of
        linear equations by successive elimination of the un-
        knowns. The augmented matrix is a and u the number of
        unknowns. The solution vector is y. If the system hasn't
        any solution or many solutions, this is indicated by go
        to stop;
    begin
    integer i, j, k, m, n;
    n := 0;
ck0:  n := n+1;
    for k := n step 1 until u do if a[k, n]≠0 then go to ck1;
    go to stop;
ck1:  if k=n then go to ck2;
    for m := n step 1 until u+1 do
    begin
      temp := a[n, m];  a[n, m] := a[k, m];  a[k, m] := temp
    end;
ck2:  for j := u+1 step−1 until n do a[n, j] := a[n, j]/a[n, n];
    for i := k+1 step 1 until u do
    for j := n+1 step 1 until u+1 do
    a[i, j] := a[i, j]−a[i, n]×a[n, j];
    if n≠u then go to ck0;
    for i := u step−1 until 1 do
    begin
      y[i] := a[i, u+1]/a[i, i];
      for k := i−1 step−1 until 1 do
      a[k, u+1] := a[k, u+1]−a[k, i]×y[i]
    end end;
```

REMARK ON ALGORITHM 107
GAUSS'S METHOD [J. W. Counts, *Comm. ACM*,
    July 1962]
P. NAUR
Regnecentralen, Copenhagen, Denmark

Algorithm 107 cannot be recommended since it does not search
for pivot and therefore will yield poor accuracy (cf. Remarks on
Algorithm 42 above).

CORRECTION TO EARLIER REMARKS ON AL-
GORITHM 42 INVERT, ALG. 107 GAUSS'S METHOD,
ALG. 120 INVERSION II, AND gjr [P. Naur, *Comm.
ACM*, Jan. 1963, 38–40.]
P. NAUR
Regnecentralen, Copenhagen, Denmark

George Forsythe, Stanford University, in a private communication has informed me of two major weaknesses in my remarks on
the above algorithms:

1) The computed inverses of rounded Hilbert matrices are compared with the exact inverses of unrounded Hilbert matrices, instead of with very accurate inverses of the rounded Hilbert
matrices.

2) In criticizing matrix inversion procedures for not searching
for pivot, the errors in inverting positive definite matrices cannot
be used since pivot searching seems to make little difference with
such matrices.

It is therefore clear that although the figures quoted in the
earlier certification are correct as they stand, they do not substantiate the claims I have made for them.

To obtain a more valid criterion, without going into the considerable trouble of obtaining the very accurate inverses of the
rounded Hilbert matrices, I have multiplied the calculated inverses by the original rounded matrices and compared the results
with the unit matrix. The largest deviation was found as follows:

	Maximum deviation from elements of the unit matrix		
Order	INVERSION II	gjr	Ratio
2	$-1.49_{10}-8$	$-1.49_{10}-8$	1.0
3	$-4.77_{10}-7$	$-8.34_{10}-7$	0.57
4	$-9.54_{10}-6$	$-3.43_{10}-5$	0.28
5	$-7.32_{10}-4$	$-4.58_{10}-4$	1.6
6	$-1.61_{10}-2$	$-1.42_{10}-2$	1.1
7	$-5.78_{10}-1$	$-5.47_{10}-1$	1.1
8	$-1.20_{10}-2$	$-1.38_{10}1$	8.7
9	$-4.91_{10}1$	$-2.22_{10}1$	2.2

This criterion supports Forsythe's criticism. In fact, on the
basis of this criterion no preference of INVERSION II or *gjr* can
be made.

The calculations were made in the GIER ALGOL system, which
has floating numbers of 29 significant bits.

ALGORITHM 108
DEFINITE EXPONENTIAL INTEGRALS A
Yuri A. Kruglyak
Kharkov State University, Kharkov, U.S.S.R., AND
Donald R. Whitman
Case Institute of Technology, Cleveland, Ohio

real procedure As (n, b); **value** n, b; **integer** n; **real** b;
comment: This procedure computes a value of integral
$A_{n-1}(1, b) = \int_1^\infty x^{n-1}\exp(-bx)\,dx$ for any given positive integer, n,
and any positive real parameter, b, by the recursion formula
$A_n(1, b) = A_0(1, b) + (n/b)A_{n-1}(1, b)$ with $A_0(1, b) = \exp(-b)/b$;
begin integer m; **real** db; **real array** a[1:n];
    a[1] := exp (−b)/b;
  if n = 1 **then go to** exit;
comment integral $a[1] = A_0(1, b)$ was evaluated;
    db := 1/b; **for** m := 2 **step** 1 **until** n **do** a[m] :=
    a[1]+db× (m−1)×a[m−1];
comment integral $a[n] = A_{n-1}(1, b)$ was evaluated;
    As := a[n] **end** As;

CERTIFICATION OF ALGORITHM 108
DEFINITE EXPONENTIAL INTEGRALS A [Yuri
  A. Kruglyak and Donald R. Whitman, *Comm. ACM 5*
  (July 1962)]
Yuri A. Kruglyak
Kharkov State University, Kharkov, U.S.S.R. and
Donald R. Whitman
Case Institute of Technology, Cleveland, Ohio

Integrals $A_n(1,b) = \int_1^\infty x^n\exp(-bx)dx$ occur in physical problems
involving spheroidal coordinates, particularly in quantum chemistry calculations. This algorithm was programmed for the Burrough's 220 computer using Burrough's Algebraic Compiler. The
program was used to compute tables of $A_n(1,b)$ in the ranges
$n = 0(1)15$, and $b = 0.01(0.01)30.14$. For example, for $n = 0(1)15$, and
$b = 0.25$ and $b = 24.0$, the results below were obtained. These are
compared with the results (columns 3 and 5) obtained by James
Miller, John M. Gerhauser, and F. A. Matsen [*Quantum Chemistry
Integrals and Tables*, University of Texas Press, 1959].

n	$b=0.25$	$b=0.25$ (Miller et al.)	$b=24.0$	$b=24.0$ (Miller et al.)
0	.31152031, 01	.31152031322856, 01	.15729727, −11	.15729727267830, −11
1	.15576015, 02	.15576015661428, 02	.16385132, −11	.16385132570656, −11
2	.12772332, 03	.12772332842371, 03	.17095154, −11	.17095154982051, −11
3	.15357950, 04	.15357951442168, 04	.17866621, −11	.17866621640586, −11
4	.24575835, 05	.24575837510601, 05	.18707497, −11	.18707497541261, −11
5	.49151976, 06	.49151986541516, 06	.19627122, −11	.19627122588926, −11
6	.11796476, 08	.11796479885167, 08	.20636507, −11	.20636507915061, −11
7	.33030132, 09	.33030143989988, 09	.21748707, −11	.21748708743056, −11
8	.10569642, 11	.10569646079911, 11	.22979295, −11	.22979296848848, −11
9	.38050711, 12	.38050725887992, 12	.24346962, −11	.24346963586148, −11
10	.15220284, 14	.15220290355200, 14	.25874294, −11	.25874295428724, −11
11	.66969248, 15	.66969277562880, 15	.27588778, −11	.27588779339328, −11
12	.32145238, 17	.32145253230182, 17	.29524115, −11	.29524116937494, −11
13	.16715523, 19	.16715531679695, 19	.31721955, −11	.31721957275639, −11
14	.93606928, 20	.93606977406291, 20	.34234200, −11	.34234202345285, −11
15	.56164156, 22	.56164186443775, 22	.37126102, −11	.37126103733633, −11

The accuracy is at least six significant figures over the entire
range. This accuracy is completely satisfactory for all quantum
chemical calculations.

ALGORITHM 109
DEFINITE EXPONENTIAL INTEGRALS B
YURI A. KRUGLYAK
Kharkov State University, Kharkov, U.S.S.R., AND
DONALD R. WHITMAN
Case Institute of Technology, Cleveland, Ohio

real procedure Bs(n, a); **value** n, a; **integer** n; **real** a;
comment This procedure computes a value of the integral
$B_{n-1}(a) = \int_{-1}^{+1} x^{n-1} \exp(-ax)dx$ for any given positive integer, n,
and any real parameter, a. If $|a| <$ alim an expansion of
$\exp(-ax)$ is used, otherwise the recursion formula $B_n(a) =$
$[(-1)^n e^a - e^{-a} + nB_{n-1}(a)]/a$ with $b_0(a) = 2 \sinh(a)/a$ is used. The
value of alim depends upon the highest n appearing in the
calculations and upon the maximum errors in the last significant
digits in the library procedures. For example, we have used
alim $= 8$ for $n_{max} = 16$ with gamma $= 1 \times 10^{-8}$. The intrinsic function mod(E_1, E_2) which requires two integer arguments, is the
conventional modulus;

```
begin      integer m; real alim, delta, gamma, r, epsilon,
             s, k, a2, omega, da, jp, jm, q1, q2; real array
             b[1:n]; if a=0 then
L1:        begin if mod(n−1, 2)=0 then
L2:        begin b[n] := 2/n;  go to exit end L2;
comment    integral b[n] = Bn−1(0) for odd n was evaluated;
           b[n] := 0;  go to exit end L1;
comment    integral b[n] = Bn−1(0) for even n was evaluated;
           if abs(a) alim then
L3:        begin delta := gamma;  if mod(n−1, 2)=0 then
L4:        begin r := 2/n;  epsilon := r×delta;  s := r;
             k := 0;  a2 := a↑2;
Even:      k := k+2;
             r := r×a2×(n+k−2)/(k×(k−1)×(n+k));
             s := s+r;  if r>epsilon then go to Even;
             b[n] := s+r;
           go to exit end L4;
comment    integral b[n]=Bn−1(a) for odd n and | a |<alim was
             evaluated;
           r := 2×a/(n+1);  omega := abs(r×delta);
             s := r;  k := 1;
           a2 := a↑2;
Odd:       k := k+2;
             r := r×a2×(n+k−2)/(k×(k−1)×(n+k));
           s := s+r;  if abs(r)>omega then go to Odd;
           b[n] := −(s+r);  go to exit end L3;
comment    integral b[n]=Bn−1(a) for even n and | a |<alim was
             evaluated;
           da := 1/a;  jp := da×exp(a);  jm := (da↑2)/jp;
             b[1] := jp−jm;
           if n=1 then go to exit;
comment    integral b[1] = B0(a) for | a |≧alim was evaluated;
           q1 := −1;  q2 := 1;  for m := 2 step 1 until n do
L5:        begin  b[m]  :=  q1×jp−jm+q2×da×b[m−1];
           q1 := −q1;  q2 := q2+1 end L5;
comment    integral b[n]=Bn−1(a) for integer n≧2 and | a |≧alim
             was evaluated;
exit:      Bs := b[n] end Bs;
```

CERTIFICATION OF ALGORITHM 109
DEFINITE EXPONENTIAL INTEGRALS B [Yuri A.
   Kruglyak, D. R. Whitman, *Comm. ACM 5* (July 1962)]
YURI A. KRUGLYAK
Kharkov State University, Kharkov, U.S.S.R., and
DONALD R. WHITMAN
Case Institute of Technology, Cleveland, Ohio

Integrals $B_n(a) = \int_{-1}^{+1} x^n \exp(-ax)dx$ occur in physical problems
involving spheroidal coordinates, particularly in quantum chemistry calculations. This algorithm was programmed for the Bur-
roughs-220 computer using a Burroughs Algebraic Compiler. The
program was used to compute tables of $B_n(a)$ in the ranges
$n=0(1)15$, and $a=0.00(0.01)32.54$. For example, for $n=0(1)15$ and
$a=0.25$, and $a=24.0$ the results below were obtained. These are
compared with the results (columns 3 and 5) obtained by James
Miller, John M. Gerhauser, and F. A. Matsen [*Quantum Chemistry
Integrals and Tables*, University of Texas Press, Austin, 1959].

n	$a=0.25$		$a=0.25$ (Miller et al.)		$a=24.0$		$a=24.0$ (Miller et al.)	
0	.20208984,	01	.20208985344653,	01	.11037134,	10	.110371342208,	10
1	−.16771064,	00	−.16771066117520,	00	−.10577253,	10	−.105772536282,	10
2	.67921322,	00	.67921324506375,	00	.10155696,	10	.101556964184,	10
3	−.10074584,	00	−.10074585827159,	00	−.97676725,	09	−.976767216847,	09
4	.40896479,	00	.40896480211998,	00	.94091887,	09	.940918885936,	09
5	−.72008754,	−01	−.72008756636929,	−01	−.90768866,	09	−.907688654174,	09
6	.29268836,	00	.29268837517905,	00	.87679129,	09	.876921258533,	09
7	−.56030292,	−01	−.56030294023170,	−01	−.84798262,	09	−.847982638338,	09
8	.22792911,	00	.22792912573392,	00	.82105258,	09	.821052542631,	09
9	−.45856272,	−01	−.45856272975462,	−01	−.79581870,	09	−.795818718590,	09
10	.18664760,	00	.18664761544688,	00	.77212229,	09	.772122289331,	09
11	−.38809718,	−01	−.38809719373731,	−01	−.74982404,	09	−.749824039467,	09
12	.15803198,	00	.15803200452627,	00	.72880141,	09	.728801402343,	09
13	−.33640562,	−01	−.33640563670387,	−01	−.70894600,	09	−.708945995807,	09
14	.13702696,	00	.13702696892367,	00	.69016158,	09	.690161591189,	09
15	−.29686662,	−01	−.29686663616401,	−01	−.67236245,	09	−.672362427583,	09

The accuracy is at least six significant figures in the ranges
mentioned above. This accuracy is enough for the majority of
quantum chemistry calculations.

ALGORITHM 110
QUANTUM MECHANICAL INTEGRALS OF
   SLATER-TYPE ORBITALS
Yuri A. Kruglyak
Kharkov State University, Kharkov, U.S.S.R., and
Donald R. Whitman
Case Institute of Technology, Cleveland, Ohio

```
real procedure  INTSOLI (n, r, za, ab, As, Bs) Result:
   (s, i1, i2, i3);  value n, r, za, zb;  integer n;  real r, za, zb;
   real array a[1:8], b[1:8], G[1:2×n];  integer array bc[1:2×n,
   1:2×n];  real procedure As, Bs;
comment  Procedure INTSOLI computes the quantum mechan-
   ical integrals  s = ⟨ψ_{n00}^{ai} | ψ_{210}^{bi}⟩         (overlap integral),
              i1 = ⟨ψ_{n00}^{ai} | Z_a*/r_{ai} | ψ_{210}^{bi}⟩  (exchange integral),
              i2 = ⟨ψ_{n00}^{ai} | Z_b*/r_{bi} | ψ_{n00}^{ai}⟩  (coulomb integral),
   and        i3 = ⟨ψ_{210}^{bi} | Z_a*/r_{ai} | ψ_{210}^{bi}⟩  (coulomb integral).
```

Here $|\psi_{nlm}^{ai}\rangle$ is a Slater-type orbital of electron i centered on atomic nucleus a. The integer n is the effective principal quantum number with values 1, 2, 3 and 4. $Z_a^* = za$ and $Z_b^* = zb$ are effective nuclear charges. r_{bi} is the distance of electron i from nucleus b. The input parameter r is the distance between the two centers a and b. All physical quantities are given in atomic units;

```
begin       integer q, t, c, m;
            real g, zsa, zsb, ks, p, pt, lilya, s, k1, exc, i1, pppt,
            k2, sue, i2, pmpt, ptmp, k3, i3;
            bc[1, 1) := bc[2, 1] := bc[2, 2] := 1;
            for q := 3 step 1 until 2×n do
L6:         begin bc[g, 1] := 1;  for t := 2 step 1 until g−1 do
            bc[q, t] := bc[q−1, t−1]+bc[q−1, t];
            bc[q, q] := 1 end L6;
```

comment   binomial coefficients $bc[q, t] = \binom{q-1}{t-1}$ were computed

using the recursion formula $\binom{q}{t} = \binom{q-1}{t-1} + \binom{q-1}{t}$;

```
procedure   As(n, b) Result: (a[n]);  value n, b;  integer n;
            real b;
```

comment   procedure *As* computes a value of integral $A_{n-1}(1, b)$ [see Algorithm 108, "Definite Exponential Integrals A," by Yuri A. Kruglyak and D. R. Whitman, *Comm. ACM* (July 1962)]. Any identifier occurring within the *As* is specified to be local to the *As*;

```
begin       integer m;  real db;  a[1] := exp(−b)/b;
            if n=1 then go to exitAs;  db := 1/b;
            for m := 2 step 1 until n do a[m] := a[1] +
            db×(m−1)×a[m−1]
exitAs:     end As;
procedure   Bs(n, a) Result: (b[n]);  value n, a;  integer n;
            real a;
```

comment   procedure *Bs* computes a value of integral $B_{n-1}(a)$ [see Algorithm 109, "Definite Exponential Integrals B" by Yuri A. Kruglyak and D. R. Whitman, *Comm. ACM* (July 1962)]. Any identifier occurring within the *Bs* is specified to be local to the *Bs*;

```
begin       integer m;  real alim, delta, gamma, r, epsilon,
            s, k, a2, omega, da, up, jm, q1, q2;
            if a=0 then begin if mod(n−1, 2)=0
            then begin b[n] := 2/n;  go to exitBs end;
            b[n] := 0;  go to exitBs end;
            if abs(a)<alim then begin delta := gamma;
```

comment   we have used alim=8 and gamma=1×10⁻⁸;

```
            if mod(n−1, 2)=0 then begin r := 2/n;
            epsilon := r×delta;
            s := r;  k := 0;  a2 := a↑2;
Even:       k := k+2;
            r := r×a2×(n+k−2)/((k×(k−1)×(n+k));
            s := s+r;  if r>epsilon then go to Even;
            b[n] := s+r;
            go to exitBs end;  r := 2×a/(n+1);
            omega := abs(r×delta);
            s := r;  k := 1;  a2 := a↑2;
Odd:        k := k+2;
            r := r×a2×(n+k−2)/(k×(k−1)×(n+k));
            s := s+r;  if abs(r)>omega then go to Odd;
            b[n] := −(s+r);  go to exitBs end;  da := 1/a;
            jp := da×exp(a);
            jm := (da↑2)/jp;  b[1] := jp−jm;
            if n=1 then go to exitBs;
            q1 := −1;  q2 := 1;  for m := 2 step 1 until n
            do begin b[m] := q1×jp−jm+q2×da×b[m−1];
            q1 := −q1;  q2 := q2+1 end
exitBs:     end Bs;
            g := 1;  for m := 1 step 1 until 2×n do g := g/m;
            G[2×n] := g;
```

comment $1/(2n)! = G[2×n]$ was evaluated;

```
            zsa := za/n;  zsb := zb/2;  ks := (r/2)↑(n+3)×
            (2×zsa)↑(n+1/2)×zsb↑(5/2)×G[2×n]↑(1/2);
            p := r×(zsa+zsb)/2;
            pt := r×(zsa−zsb)/2;
            for c := 1 step 1 until n+3 do
ABSI:       begin As(c, p) Result: (a[c]);
            Bs(c, pt) Result: (b[c])
            end ABSI;
S:          lilya := 0;  for m := 0 step 1 until n do lilya :=
            lilya+bc[n+1, m+1]×(a[n−m+2]×(b[m+1]+
            b[m+3])−b[m+2]×(a[n−m+1]+a[n−m+3]));
            s := ks×lilya;
I1:         k1 := ks×2×za/r;  exc := 0;
            for m := 0 step 1 until n−1 do
            exc := exc+bc[n, m+1]×(a[n−m+1]×(b[m+1]+
            b[m+3])−b[m+2]×(a[n−m]+a[n−m+2]));
            i1 := k1×exc;
I2:         pppt := p+pt;  k2 := (r/2)↑(2×n)×(2×zsa)↑
            (2×n+1)×zsb×G[2×n];
            for c := 1 step 1 until 2×n do
BA2:        begin As(c, pppt) Result: (a[c]);
            Bs(c, pppt) Result: (b[c])
            end BA2;  sue := 0;
            for m := 0 step 1 until 2×n−1 do
            sue := sue+bc[2×n, m+1]×a[2×n−m]×b[m+1];
            i2 := k2×sue;
I3:         pmpt := p−pt;  ptmp := −pmpt;
            k3 := (r/2)↑4×2×za×zsb↑5;
            for c := 1 step 1 until 4 do
AB3:        begin As(c, pmpt) Result: (a[c]);
            Bs(c, ptmp) Result: (b[c])
            end AB3;  i3 := k3×(a[2]×(b[1]+2×b[3])−b[2]×
            (a[1]+2×a[3])+a[4]×b[3]−a[3]×b[4]) end
            INTSOLI;
```

CERTIFICATION OF ALGORITHM 110
QUANTUM MECHANICAL INTEGRALS OF
SLATER-TYPE ORBITALS [Yuri A. Kruglyak and
Donald R. Whitman, *Comm. ACM 5* (July 1962)]

YURI A. KRUGLYAK
Kharkov State University, Kharkov, U.S.S.R. and
DONALD R. WHITMAN
Case Institute of Technology, Cleveland, Ohio

This procedure was written and tested in the Burroughs 220 version of the ALGOL language in the spring of 1961 at Case Institute of Technology. The program was used to compute tables of quatnum mechanical integrals s, $i1$, $i2$, and $i3$ in the ranges: $r(\text{Å}) = 0.64(0.02)1.40(0.10)3.10$; $Z_b^* = 0.25(0.50)3.75$, 3.90, 4.25, 4.55, 4.75, 5.20, 5.25; $Z_a^* = 0.7$, 1.0 for $n=1$; 1.3(1.0)3.3 for $n=2$; 0.2, 2.2(1.0)4.2 for $n=3$; and 0.2, 2.2, 3.2 for $n=4$. The table at the right shows typical results compared with values from *Integraltafeln zur Quantenchemie* by H. Preuss (Springer-Verlag, 1957), Zweiter Band. Accuracy is at least six significant figures in the ranges mentioned above. This is ample for the overwhelming majority of quantum chemistry calculations.

Certification of INTSOLI

Input				K. and W. Result		Preuss' Result		Table No.
n	r	za	zb			Notation		
1	5	0.5	0.2	s	0.14841691	0.5 0.1 [1a 3b]	0.148417	23
1	1	4.5	8.0	s	0.35203437	4.5 4.0 [1a 3b]	0.352034	30
2	1	20	20	s	$0.25032133 \times 10^{-1}$	10 10 [2a 3b]	0.250321×10^{-1}	40
1	5	0.5	0.2	$i1$	0.22058816	0.5 0.1 $5[a^{-1} \mid$ 1a 3b]	0.220588	59
1	1	4.5	8.0	$i1$	0.96587055	4.5 4.0 $1[a^{-1} \mid$ 1a 3b]	0.965871	66
2	1	20	20	$i1$	$0.58102500 \times 10^{-1}$	10 10 $1[a^{-1} \mid$ 2a 3b]	0.581025×10^{-1}	76
1	1	0.5	0.2	$i2$	0.44818080	0.5 0.5 $1[b^{-1} \mid$ 1a 1a]	0.448181	41
1	5	0.5	0.2	$i2$	0.97641725	0.5 0.5 $5[b^{-1} \mid$ 1a 1a]	0.976417	45
2	1	20	20	$i2$	0.99999530	10 10 $1[b^{-1} \mid$ 2a 2a]	0.100000×10^{1}	58
1	1	10	1	$i3$	0.26217432	0.5 0.5 $1[b^{-1} \mid$ 3a 3a]	0.262174	41
1	1	10	10	$i3$	0.11093011×10^{1}	5 5 $1[b^{-1} \mid$ 3a 3a]	0.110929×10^{1}	50
1	1	10	20	$i3$	0.10300137×10^{1}	10 10 $1[b^{-1} \mid$ 3a 3a]	0.103001×10^{1}	58

ALGORITHM 111

MOLECULAR-ORBITAL CALCULATION OF MOLECULAR INTERACTIONS

Yuri A. Kruglyak

Kharkov State University, Kharkov, U.S.S.R., and

Donald R. Whitman

Case Institute of Technology, Cleveland, Ohio

real procedure SOLI(n, za, zb, rem, coef2, r1, E1, dr1, drk1, acc, acc1, rk2, rk1, dr2, asy, rk3, dr3, As, Bs, RESULT) Result: (ra, Ep, Em, ca, ca2, cb, cb2, DEa, DEb, s, i1, i2, i3, haa, hab, hbb); **value** n, za, zb, rem, coef2, r1, E1, dr1, rk1, drk1, acc, acc1, rk2, dr2, asy, rk3, dr3; **integer** n, rem; **real** otherwise; **real array** a[1:8], b[1:8], G[1:8], zsap [1:9], rp[1:8], ans[1:rem]; **real procedure** As, Bs, RESULT;

comment This procedure calculates a one-electron approximation to the energy of interaction of molecular species by the use of the molecular-orbital (MO) method with a linear combination of Slater-type orbitals (LCSTO). The wave function used is $|\phi\rangle = c_a |\psi_{n00}^a\rangle + c_b |\psi_{210}^t\rangle$, where $|\psi_{n1m}^a\rangle$ is a STO centered on nucleus a. The effective principal quantum number n takes the integral value 1, 2, 3 or 4. The Hamiltonian used is: $\mathcal{H}_{ab} = -\Delta/2 - Z_a^*/r_a - Z_b^*/r_b + Z_a^* Z_b^*/R_{ab}$. Here Z_a^* and Z_b^* are effective nuclear charges, r_a and r_b the distances of the electron from nucleus a and b, and R_{ab} is the distance between nuclei a and b. The calculations are in atomic units, while the output ra is in Angstroms and DEa and DEb are in kcal/gm-ion. Abbreviations of the following type are used: $Za = Z_a^*$, $ra = R_{ab}(\text{Å})$, $haa = \langle\psi_{n00}^a | \mathcal{H}_{ab} | \psi_{n00}^a\rangle$, $DEa = D(a, b+\text{electron})$, $e1 = \langle\psi_{n00}^a | -\Delta/2 - Z_a^*/ra | \psi_{n00}^a\rangle$. The values of coef1 and coef2 are 627.71 (kcal/gm-ion) and 0.5291Å, respectively. $r1$, $E1$, $dr1$, $dr2$, $dr3$, acc, $acc1$, and asy are control parameters. The accuracy of the calculations $(acc, acc1)$ is 1×10^{-5}. The initial values of R_{ab} and E_- are conveniently: $r1=0.4(\text{Å})$, and $E1=100$(a.u.). The steps are: $dr1=0.1$, $dr2=0.4$, and $dr3=0.01$, all in Angstroms. asy is -1×10^{-3} (a.u.);

begin **integer** q, t, c, m, f; **real** otherwise;

procedure As(n, b) Result: (a[n]); **value** n, b; **integer** n; **real** b;

comment any identifier occurring within the As is specified to be local to the As;

begin **integer** m; **real** db; a[1] := exp(−b)/b; **if** n=1 **then go to** exitAs; db := 1/b; **for** m := 2 **step** 1 **until** n **do** a[m] := a[1]+db× (m−1)×a[m−1]

exitAs: **end** As;

procedure Bs(n, a) Result: (b[n]); **value** n, a; **integer** n; **real** a;

comment any identifier occurring within the Bs is specified to be local to the Bs;

begin **integer** m; **real** otherwise; **if** a=0 **then begin if** mod(n−1, 2)=0 **then begin** b[n] := 2/n; **go to** exitBs **end**; b[n] := 0; **go to** exitBs **end**; **if** abs(a)<alim **then begin** delta := gamma; **if** mod(n−1, 2)=0 **then begin** r := 2/n; epsilon := r×delta; s := r; k := 0; a2 := a↑2;

Even: k := k+2; r := r×a2×(n+k−2)/(k(k−1)(n+k)); s := s+r; **if** r>epsilon **then go to** Even; b[n] := s+r; **go to** exitBs **end**; r := 2×a/(n+1); omega := abs(r×delta);

Odd: s := r; k:=1; a2 := a↑2; k := k+2; r := r×a2×(n+k−2)/(k(k−1)(n+k)); s := s+r; **if** abs(r)>omega **then go to** Odd; b[n] := −(s+r); **go to** exitBs **end**; da := 1/a; jp := da×exp(a); jm := (da↑2)/jp; b[1] := jp−jm; **if** n=1 **then go to** exitBs; q1 := −1; q2 := 1; **for** m := 2 **step** 1 **until** n **do begin** b[m] := q1×jp−jm+ q2×da×b[m−1]; q1 := −q1; q2 := q2+1 **end**

exitBs: **end** Bs;

procedure Result(coef1); **real** coef1;

comment RESULT computes Ep, Em, ca, $ca2$, cb, $cb2$, DEa, DEb, s, $i1$, $i2$, $i3$, nn, haa, hab, hbb. Important: RESULT and any identifier occurring within the RESULT enter SOLI as nonlocal entities;

begin r := ra×br; rp[1] := r; **for** c := 2 **step** 1 **until** n+4 **do** rp[c] := rp[c−1]×rp[1]; p := r×sum; pt := r×dif; ks := rp[n+3]×zss; **for** c:= 1 **step** 1 **until** n+3 **do begin** As(c, p) Result: (a[c]); Bs(c, pt) Result: (b[c]) **end**; lilya := 0; **for** m := 0 **step** 1 **until** n **do** lilya := lilya+bc[n+1, m+1]× (a[n−m+2]× (b[m+1]+b[m+3])−b[m+2]× (a[n−m+1] +a[n−m+3])); s := ks×lilya: 1i12 := 2×s; k1 := ks×za/r; exc := 0; **for** m := 0 **step** 1 **until** n−1 **do** exc := exc+bc[n, m+1]× (a[n− m+1]× (b[m+1]+b[m+3])−b[m+2]× (a[n−m]+ a[n−m+2])); i1 := k1×exc; pppt := p+pt; k2 := rp[2×n]×zsbd; **for** c := 1 **step** 1 **until** 2×n **do begin** As(c, pppt) Result: (a[c]); Bs(c, pppt) Result: (b[c]) **end**; sue := 0; **for** m := 0 **step** 1 **until** 2×n−1 **do** sue := sue+bc[2×n, m+1]×a[2×n−m]×b[m+1]; i2 := k2×sue; pmpt := p−pt; ptmp := −pmpt; k3 := rp[4]×z5; **for** c := 1 **step** 1 **until** 4 **do begin** As(c, pmpt) Result: (a[c]); Bs(c, ptmp) Result: (b[c]) **end**; i3 := k3× (a[2]× (b[1]+2×b[3])−b[2]× (a[1]+2× a[3])+a[4]×b[3]−a[3]×b[4]);

comment Two-center integrals s, $i1$, $i2$, and $i3$ were computed [see Algorithm 110, "Quantum Mechanical Integrals of Slater-Type Orbitals," by Yuri A. Kruglyak and D. R. Whitman, *Comm. ACM* (July 1962)]; nn := zz/(2×r); e2pnn := e2+nn; haa := e1−e2+nn; hbb := e2pnn−i3; hab := e2pnn×s−i1; den := 2−s×1i12; bsr := haa+hbb−hab×1i12; root := sqrt(bsr↑2−2×den× (haa×hbb−hab↑2)); Ep := (bsr+root)/den; Em := (bsr−root)/den; ans[f] := Em; DEa := coef1× (e2−Em); DEb := coef1× (e1−Em); Emhaa := Em−haa; Emhbb := Em−hbb; ES := Em×s; habmES := hab−ES; caDcb1 := habmES/Emhaa; cbDca2 := habmES/Emhbb; **if** abs(Emhaa)>abs(Emhbb) **then begin** col := caDcb1↑2; cb2 := 1/(1+1i12×caDcb1+col); ca2 := cb2×col; ca := sqrt(ca2); cb := ca/caDcb1 **go to** NATA **end**; co2 := cbDca2↑2; ca2 := 1/(1+1i12×cbDca2+co2); cb2 := ca2×co2; ca := sqrt(ca2); cb := ca×cbDca2

NATA: **end** RESULT;

Begin of program: bc[1, 1] := bc[2, 1] := bc[2, 2] := 1; **for** q := 3 **step** 1 **until** 8 **do begin** bc[q, 1] := 1; **for** t := 2

```
      step 1 until q−1 do bc[q, t] := bc[q−1, t−1]+bc[q−1, t];
      bc[q, q] := 1 end;
IZM:      g := 1;  for m := 1 step 1 until 2×n do g := g/m;
          G [2×n]:=g;  zsa := za/n;  zsap[1] := zsa×2;
          for c := 2 step 1
          until  2×n+1  do  xsap[c]  :=  zsap[c−1]×zsap[1];
          D   :=   zsap[2×n+1]×G[2×n];  DS   :=   sqrt(D);
          e1   :=   −zsap[2]×0.125×(4×n−3)/(2×n−1);
          zsb := zb×0.5;  sum := zsa+zsb;  dif := zsa−zsb;
          zsb5  :=  zsb↑5;  zss   :=  DS×sqrt(zsb5);
          zsbd := zsb×D;  z5 := 2×za×zsb5;  zz := za×zb;
          e2 :=  −(zsb↑2)/2;  br :=  0.5/coef2;  f :=  1;
          ans[1] := E1;  ra := r1;
KOM:      ra := ra+dr1;  if ra>rk1 then ra := ra+drk1;
          f := f+1;  RESULT (coef1);  if ans [f]−ans[f−1]≦
          acc then begin if ra>rk2 then go to IZM;
          go to KOM end;  ansf := ans[f];  d1 := ra;
CLEV:     ra := ra+dr2;  RESULT (coef1);  if e1<e2 then
             begin if Em−e1≦asy∧ra<rk3 then go to CLEV;
             go to KHAR end;  if e1≧e2 then begin if Em−e2≦
             asy∧ra<rk3 then go to CLEV;  go to KHAR end;
KHAR:     ra := d1;  ans[f−1] := ansf;
CASE:     ra := ra−dr3;  f := f+1;  RESULT (coef1);
          if ans[f]−ans[f−1]≦acc1 then go to CASE;
          go to IZM end SOLI;
```

ALGORITHM 112
POSITION OF POINT RELATIVE TO POLYGON
M. Shimrat
University of Alberta, Calgary, Alberta, Canada

Boolean procedure *POINT IN POLYGON* $(n, x, y, x0, y0)$;
value $n, x0, y0$; **integer** n; **array** x, y; **real** $x0, y0$;
comment if the points $(x[i], y[i])$ $(i = 1, 2, \cdots, n)$ are—in
    this cyclic order—the vertices of a simple closed polygon and
    $(x0, y0)$ is a point not on any side of the polygon, then the pro-
    cedure determines, by setting "point in polygon" to **true**,
    whether $(x0, y0)$ lies in the interior of the polygon;
begin integer i; **Boolean** b;
  $x[n + 1] := x[1]$; $y[n + 1] := y[1]$; $b :=$ **true**;
  for $i := 1$ **step** 1 **until** n **do**
    if $(y < y[i] \equiv y > y[i + 1]) \wedge$
      $x0 - x[i] - (y0 - y[i]) \times (x[i + 1] - x[i])/(y[i + 1] - y[i]) < 0$
    then $b := \neg\, b$;
  POINT IN POLYGON $:= \neg\, b$;
end POINT IN POLYGON

CERTIFICATION OF ALGORITHM 112
POSITION OF POINT RELATIVE TO POLYGON
  [M. Shimrat, *Comm. ACM*, Aug. 1962]
Richard Hacker
The Boeing Co., Seattle Wash.

    The Boolean procedure *POINT IN POLYGON* was programmed
in Fortran for the IBM 7090. The algorithm gave satisfactory
results except for a case such as the following:

  Let the polygon points be: $(0, 0)$, $(1, 0)$, $(2, 1)$, $(1, 2)$, $(0, 2)$.

In this case the procedure would not detect that the point $(1, 1)$
is in the polygon. However, the correct result was obtained by
changing:

        if $(y < y[i] \equiv y > y[i+1]) \wedge$

to read:

        if $(y0 \leqq y[i] \equiv y0 > y[i+1]) \wedge$

ALGORITHM 113
TREESORT
Robert W. Floyd
Computer Associates, Inc., Woburn, Mass.

procedure *TREESORT* (*UNSORTED*, *n*, *SORTED*, *k*); **value**
 n, *k*;

integer *n*, *k*; **array** *UNSORTED*, *SORTED*;

comment *TREESORT* sorts the smallest *k* elements of the *n*-
 component array *UNSORTED* into the *k*-component array
 SORTED (the two arrays may be the same). The number of
 operations is on the order of $2 \times n + k \times \log_2(n)$. The number
 of auxiliary storage cells required is on the order of $2 \times n$. It is
 assumed that procedures are available for finding the minimum
 of two quantities, for packing one real number and one integer
 into a word, and for obtaining the left and right half of a packed
 word. The value of infinity is assumed to be larger than that of
 any element of *UNSORTED*;

begin integer *i*, *j*; **array** $m[1:2 \times n - 1]$;

for $i := 1$ **step** 1 **until** *n* **do** $m[n + i - 1] := pack\ (UNSORTED$
 $[i], n + i - 1)$;

for $i := n - 1$ **step** $- 1$ **until** 1 **do** $m[i] := minimum\ (m[2 \times i],$
 $m[2 \times i + 1])$;

for $j := 1$ **step** 1 **until** *k* **do**

  begin *SORTED* $[j] := left\ half\ (m[1])$; $i := right\ half\ (m[1])$;
   $m[i] := infinity$;

  for $i := i \div 2$ **while** $i > 0$ **do** $m[i] := minimum\ (m[2 \times i], m[2 \times$
   $i + 1])$

  end

end TREESORT

ALGORITHM 114

GENERATION OF PARTITIONS WITH CON-
STRAINTS

Frank Stockmal

System Development Corp., Santa Monica, Calif.

```
procedure CP GENERATOR (N, K, H, p, F, Z);  integer
  N, K, H;  integer array p;  Boolean F, Z;
comment CP GENERATOR generates a partition of N into K
  parts, no part greater than H. Each partition is represented by
  the array of parts p[1] thru p[K], where p[1] ≧ p[2] ≧ ··· ≧ p[K].
  Initial entry is made with F = true and Z = true if parts = 0
  are allowable, or F = true and Z = false if only nonzero parts
  are desired. Upon initial entry, procedure ignores the input
  array p, sets F = false, and generates the initial parti-
  tion. Subsequent calls made with F = false will cause
  procedure to operate upon the input partition to produce
  another partition if one exists, so that all possible unpermuted
  partitions with the specified constraints will be produced if CP
  GENERATOR is allowed to operate upon its previous output.
  When this scheme is followed, and initial entry is made with
  F = true, Z = true, K = N, H = N, all possible un-
  permuted partitions of N will be produced. Upon generating
  the last partition, procedure resets F to true. The input param-
  eters are restricted as follows: K ≧ 1, H ≧ 1, p[1] ≧ p[2]
  ≧ ··· ≧ p[K]. For Z = true, N is restricted to the range
  0 ≦ N ≦ KH, and for Z = false, K ≦ N ≦ KH. A call should
  not be made with p[1] − p[K] < 2 and F = false;
begin integer a, b, i, j, q, r;
  if F then go to first;
  a := p[1] − p[2] − 2;  j := 2;
test:  if p[1] − p[j] ≧ 2 then go to divide;
  a := a − 1 + j × (p[j] − p[j + 1]);  j := j + 1;  go to test;
first:  if Z then go to alpha;
  a := N − K;  p[K] := 0;  go to beta;
alpha:  a := N;  p[K] := −1;
beta:  F := false;  j := K;
divide:  b := H − 1 − p[j];  q := entier (a/b);  r := a − b × q;
  for i := 1 step 1 until q do p[i] := H;
  if q = K then go to last;
  for i := q + 1 step 1 until j do p[i] := 1 + p[j];
  p[q + 1] := r + p[q + 1];
  if p[1] − p[K] ≧ 2 then go to exit;
last:  F := true;
exit:  end CP GENERATOR
```

ALGORITHM 115
PERM

H. F. TROTTER
Princeton University, Princeton, N. J.

```
procedure PERM (x, n);  value n;
integer n;  array x;
```

comment This algorithm was inspired by the procedure
PERMUTE of Peck and Schrack (Algorithm 86, *Comm. ACM*
Apr. 1962) and performs the same function. Each call of *PERM*
changes the order of the first n components of x, and $n!$ successive calls will generate all $n!$ permutations. A nonlocal Boolean
variable '*first*' is assumed, which must be **true** when *PERM* is
first called, to cause proper initialization. The first call of *PERM*
makes '*first*' **false**, and it remains so (unless changed by the
external program) until the exit from the $(n!)$th call of *PERM*.
At that time x is restored to its original order and '*first*' is made
true.

The excuse for adding *PERM* to the growing pile of permutation generators is that, at the expense of some extra **own** storage,
it cuts the manipulation of x to the theoretical minimum of $n!$
transpositions, and appears to offer an advantage in speed. It
also has the (probably useless) property that the permutations
it generates are alternately odd and even;

```
begin own integer array p, d[2: n];  integer k, q;  real t;
if first then initialize:
begin for k := 2 step 1 until n do
  begin p[k] := 0;  d[k] := 1 end;
  first := false
end initialize;
k := 0;
INDEX:  p[n] := q := p[n] + d[n];
  if q = n then
  begin d[n] := −1;  go to LOOP end;
  if q ≠ 0 then go to TRANSPOSE;
  d[n] := 1;  k := k + 1;
  LOOP:  if n > 2 then begin
    comment  Note that n was called by value;
    n := n − 1;  go to INDEX end LOOP;
Final exit:  q := 1;  first := true;
TRANSPOSE:  q := q + k;  t := x[q];
  x[q] := x[q + 1];  x[q + 1] := t
end PERM:
```


CERTIFICATION OF ALGORITHM 115
PERM [H. F. Trotter, *Comm. ACM* (Aug. 1962)]

G. F. SCHRACK
University of Alberta, Calgary, Alb., Canada

PERM was translated into FORTRAN for the IBM 1620 and it
performed satisfactorily. Timing tests were carried out under the
same conditions as for PERMUTATION (Algorithm 71) and
PERMUTE (Algorithm 86).

PERM is indeed the fastest permutation generator so far en-
countered. For $n = 8$, PERM is 25% faster than PERMUTE
(989 against 1316 sec.). The values for r_n are (for a definition of
r_n, see Certification of Algorithm 71, *Comm. ACM*, Apr. 1962):

n	6	7	8
r_n	.92	.95	.98

CERTIFICATION OF ALGORITHM 115
PERM [H. F. Trotter, *Comm. ACM*, Aug. 1962]

E. S. PHILLIPS
Michigan State University, East Lansing, Mich.

PERM was translated into FORTRAN for the CDC 160-A, and
it performed correctly. For $n = 8$, this method requires 2822
seconds. For comparison, Algorithm 86, *PERMUTE*, was translated and run on the same machine, requiring 3710 seconds as
opposed to 1316 when run on an IBM 1620.

ALGORITHM 116
COMPLEX DIVISION
Robert L. Smith
Stanford University, Stanford, Calif.

procedure *complexdiv* (a, b, c, d) results: (e, f);
value a, b, c, d; **real** a, b, c, d;
comment *complexdiv* yields the complex quotient of $a + ib$
   divided by $c + id$. The method used here tends to avoid arithmetic overflow or underflow. Such spills could otherwise occur
   when squaring the component parts of the denominator if the
   usual method were used;
begin real r, den;
   if abs (c) $\geqq$ abs (d) **then**
   begin $r := d/c$;
      $den := c + r \times d$;
      $e := (a + b \times r)/den$;
      $f := (b - a \times r)/den$;
   end
   else
   begin $r := c/d$;
      $den := d + r \times c$;
      $e := (a \times r + b)/den$;
      $f := (b \times r - a)/den$;
   end
end complexdiv

ALGORITHM 117
MAGIC SQUARE (EVEN ORDER)
D. M. COLLISON
Elliott Brothers (London) Limited, Borehamwood, Herts.,
England

```
procedure magiceven (n, x);  value n;  integer array x;  in-
    teger n;
comment  the method of Devedec for even n is described in
    "Mathematical Recreations" by M. Kraitchik, pp. 150-2. Enter
    with side of square n to produce a magic square of the integers
    1 − n ↑ 2 in x, where n ≧ 4;
begin integer a, b, n2, nn;  Boolean p, q, r;
    n2 := n ÷ 2;  nn := n × n;
    begin
    procedure alpha (p, q, a, h);  value p, q, a, h;  integer p, q, a;
        Boolean h;
    Comment pattern 0/0/0/ ··· ;
        begin integer r;
            for r := p step 1 until q do begin
                x[r, a] := if h then (a × n − n + r) else (nn − a × n +
                1 + n − r);  h := ¬h end;
    end alpha;
    procedure beta (p, q, a, h);  value p, q, a, h;  integer p, q, a;
        Boolean h;
    comment pattern 1 − 1 − 1 − ··· ;
        begin integer r;
            for r := p step 1 until q do begin
                x[r, a] := if h then [nn − a × n + r) else (a × n + 1 − r);
                h := ¬ h end;
    end beta;
    procedure gamma (p, q, a, h);  value p, q, a, h;  integer p, q, a;
        Boolean h;
    comment pattern /−/−/− ··· ;
        begin integer r;
            for r := p step 1 until q do begin
                x[r, a] := if h then (nn − a × n + n − r + 1) else (a × n
                + 1 − r);  h := ¬ h end;
    end gamma;
    comment program begins;
    p := q := (n − (n ÷ 4) × 4 = 0); r := true;
    for a := 1 step 1 until (n2 − 1) do begin
        beta (1, a − 1, a, r);  alpha (a, n2 − 1, a, true);
        x[n2, a] := if q then (nn − a × n + n2 + 1) else (nn − a ×
            n + n2);
        alpha (n2 + 1, n, a, ¬ q);
        q := ¬ q;  r := ¬ r end;
    alpha (1, n2 − 1, n2, ¬ p);  alpha (n2 + 2, n, n2, false);
    gamma (1, n2 − 1, n2 + 1, p);  gamma (n2 + 2, n, n2 + 1, true);
    q := p;  r := true;
    for a := (n2 + 2) step 1 until n do begin
        beta (1, n − a, a, q);  x[n − a + 1, a] := a × n − a + 1;
        beta (n − a + 2, n2 − 1, a, true);
        if r then for b := n2, n2 + 1 do x[b, a] := nn − a × n +
            n − b + 1
        else begin x[n2, a] := nn − a × n + n2;
            x[n2 + 1, a] := a × n − n2 + 1 end;
        beta (n2 + 2, a − 1, a, ¬ r);  alpha (a, n, a, true);
        q := ¬ q;  r := ¬ r end;
```

```
    for a := n2, n2 + 1 do for b := n2, n2 + 1 do
        x[b, a] := if p then (a × n − n + b) else (nn − a × n + n −
            b + 1);
    if ¬ p then begin
        for a := n2, n2 + 1 do x[n2 − 1, a] := a × n − n2 + 2;
        for b := n2, n2 + 1 do x[b, n2 + 2] := n × n2 − 2 × n + b end;
    end end magiceven
```

CERTIFICATION OF ALGORITHMS 117 AND 118
MAGIC SQUARE (ODD AND, EVEN ORDERS)
[D. M. Collison, *Comm. ACM*, Aug. 1962]
D. M. COLLISON
Elliott Bros. (London) Ltd., Borehamwood, Herts.,
England

Both algorithms were checked and timed, using a special ALGOL
program, with the Elliott ALGOL translator on the National-
Elliott 803. The procedure for odd orders was the slower:

Procedure	Size of Square	Time
Odd order	9	10 sec.
	19	45 sec.
Even order	10	7 sec.
	20	23 sec.

Because of the different methods used and the length of the even
order procedure it was decided not to combine the two. The
smallest square of even order generated is given below:—

13	3	2	16
8	10	11	5
12	6	7	9
1	15	14	4

CERTIFICATION OF ALGORITHMS 117 AND 118
MAGIC SQUARES (EVEN AND ODD ORDERS)
[D. M. Collison, [*Comm. ACM*, Aug. 1962]
P. NAUR
Regnecentralen, Copenhagen, Denmark

MAGICEVEN needed the following correction: Within the
body of **procedure** beta a left square bracket: ... **then** [nn ...
should be changed to a left parenthesis: ... **then** (nn ...

With this correction it has run successfully in the GIER ALGOL
system. The squares of even orders from 4 to 20 were generated
and checked for magicity in rows and columns, but not in
diagonals.

The algorithm contains 11 pairs of superfluous parentheses (10
of which are in conditional expressions) and if the assignments to
n2 and nn are moved to the place just following "**end gamma;**"
the inner block becomes unnecessary.

MAGICODD ran without correction in GIER ALGOL and produced a few reasonable-looking squares.

Run times are as follows:

Procedure	Size of square	Time
Magicodd	9	0.6 sec
	19	2.5 sec
Magiceven	10	0.9 sec
	20	2.3 sec

CERTIFICATIONS OF ALGORITHMS 117 and 118
MAGIC SQUARE (ODD AND EVEN ORDERS)
[D. M. Collison, *Comm. ACM*, Aug. 1962]

K. M. BOSWORTH
I.C.T. Ltd., Blyth Road, Hayes, Middlesex, England

The statement within the **Booleon procedure** *beta* should be changed from

$$x[r,a] \; := \; \textbf{if } h \textbf{ then } [nn-a\times n+r) \textbf{ else } (a\times n+1-r);$$

to

$$x[r,a] \; := \; \textbf{if } h \textbf{ then } (nn-a\times n+r) \textbf{ else } (a\times n+1-r);$$

The procedures were then tested on magic squares of order 3 to 17 inclusive without fault.

ALGORITHM 118
MAGIC SQUARE (ODD ORDER)
D. M. COLLISON
Elliott Brothers (London) Limited, Borehamwood, Herts.,
England

procedure *magicodd* (*n*, *x*); **value** *n*; **integer** *n*; **integer
array** *x*;

comment for given side *n* the procedure generates a magic
square of the integers $1 - n \uparrow 2$. For the method of De la
Loubère, see M. Kraitchik, "Mathematical Recreations," p.
149. *n* must be odd and $n \geq 3$;

begin integer *i*, *j*, *k*;
  for $i := 1$ **step** 1 **until** *n* **do**
    for $j := 1$ **step** 1 **until** *n* **do** $x[i, j] := 0$;
  $i := (n + 1) \div 2$; $j := n$;
  for $k := 1$ **step** 1 **until** $n \times n$ **do begin**
    if $x[i, j] \neq 0$ **then begin** $i := i - 1$; $j := j - 2$;
      if $i < 1$ **then** $i := i + n$; **if** $j < 1$ **then** $j := j + n$ **end**;
    $x[i, j] := k$;
    $i := i + 1$; **if** $i > n$ **then** $i := i - n$;
    $j := j + 1$; **if** $j > n$ **then** $j := j - n$;
    end;
end magicodd

CERTIFICATION OF ALGORITHMS 117 AND 118
MAGIC SQUARE (ODD AND, EVEN ORDERS)
[D. M. Collison, *Comm. ACM*, Aug. 1962]
D. M. COLLISON
Elliott Bros. (London) Ltd., Borehamwood, Herts.,
England

Both algorithms were checked and timed, using a special ALGOL
program, with the Elliott ALGOL translator on the National-
Elliott 803. The procedure for odd orders was the slower:

Procedure	Size of Square	Time
Odd order	9	10 sec.
	19	45 sec.
Even order	10	7 sec.
	20	23 sec.

Because of the different methods used and the length of the even
order procedure it was decided not to combine the two. The
smallest square of even order generated is given below:—

13	3	2	16
8	10	11	5
12	6	7	9
1	15	14	4

CERTIFICATION OF ALGORITHMS 117 AND 118
MAGIC SQUARES (EVEN AND ODD ORDERS)
[D. M. Collison, [*Comm. ACM*, Aug. 1962]
P. NAUR
Regnecentralen, Copenhagen, Denmark

MAGICEVEN needed the following correction: Within the
body of **procedure** beta a left square bracket: . . . **then** [*nn* . . .

should be changed to a left parenthesis: . . . **then** (*nn* . . .

With this correction it has run successfully in the GIER ALGOL
system. The squares of even orders from 4 to 20 were generated
and checked for magicity in rows and columns, but not in
diagonals.

The algorithm contains 11 pairs of superfluous parentheses (10
of which are in conditional expressions) and if the assignments to
n2 and *nn* are moved to the place just following "**end** gamma;"
the inner block becomes unnecessary.

MAGICODD ran without correction in GIER ALGOL and produced a few reasonable-looking squares.

Run times are as follows:

Procedure	Size of square	Time
Magicodd	9	0.6 sec
	19	2.5 sec
Magiceven	10	0.9 sec
	20	2.3 sec

CERTIFICATION OF ALGORITHM 118
MAGIC SQUARE (ODD ORDER) [D. M. Collison,
Comm. ACM, Aug. 1962]
HENRY C. THACHER, JR.*
Reactor Engineering Div., Argonne National Lab.,
Argonne, Ill.
* Work supported by the U. S. Atomic Energy Commission.

The body of the procedure *magicodd* was tested on the LGP-30
using the Dartmouth ALGOL 60 translator. No syntactical errors
were found. The procedure generated odd-order magic squares
satisfactorily. For orders up to 9, times were as follows (including
output on the Flexowriter):

Order	Time (sec)
3	171
5	422
7	804
9	1285

The 3×3 square was:

4	3	8
9	5	1
2	7	6

CERTIFICATIONS OF ALGORITHMS 117 and 118
MAGIC SQUARE (ODD AND EVEN ORDERS)
[D. M. Collison, *Comm. ACM*, Aug. 1962]
K. M. BOSWORTH
I.C.T. Ltd., Blyth Road, Hayes, Middlesex, England

The statement within the **Booleon procedure** *beta* should be
changed from

  $x[r,a] :=$ **if** *h* **then** [$nn - a \times n + r$) **else** $(a \times n + 1 - r)$;
to
  $x[r,a] :=$ **if** *h* **then** $(nn - a \times n + r)$ **else** $(a \times n + 1 - r)$;

The procedures were then tested on magic squares of order
3 to 17 inclusive without fault.

ALGORITHM 119
EVALUATION OF A PERT NETWORK
BURTON EISENMAN AND MARTIN SHAPIRO
United Nuclear Corp., White Plains, N. Y.

```
procedure  pert (nmax, i, j, te, st, emax, l, es, at);
comment  An algorithm describing an iterative procedure for
    evaluating a PERT network that permits the use of arbitrarily
    ordered activities and event identifiers such that an upper
    triangular matrix type of solution is unnecessary.
```

It has been observed by investigations of PERT networks, that an $N \times N$ matrix whose rows are designated as predecessor and whose columns are designated as successor events, has an entry in the (i, j)-element representing the activity time required in going from event i to event j. By elementary transformations, the matrix is transformed generally into an upper triangular matrix. The resultant upper triangular matrix is well ordered (i.e. any activity time appearing in a column is not dependent upon those activity times which appear in columns to the right of it).

This precise manipulation generally demands considerable running time. By direct evaluation not requiring a collection of elementary transformations, it is possible to evaluate the network with considerable reduction of running time;

```
integer  nmax, emax;
real  st;
integer array  i, j, l;
real array  te, es, at;
comment  Given the total number of activities, nmax, the pre-
    ceding and succeeding event identifiers, i_n and j_n, the cor-
    responding expected time, te, for each activity, and the starting
    time, st, of the network, this procedure computes the early start
    and late finish times, es_e and at_e, for each event, l_e, in the net-
    work;
begin
procedure  scan (e, t, l);
integer  e, t;
integer array  l:
comment  Given the number of events, e−1, contained thus far
    in vector array, l, and an event identifier i_n or j_n, stored in t,
    this procedure scans the existing array, l, to determine whether
    the event should be added to the list or not. If it is to be added,
    it becomes l_e and e replaces the event identifier. If it is not
    added, k replaces the event identifier.;
begin
integer  k;
        if e = 1 then go to add;
        for k := e−1 step −1 until 1 do
begin   if t = l[k] then
begin   t := k;
        go to out
end
end;
add:    l[e] := t;
        t := e;
        e := e + 1;
out:
end scan;
```

```
integer  n, e, s, t, k;
real     a, x;
         e := 1;
         for n := 1 step 1 until nmax do
begin    t := j[n];
         scan (e, t, l);
         j[n] := t;
         t := i[n];
         scan (e, t, l);
         i[n] := t
end;
comment  By means of the switch, s, we will either compute the
    activity times, at_e , and transfer the values to the early start
    vector, es_e , or we will compute at_e without any transfer process,
    in which case the late finish times will be obtained.;
         emax := e − 1;
         s := 1;
         a := st;
s1:      k := emax;
         for e := 1 step 1 until emax do
         at[e] := a;
s2:      for n := 1 step 1 until nmax do
begin    if l[i[n]] > 0 then
begin    switch s := b1, b2;
b1:      x := abs (at[i[n]]) + te[n];
         if x > abs (at[j[n]]) then go to l1;
         go to l2;
b2:      x := abs (at[i[n]]) − te[n];
         if x < abs (at[j[n]]) then
l1:      at[j[n]] := − x;
l2:
end
end;

         for e := 1 step 1 until emax do
begin    if l[e] < 0 then
begin    if at[e] < 0 then
begin    l[e] := abs (l[e]);
         k := k + 1;
s3:      at[e] := abs (at[e]);
         go to l3
end;

         go to l3
end;

         if at[e] ≧ 0 then
begin    l[e] := − l[e];
         k := k − 1;
         go to l3
end;

         go to s3;
l3:
end;

         if k ≠ 0 then go to s2;
         switch s := g1, g2;
g1:      s := 2;
         for n := 1 step 1 until nmax do
begin    t := i[n];
         i[n] := j[n];
         j[n] := t
end;
```

```
            a := 0;
            for e := 1 step 1 until emax do
begin       es[e] := at[e];
            l[e] := abs (l[e]);
            if at[e] > a then
            a := at[e]
end;
            go to s1;
g2:         for e := 1 step 1 until emax do
            l[e] := abs (l[e]);
end pert
```

CERTIFICATION OF ALGORITHM 119 [H]
EVALUATION OF A PERT NETWORK [Burton Eisenman and Martin Shapiro, *Comm. ACM 5* (Aug. 1962), 436]

L. Stephen Coles (Recd. 10 Nov. 1964 and 7 Dec. 1964)
Carnegie Institute of Technology, Pittsburgh, Pa.

The procedure was tested on a CDC-G20, using the Algol compiler developed by Carnegie Tech. Before compilation was possible, the following modifications were required in order to make it a correct Algol 60 procedure.

1. Insert after the end of *scan*

    switch $sw2 := g1, g2$;

2. Modify **comment** By means of the switch, s, $\cdots$

to read

    comment By means of the switches, $sw1$ and $sw2$, $\cdots$

3. Modify **begin switch** $s := b1, b2$;

to read

    begin switch $sw1 := b1, b2$; **go to** $sw1 [s]$;

4. Modify **switch** $s := g1, g2$;

to read

    go to $sw2 [s]$;

With these changes the procedure was operated successfully on a number of small test problems.

ALGORITHM 120
MATRIX INVERSION II
RICHARD GEORGE*
Particle Accelerator Division Argonne National Laboratory Argonne, Illinois
* Work supported by the U. S. Atomic Energy Commission.

procedure *INVERSION II* (*n, a, epsilon, ALARM, delta*);
comment This is a revision of Algorithm 58. It accomplishes inversion of the matrix *a*, with the result stored in matrix *a*. The order of the matrix is *n*. If in the process of calculating, any pivot element has an absolute value less than *epsilon*, there will be a jump to the non-local label *ALARM*. The variable *delta* will contain the value of the determinant of the original matrix on normal exit, zero or a very small number on exit to *ALARM*.;
value *n*;
array *a*;
real *epsilon, delta*;
integer *n*;
begin
```
    array b, c[1:n];  real w, y;
    integer array z[1:n];  integer i, j, k, l, p;
    delta := 1.0;
    for j := 1 step 1 until n do
      z[j] := j;
    for i := 1 step 1 until n do
      begin
        k := i;  y := a[i, i];  l := i-1;  p := i+1;
        for j := p step 1 until n do
          begin
            w := a[i, j];
          if abs(w) > abs(y) then
            begin
              k := j;
              y := w
            end;
        end;
    delta := delta × y;
    if abs(y) < epsilon then go to ALARM;
    y := 1.0 / y;
    for j := 1 step 1 until n do
      begin
        c[j] := a[j, k];
        a[j, k] := a[j, i];
        a[j, i] := − c[j] × y;
        b[j] := a[i, j] := a[i, j] × y
      end;
    a[i, i] := y;
    j := z[i];
    z[i] := z[k];
    z[k] := j;
    for k := 1 step 1 until l, p step 1 until n do
      for j := 1 step 1 until l, p step 1 until n do
        a[k, j] := a[k, j] − b[j] × c[k]
      end;
    for i := 1 step 1 until n do
      begin
REPEAT: k := z [i];
        if k=i then go to ADVANCE;
        for j := 1 step 1 until n do
          begin
            w := a [i, j];
            a [i, j] := a [k, j];
            a [k, j] := w
          end;
        p := z [i];
        z [i] := z [k];
        z [k] := p;
        delta := − delta;
        go to REPEAT;
ADVANCE:  end;
end
```

CERTIFICATION OF ALGORITHMS 120 AND MATRIX INVERSION BY GAUSS-JORDAN INVERSION II [R. George, *Comm. ACM* Aug. 1962] and gjr [by H. Rutishauser, quoted by H. R. Schwarz, *Comm. ACM* Febr. 1962]
P. NAUR
Regnecentralen, Copenhagen, Denmark

These two procedures were compared using the GIER ALGOL system (30 bits for the normalized mantissa including sign). The following changes (in part dictated by the requirements of the compiler) were included:

INVERSION II: (1) Epsilon was included in the value part. (2) The specification **label** *ALARM* was added.

gjr: (1) The value part: **value** *n, eps* was inserted. (2) The second *a* in the formal parameter part was taken out.

With these changes both procedures ran smoothly through the compiler. In order to obtain a comparison each of them was tested as follows: With a given, rather large value of epsilon the procedure was called to invert a segment of the Hilbert matrix. Upon alarm exit, the value of epsilon was divided by 10 and a fresh call was made. In this way an estimate of the largest permissible epsilon was obtained. When the inverse had been obtained, that element of it which was most in error was found through a comparison with the accurate inverse as calculated by means of INVHILBERT (Algorithm 50, see certification above). A relative error was obtained through division by the largest element of the accurate inverse.

This process was carried out for segments of the Hilbert matrix of orders 2 through 15. For orders above 9, the results of the inversion are dominated by errors. Below 9 we obtained the following output:

Inversion by INVERSION II

Order	eps	Determinant	Subscr.	Maximum error Error	Maximum error Relative
2	$_{10}-2$	$8.3333333_{10}-2$	2,2	$2.98_{10}-8$	$2.48_{10}-9$
3	$_{10}-3$	$4.6296284_{10}-4$	2,2	$5.01_{10}-5$	$2.61_{10}-7$
4	$_{10}-4$	$1.6534314_{10}-7$	3,3	$3.06_{10}-2$	$4.72_{10}-6$
5	$_{10}-5$	$3.7490001_{10}-12$	4,4	$1.38_{10}1$	$7.72_{10}-5$
6	$_{10}-7$	$5.3601875_{10}-18$	5,5	$5.78_{10}3$	$1.31_{10}-3$
7	$_{10}-8$	$4.8485529_{10}-25$	5,5	$3.70_{10}5$	$2.77_{10}-3$
8	$_{10}-10$	$1.5221000_{10}-33$	6,6	$3.33_{10}9$	$7.84_{10}-1$

Similarly we got for gjr, and the ratio of errors of the two procedures:

| Order | eps | Inversion by gjr | | Ratio of errors |
| | | Maximum error | Maximum error | INVERSION II |
		Subscr. Error	Relative	to gjr
2	$10-2$	2,1 $2.98_{10}-8$	$2.48_{10}-9$	1.0
3	$10-3$	2,2 $2.86_{10}-6$	$1.49_{10}-8$	18
4	$10-4$	4,3 $1.07_{10}-4$	$1.65_{10}-8$	290
5	$10-6$	1,1 2.48	$1.39_{10}-5$	5.6
6	$10-7$	5,5 $4.05_{10}3$	$9.18_{10}-4$	1.4
7	$10-8$	5,5 $4.32_{10}6$	$3.24_{10}-2$	.086
8	$10-10$	7,7 $5.55_{10}7$	$1.31_{10}-2$	60

Although the superiority of gjr, which searches for the pivot in both columns and rows, over INVERSION II, which only searches in the next column, is well brought out in the last column of the second table the behavior for $n = 7$ is curious and ought to be confirmed elsewhere.

As a further test both procedures were used to invert the matrices produced by Algorithm 52, TESTMATRIX (see certification above). Again, the error of the inverse was found by a comparison with the known inverse. The comparison of the two procedures was made for orders 2 through 23 and revealed a surprisingly small difference of accuracy. Typical output was as follows:

| Order | Location and size of max. error | | | | Ratio of errors |
| | INVERSION II | | gjr | | INVERSION II |
	Subscr.	Error	Subscr.	Error	to gjr
5	5,5	$8.94_{10}-8$	5,5	$8.94_{10}-8$	1.00
10	10,10	$3.76_{10}-6$	10,10	$3.52_{10}-6$	1.07
15	15,15	$2.12_{10}-5$	15,15	$1.78_{10}-5$	1.19
20	20,20	$6.81_{10}-5$	20,20	$6.71_{10}-5$	1.02

The relative errors of the determinants calculated by INVERSION II increased slowly with n, reaching $2.3_{10}-7$ for $n = 24$.

Typical execution times were found as follows:

Order	INVERSION II	gjr
5	2 seconds	3 seconds
10	5 "	8 "
15	16 "	17 "
20	53 "	57 "

However, it should be noted that owing to the automatic segmentation of the program into drum tracks in GIER ALGOL the execution time may vary somewhat from one program in which a procedure is used to another. The above times do not, in fact, refer to the same program.

CORRECTION TO EARLIER REMARKS ON ALGORITHM 42 INVERT, ALG. 107 GAUSS'S METHOD, ALG. 120 INVERSION II, AND gjr [P. Naur, *Comm. ACM*, Jan. 1963, 38–40.]

P. NAUR

Regnecentralen, Copenhagen, Denmark

George Forsythe, Stanford University, in a private communication has informed me of two major weaknesses in my remarks on the above algorithms:

1) The computed inverses of rounded Hilbert matrices are compared with the exact inverses of unrounded Hilbert matrices, instead of with very accurate inverses of the rounded Hilbert matrices.

2) In criticizing matrix inversion procedures for not searching for pivot, the errors in inverting positive definite matrices cannot be used since pivot searching seems to make little difference with such matrices.

It is therefore clear that although the figures quoted in the earlier certification are correct as they stand, they do not substantiate the claims I have made for them.

To obtain a more valid criterion, without going into the considerable trouble of obtaining the very accurate inverses of the rounded Hilbert matrices, I have multiplied the calculated inverses by the original rounded matrices and compared the results with the unit matrix. The largest deviation was found as follows:

| Order | Maximum deviation from elements of the unit matrix | | |
	INVERSION II	gjr	Ratio
2	$-1.49_{10}-8$	$-1.49_{10}-8$	1.0
3	$-4.77_{10}-7$	$-8.34_{10}-7$	0.57
4	$-9.54_{10}-6$	$-3.43_{10}-5$	0.28
5	$-7.32_{10}-4$	$-4.58_{10}-4$	1.6
6	$-1.61_{10}-2$	$-1.42_{10}-2$	1.1
7	$-5.78_{10}-1$	$-5.47_{10}-1$	1.1
8	$-1.20_{10}-2$	$-1.38_{10}1$	8.7
9	$-4.91_{10}1$	$-2.22_{10}1$	2.2

This criterion supports Forsythe's criticism. In fact, on the basis of this criterion no preference of INVERSION II or gjr can be made.

The calculations were made in the GIER ALGOL system, which has floating numbers of 29 significant bits.

ALGORITHM 121
NORMDEV
DAVID SHAFER
University of Chicago, Chicago, Ill.

```
procedure  NormDev(Random,A,x);
  procedure  Random;  real A,x;
  comment  'NormDev' uses (1) a procedure 'Random(y)' as-
    sumed to produce a random number, 0 < y < 1, and (2) the
    constant  A = sqrt(2/pi) × integral [0:1] exp(−x↑2/2)dx, to
    produce a positive normal deviate 'x';
  begin real y;
  Random(x);  if x > A then go to large;
  x := x/A;
  1:  Random(y);  if y < exp(−x↑2/2) then go to EndND;
      Random(x);  go to 1;
  large:  x := (x − A)/(1 − A);
  2:  x := sqrt(1 − 2 × log(x));
      Random(y);  if y < 1/x then go to EndND;
      Random(x);  go to 2;
  EndND:  end
```


CERTIFICATION OF ALGORITHM 121 [G5]
NORMDEV
[David Shafer, *Comm. ACM 5* (Sept. 1962), 482]
M. C. PIKE (Recd. 3 May 1965)
Statistical Research Unit of the Medical Research Coun-
  cil, U. College Hospital Medical School, London.

Algorithm 121 has the following error: The line
$$2: x := sqrt (1 − 2 × log (x));$$
should read
$$2: x := sqrt (1 − 2 × ln (x));$$
With this correction *NormDev* has been run successfully on the
ICT Atlas computer with the Atlas ALGOL compiler.

ALGORITHM 122
TRIDIAGONAL MATRIX
GERARD F. DIETZEL
Burroughs Corp., Pasadena, Calif.

```
procedure TRIDIAG (n,A,U);
integer n;  array A,U;
comment  This procedure reduces a real symmetric matrix A of
    order n to tridiagonal form (UT)AU (UT = transpose of U) by
    a sequence of at most (n-1)(n-2)/2 binary orthogonal trans-
    formations. Also, the matrix U is calculated. [Cf. W. Givens,
    "Numerical computation of the characteristic values of a real
    symmetric matrix," Report ORNL1574 (1954), Oak Ridge Nat.
    Lab., Tenn., and D. E. Johansen, "A modified Givens method
    for the eigenvalue evaluation of large matrices," J. ACM 8, 3
    (1961)];
begin  real fact,c1,c2,loc1,loc2,temp;  integer i,j,j1,j2,j3,j4,n1;
    comment  Set array U = identity matrix of order n;
    for i := 1 step 1 until n do
    begin
        for j := i+1 step 1 until n do U[i,j] := U[j,i]  = 0;
        U[i,i] := 1.0
    end;
    comment  The reduction of the matrix A begins here. Only the
        upper triangular elements of A are used in the computation;
    n1 := n - 2;
    for i := 1 step 1 until n1 do
    begin
        j1 := i + 1;  j2 := i + 2;
        for j := j2 step 1 until n do
        begin
            if A[i,j] = 0 then go to lab;
            fact := 1 / sqrt(A[i,j1]↑2 + A[i,j]↑2);
            c1 := fact × A[i,j1];  c2 := fact × A[i,j];
            loc1 := A[j1,j1];  loc2 := A[j1,j];
            A[j1,j1] := c1↑2 × loc1 + 2.0 × c1 × c2 × loc2 + c2↑2 ×
                A[j,j];
            A[j1,j] := −c1 × c2 × loc1 + (c1↑2 − c2↑2) × loc2 + c1 ×
                c2 × A[j,j];
            A[j,j] := c2↑2 × loc1 − 2.0 × c1 × c2 × loc2 + c1↑2 ×
                A[j,j];
            j3 := j + 1;
            for k := j3 step 1 until n do
            begin
                temp := A[j1,k];
                A[j1,k] := c1 × temp + c2 × A[j,k];
                A[j,k] := −c2 × temp + c1 × A[j,k]
            end;
            j4 := j − 1;
            for k := j2 step 1 until j4 do
            begin
                temp := A[j1,k];
                A[j1,k] := c1 × temp + c2 × A[k,j];
                A[k,j] := −c2 × temp + c1 × A[k,j]
            end;
            A[i,j1] := c1 × A[i,j1] + c2 × A[i,j];
            A[i,j] := 0;
            for k := 1 step 1 until n do
            begin
                temp := U[k,j1];
                U[k,j1] := c1 × temp + c2 × U[k,j];
                U[k,j] := −c2 × temp + c1 × U[k,j]
            end;
lab:    end
    end;
    for i := 1 step 1 until n do
        for j := i+1 step 1 until n do
            A[j,i] := A[i,j]
end  TRIDIAG
```

CERTIFICATION OF ALGORITHM 122
TRIDIAGONAL MATRIX [Gerard F. Dietzel, *Comm.*
ACM 5 (Sept. 1962), 482]
PETER NAUR (Recd 27 Sept. 63)
Regnecentralen, Copenhagen, Denmark

TRIDIAG needed the following corrections:
1. Insert k among the local integers to read:
    integer i, j, $j1$, $j2$, $j3$, $j4$, $n1$, k;
2. At the end of line 5 of the procedure body, insert the colon to
    read $U[j, i] := 0$;
3. Change the round parenthesis to a square bracket following
    for $k := j3 \cdots$ to read $temp := A[j1, k]$;

With these corrections the algorithm worked satisfactorily with
the GIER ALGOL system. As a test it was tried with the following
matrix:

$$HBH\ TESTMATRIX[j, i] = HBH\ TESTMATRIX[i, j]$$

$$= n + 1 - j \qquad\qquad (j \geq i)$$

(cf. the Certification of Alg. 85, *Comm. ACM 6* (Aug. 1963), 447).
As a check the resulting matrix was rotated back again, using the
resulting U-matrix, and the largest deviation of any element from
the original was found.

For comparison the figures obtained by using the algorithms
given by Wilkinson in *Numerische Mauhematik 4* (1962), 354–376,
may be used. Wilkinson's algorithms use Householder's method of
obtaining the tridiagonal form. It should be noted that the deviations given in the table below for Householder's method refer to
the final result of obtaining the eigenvalues and vectors, and not
only the tridiagonal form, and thus include error contributions
from a rather longer chain of calculations than the ones given for
TRIDIAG. The times, however, only refer to the tridiagonalisation process in both cases.

	$n=5$	$n=10$	$n=15$
Largest deviation			
TRIDIAG,	$1.4_{10} - 7$	$7.0_{10} - 7$	$2.4_{10} - 6$
householder tridiagonalisation		$1.4_{10} - 7$	$1.3_{10} - 6$
Time of execution, in GIER ALGOL, seconds			
TRIDIAG	2	7	34
householder tridiagonalisation	1	4	10

These figures clearly demonstrate the superiority of the Householder process. Since, in addition, the Householder method in the
form given by Wilkinson uses much less storage for variables,
Algorithm 122 cannot be recommended.

ALGORITHM 123
REAL ERROR FUNCTION, ERF(x)

MARTIN CRAWFORD AND ROBERT TECHO
Georgia Institute of Technology, Atlanta, Ga.

real procedure *Erf*(x); **real** x;
comment $\Phi(x) = \mathrm{Erf}(x) = (2/\sqrt{\pi})\int_0^x e^{-u^2}\,du$ can be computed by using the recursive relation for derivatives with $\Phi^1(x) = (2/\sqrt{\pi})e^{-x^2}$, where $\Phi^{(n)}(x) = -2x\Phi^{(n-1)}(x) - 2(n-2)\Phi^{(n-2)}(x)$, for $n = 2, 3, \cdots$. The Taylor's series expansions of $\Phi(a_k)$ are taken about $k+1$ points on the interval $0 < a_k \leqq x$ and summed to get $\Phi(x)$;

```
begin  real A, U, V, W, Y, Z, T;  integer N;
       Z := 0;  1:  if x ≠ 0 then
begin  if 0.5 < abs (x) then A := − sign (x) × 0.5
       else A := − x;
       U := V := 1.12837917 × exp(−x↑2);  Y := T := −V ×
         A;  N := 1;
2:     if abs(T) ≧ ₁₀− 10 then
begin  N := N + 1;  W := −2 × x × V − 2 × U × (N−2);
         T := T × W × A/(V × N);
       U := V;  V := W;  Y := Y + T;  go to 2 end;
       Z := Z + Y;  x := x + A;  go to 1 end;
       Erf := Z end Erf
```

CERTIFICATION OF ALGORITHM 123
REAL ERROR FUNCTION, ERF (x) [Martin Crawford and Robert Techo, *Comm. ACM*, Sept. 1962]

HENRY C. THACHER, JR.*
Argonne National Laboratory, Argonne, Ill.

* Work supported by the U. S. Atomic Energy Commission.

The body of *Erf*(x) was tested using the Dartmouth SCALP compiler for the LGP-30. For $x = 0(0.01)0.3$, the results agreed with tabulated values to 8 in the 7th decimal place, and for $x = 0.4(0.2)1.6$ the error was less than 1 in the 6th decimal. These results are compatible with the roundoff error in the arithmetic used. The computing time increased rapidly (by a factor of more than 10) as x increased from 0.01 to 1.6.

The following comments should be considered by users of the algorithm:

1. The parameter x should be called by value, both to allow the use of expressions, and also to avoid destruction of the actual parameter.
2. The constant ₁₀−10 in statement 2 determines the accuracy of the computation. Its value should be adjusted to the arithmetic being used, and the accuracy required. A machine-independent test could be made by substituting **if** $Y - T = Y$ **then** $\cdots$.
3. For large x, the error function is more efficiently calculated from the Laplace continued fraction for *erfc*(x). Algorithm 180 is based on this method.

REMARK ON ALGORITHM 123
ERF(x) [Martin Crawford and Robert Techo, *Comm. ACM*, Sept. 1962]

D. IBBETSON
Elliott Brothers (London) Ltd.
Elstree Way, Borehamwood, Herts., England

(1) The specification **value** x; was added to allow x to be an expression and to prevent side effects.

(2) The algorithm was then modified to give the Gaussian integral $(1/\sqrt{2\pi})\int_{-\infty}^{x} \exp(-\frac{1}{2}u^2)\,du$ by

(a) changing its name to *Gauss* (x),

(b) inserting $x := x*0.70710678$; immediately before $Z := 0$; , and

(c) changing the final statement to
Gauss := $(Z+1)/2$ **end** *Gauss*

(3) The algorithm with the above changes was tested on a National Elliott 803 computer using the Elliott-ALGOL translator with ₁₀−8 substituted for ₁₀−10. It was found to produce wrong answers when $x = \pm 1$ (corresponding to $Erf(\pm 1/\sqrt{2})$) giving 0.5 ± 0.3467899 instead of 0.5 ± 0.3413447.

REMARK ON ALGORITHM 123
ERF(x) [Martin Crawford and Robert Techo, *Comm. ACM 5* (Sept. 1962), 483; *6* (June 1963), 316; *6* (Oct. 1963), 618]

STEPHEN P. BARTON AND JOHN F. WAGNER (Recd 2 Dec. 63)
General Telephone and Electronics Laboratories, Bayside, New York

This algorithm may err when the Taylor series expands about a root of the nth-order Hermite polynomial; one such error has already been noted [Remark on Algorithm 123, D. Ibbetson, *Comm. ACM 6* (Oct. 1963), 618]. The difficulty springs from the Taylor-series truncation criterion, which assumes that the magnitude of successive terms in the Taylor series decreases. This is not always so, as may be seen by relating

$$\Phi^{(n)}(x) \equiv \frac{2}{\sqrt{\pi}} \frac{d^{n-1}}{dx^{n-1}} (e^{-x^2}), \qquad (n \geqq 1)$$

to the Hermite polynomial $H_n(x)$, which can be defined as

$$H_n(x) \equiv (-1)^n e^{x^2} \frac{d^n}{dx^n} (e^{-x^2}).$$

Therefore

$$\Phi^{(n)}(x) = \frac{2}{\sqrt{\pi}} (-1)^{n-1} e^{-x^2} H_{n-1}(x).$$

As a result, $\Phi^{(n)}(x)$ vanishes when x is a root of $H_{n-1}(x)$ and the Taylor series may be terminated prematurely.

The algorithm was translated into FORTRAN II and run on a Scientific Data Systems 910 computer (39-bit mantissa) with the following changes:

(1) The argument was decremented by 0.25 rather than 0.5.

(2) The truncation criterion for *abs* (T) was 10^{-12} rather than 10^{-10}.

Errors, detected for $x = 1/\sqrt{2}$ and $x = 2.652$, were traced to the above described premature truncation of the relevant Taylor series. These arguments correspond to the roots of $H_2(x)$ and $H_7(x)$.

The program was therefore modified to sum a fixed number of terms, with special attention to the difficulties that might arise when expanding about roots of $H_n(x)$. In particular, in Algorithm 123, line 9, the coefficient, $A^n/n!$, of the nth term in the Taylor expansion, is obtained via the intermediate step of dividing the $(n-1)$-term, T, by the $(n-1)$-derivative, V. The possibility of dividing by $V = 0$ when the Taylor expansion takes place about roots of $H_{n-2}(x)$ was avoided by modifying the program to compute coefficients directly from the recursion relation,

$$A^n/n! = [A^{n-1}/(n-1)!][A/n].$$

In selecting the number of terms to be included in each Taylor series, consideration should also be given to the size of the standard decrement (specified as 0.5 in line 3 of Algorithm 123), for it is the combination of these two parameters which largely determines the accuracy and running time. A brief survey suggested that at least 10-digit accuracy could be obtained if a decrement of 0.4 were employed with 16 terms in each Taylor series; this resulted in an average running time of about 3.5 seconds per computation for arguments in the range $0 \leq x \leq 5.0$.

REFERENCE: H. MARGENAU and G. M. MURPHY, *The Mathematics of Physics and Chemistry*, pp. 119, 122. D. van Nostrand, 1943.

REMARKS ON:
ALGORITHM 123 [S15]
REAL ERROR FUNCTION, ERF(x)
[Martin Crawford and Robert Techo *Comm. ACM 5* (Sept. 1962), 483]

ALGORITHM 180 [S15]
ERROR FUNCTION—LARGE X
[Henry C. Thacher Jr. *Comm. ACM 6* (June 1963), 314]

ALGORITHM 181 [S15]
COMPLEMENTARY ERROR FUNCTION—
LARGE X
[Henry C. Thacher Jr. *Comm. ACM 6* (June 1963), 315]

ALGORITHM 209 [S15]
GAUSS
[D. Ibbetson. *Comm. ACM 6* (Oct. 1963), 616]

ALGORITHM 226 [S15]
NORMAL DISTRIBUTION FUNCTION
[S. J. Cyvin. *Comm. ACM 7* (May 1964), 295]

ALGORITHM 272 [S15]
PROCEDURE FOR THE NORMAL DISTRIBUTION
FUNCTIONS
[M. D. MacLaren. *Comm. ACM 8* (Dec. 1965), 789]

ALGORITHM 304 [S15]
NORMAL CURVE INTEGRAL
[I. D. Hill and S. A. Joyce. *Comm. ACM 10* (June 1967), 374]

I. D. HILL AND S. A. JOYCE (Recd. 21 Nov. 1966)
Medical Research Council,
Statistical Research Unit, 115 Gower Street, London W.C.1., England

These algorithms were tested on the ICT Atlas computer using the Atlas ALGOL compiler. The following amendments were made and results found:

ALGORITHM 123
 (i) **value** x; was inserted.
 (ii) $abs(T) \leqslant {}_{10}-10$ was changed to $Y - T = Y$
  both these amendments being as suggested in [1].
 (iii) The labels 1 and 2 were changed to $L1$ and $L2$, the **go to** statements being similarly amended.
 (iv) The constant was lengthened to 1.12837916710.
 (v) The extra statement $x := 0.707106781187 \times x$ was made the first statement of the algorithm, so as to derive the normal integral instead of the error function.

The results were accurate to 10 decimal places at all points tested except $x = 1.0$ where only 2 decimal accuracy was found, as noted in [2]. There seems to be no simple way of overcoming the difficulty [3], and any search for a method of doing so would hardly be worthwhile, as the algorithm is slower than Algorithm 304 without being any more accurate.

ALGORITHM 180
 (i) $T := -0.56418958/x/exp(v)$ was changed to
  $T := -0.564189583548 \times exp(-v)/x$. This is faster and also has the advantage, when v is very large, of merely giving 0 as the answer instead of causing overflow.
 (ii) The extra statement $x := 0.707106781187 \times x$ was made as in (v) of Algorithm 123.
 (iii) **for** $m := m + 1$ was changed to **for** $m := m + 2$. $m+1$ is a misprint, and gives incorrect answers.
 The greatest error observed was 2 in the 11th decimal place.

ALGORITHM 181
 (i) Similar to (i) of Algorithm 180 (except for the minus sign).
 (ii) Similar to (ii) of Algorithm 180.
 (iii) m was declared as **real** instead of **integer**, as an alternative to the amendment suggested in [4].
 The results were accurate to 9 significant figures for $x < 8$, but to only 8 significant figures for $x = 10$ and $x = 20$.

ALGORITHM 209
No modification was made. The results were accurate to 7 decimal places.

ALGORITHM 226
 (i) $10 \uparrow m/(480 \times sqrt(2 \times 3.14159265))$ was changed to $10 \uparrow m \times 0.000831129750836$.
 (ii) **for** $i := 1$ **step** 1 **until** $2 \times n$ **do** was changed to $m := 2 \times n$; **for** $i := 1$ **step** 1 **until** m **do**.
 (iii) $-(i \times b/n) \uparrow 2/8$ was changed to $-(i \times b/n) \uparrow 2 \times 0.125$.
 (iv) **if** $i = 2 \times n - 1$ was changed to **if** $i = m - 1$
 (v) $b/(6 \times n \times sqrt(2 \times 3.14159265))$ was changed to $b/(15.0397696478 \times n)$.

Tests were made with $m = 7$ and $m = 11$ with the following results:

x	Number of significant figures correct		Number of decimal places correct	
	$m = 7$	$m = 11$	$m = 7$	$m = 11$
−0.5	7	11	7	11
−1.0	7	10	7	10
−1.5	7	10	8	10
−2.0	7	9	8	10
−2.5	6	9	8	11
−3.0	6	7	8	9
−4.0	5	7	10	11
−6.0	2	1	12	10
−8.0	0	0	11	9

Perhaps the comment with this algorithm should have referred to decimal places and not significant figures. To ask for 11 significant figures is stretching the machine's ability to the limit, and where 10 significant figures are correct, this may be regarded as acceptable.

ALGORITHM 272
The constant .99999999 was lengthened to .9999999999.

The accuracy was 8 decimal places at most of the points tested, but was only 5 decimal places at $x = 0.8$.

ALGORITHM 304
No modification was made. The errors in the 11th significant figure were:

$abs(x)$	$x > 0 \equiv upper$	$x > 0 \not\equiv upper$
0.5	1	1
1.0	1	2
1.5	21ª(5)	2
2.0	25ª(0)	4
3.0	0	0
4.0	2	3
6.0	6	0
8.0	14	0
10.0	23	0
20.0	35	0

ª Due to the subtraction error mentioned in the comment section of the algorithm. Changing the constant 2.32 to 1.28 resulted in the figures shown in brackets.

To test the claim that the algorithm works virtually to the accuracy of the machine, it was translated into double-length instructions of Mercury Autocode and run on the Atlas using the EXCHLF compiler (the constant being lengthened to 0.398942280401432677939946). The results were compared with hand calculations using Table II of [5]. The errors in the 22nd significant figure were:

$abs(x)$	$x > 0 \equiv upper$	$x > 0 \not\equiv upper$
1.0	2	3
2.0	7	1
4.0	2	0
8.0	8	0

Timings. Timings of these algorithms were made in terms of the Atlas "Instruction Count," while evaluating the function 100 times. The figures are not directly applicable to any other computer, but the relative times are likely to be much the same on other machines.

INSTRUCTION COUNT FOR 100 EVALUATIONS

$abs(x)$	Algorithm number							
	123	180	181	209	226 $m = 7$	272	304ª	304ᵇ
0.5	58			8	97	24	25	24
1.0	65ᶜ			8	176	24	29	29
1.5	164	128	127	9	273	25	35	35
2.0	194	78	90	8	387	24	39	39
2.5	252	54	68	10	515	24	131	44
3.0		42	51	9	628	25	97	50
4.0		27	39	9	900ᵈ	25	67	44
6.0		15	30	6	1400ᵈ	16	49	23
8.0		9	28	7	2100ᵈ	18	44	11
10.0		10	25	5	2700ᵈ	16	38	11
20.0		9	22	5	6500ᵈ	16	32	11
30.0		9	9	5	10900ᵈ	16	11	11

ª Readings refer to $x > 0 \equiv upper$.
ᵇ Readings refer to $x > 0 \not\equiv upper$.
ᶜ Time to produce incorrect answer. A count of 120 would fit a smooth curve with surrounding values.
ᵈ 100 times Instruction Count for 1 evaluation.

Opinion. There are advantages in having two algorithms available for normal curve tail areas. One should be very fast and reasonably accurate, the other very accurate and reasonably fast. We conclude that Algorithm 209 is the best for the first requirement, and Algorithm 304 for the second.

Algorithms 180 and 181 are faster than Algorithm 304 and may be preferred for this reason, but the method used shows itself in Algorithm 181 to be not quite as accurate, and the introduction of this method solely for the circumstances in which Algorithm 180 is applicable hardly seems worth while.

Acknowledgment. Thanks are due to Miss I. Allen for her help with the double-length hand calculations.

REFERENCES:

1. THACHER, HENRY C. JR. Certification of Algorithm 123. *Comm. ACM 6* (June 1963), 316.

2. IBBETSON, D. Remark on Algorithm 123. *Comm. ACM 6* (Oct. 1963), 618.

3. BARTON, STEPHEN P., AND WAGNER, JOHN F. Remark on Algorithm 123. *Comm. ACM 7* (Mar. 1964), 145.

4. CLAUSEN, I., AND HANSSON, L. Certification of Algorithm 181. *Comm. ACM 7* (Dec. 1964), 702.

5. SHEPPARD, W. F. *The Probability Integral*. British Association Mathematical Tables VII, Cambridge U. Press, Cambridge, England, 1939.

ALGORITHM 124
HANKEL FUNCTION
LUIS J. SCHAEFER
Purdue University, West Lafayette, Ind.

procedure $HANKEL(N,X,H)$; **value** N,X; **integer** N;
real X; **array** H;
comment This procedure evaluates the complex valued hankel function of the first kind for real argument X and integral order N and assigns it to H. The individual Bessel- and Neuman-function series are not evaluated separately. Both the real and imaginary parts are generated from the same terms;
begin **real** K, P, R, A, S, T, D, L; **integer** Q;
$A := R := 1$; $H[1] := H[2] := S := 0$;
for $Q := 1$ **step** 1 **until** N **do begin** $R := R \times Q$; $S := S + 1/Q$ **end**; $D := R/N$;
$R := 1/R$; $K := X \times X/4$; $P := (X/2)\uparrow N$; $T := ln(K) + 1.1544313298631$;
for $Q := 0, Q+1$ **while** $Q \leq N \vee L \neq H[2]$ **do**
begin $L := H[2]$; $H[1] := H[1] + A \times K \times R$;
$\quad H[2] := H[2] + A \times (R \times K \times (T-S) - ($**if** $Q < N$ **then** D/P **else** $0))$;
$\quad A := A \times K/Q$; $R := -R/(Q+N)$; $S := S+1/Q + 1/(Q+N)$;
$\quad$**if** $Q < N$ **then** $D := D/(N-Q)$
end; $H[2] := H[2] \times .31830989$
end

A := R := 1; $H[1] := H[2] := S := 0$;
if $N = 0$ **then begin** $R := 1$; $S := D := 0$ **end**
else
begin for $Q := 1$ **step** 1 **until** N **do**
$\quad$**begin** $R := R \times Q$; $S := S + 1/Q$ **end**; $D := R/N$
end;
$R := 1/R$; $K := X \times X/4$; $P := K \uparrow N$; $T := ln(K) + 1.1544313298631$;
comment The last constant is $2 \times gamma$, Euler's constant;
for $Q := 0, Q+1$ **while** $Q \leq N \vee L \neq H[2]$ **do**
begin $L := H[2]$; $H[1] := H[1] + A \times R$;
$\quad H[2] := H[2] + A \times (R \times (T-S) - ($**if** $q < N$ **then** D/P **else** $0))$;
$\quad A := A \times K/(Q+1)$; $R := -R/(Q+N+1)$;
$\quad S := S + 1/(Q+1) + 1/(Q+N+1)$;
$\quad$**if** $Q+1 < N$ **then** $D := D/(N-Q-1)$;
end;
$P := (X/2) \uparrow N$; $H[1] := H[1] \times P$; $H[2] := 0.318309886184 \times H[2] \times P$;
comment The multiplicative constant is $1/Pi$;
exit:
end $HANKEL$

CERTIFICATION OF ALGORITHM 124 [S17]
HANKEL FUNCTION [Luis J. Schaeffer, *Comm. ACM 5* (Sept. 1962), 483]
GEORGE A. REILLY (Recd. 5 Oct. 1964 and 4 Nov. 1964)
Westinghouse Research Laboratories, Pittsburgh, Pa.

This procedure, after modification, was run on the B-5000 using B-5000 ALGOL. Values obtained checked with US National Bureau of Standards *Handbook of Mathematical Functions*, Applied Mathematics Series 55, US Government Printing Office, Washington, D.C. 1964.

For $N = 0$, 1 and 2, accuracy was to 10 decimals for $X < 8.0$. It deteriorated to 6 decimals for $8 < X < 17.5$. For $3 \leq N \leq 9$ accuracy was to the 5 decimals of the tables.

Some changes proved necessary to make the algorithm run. Since the algorithm is short and the changes are involved, the algorithm is restated here. Note that a test for a zero argument X is included in the body of the procedure since $H[2]$ ought to be minus infinity when $X = 0$.

procedure $HANKEL (N, X, H)$; **value** N, X; **integer** N;
real X; **array** H;
begin real K, P, R, A, S, T, D, L; **integer** Q;
if $X = 0$ **then**
begin comment In this case $H[2]$ is minus infinity. M denotes the largest number which can be represented in the machine. The numerical value of M is to be written into the procedure:
$\quad H[2] := -M$;
$\quad H[1] :=$ **if** $N = 0$ **then** 1 **else** 0;
go to *exit*
end;

ALGORITHM 125
WEIGHTCOEFF
H. RUTISHAUSER
Eidg. Technische Hochschule, Zurich, Switzerland

procedure *weightcoeff* (n,q,e,eps,w,x); **value** n; **real** eps;
  integer n; **array** q,e,w,x;
comment Computes abscissae x_i and weight coefficients w_i for a Gaussian quadrature method $\int_0^b w(x)f(x)\,dx \approx \sum_{i=1}^n w_i f(x_i)$, where $\int_0^b w(x)\,dx = 1$ and $w(x) \geq 0$. The method requires the order n, a tolerance eps and the $2n-1$ first coefficients of the continued fraction

$$\int_0^b \frac{w(x)}{z-x}\,dx = \frac{1}{|z} - \frac{q_1}{|1} - \frac{e_1}{|z} - \frac{q_2}{|1} - \frac{e_2}{|z} - \cdots$$

to be given, the latter as two arrays $q[1:n]$ and $e[1:n-1]$ all components of which are automatically positive by virtue of the condition $w(x) \geq 0$. The method works as well if the upper bound b is actually infinity (note that b does not appear directly as parameter!) or if the density $w(x)\,dx$ is replaced by $d\alpha(x)$ with a monotonically increasing $\alpha(x)$ with at least n points of variation. The tolerance eps should be given in accordance to the machine accuracy, e.g. as $10-10$ for a computer with a ten-digit mantissa. The result is delivered as two arrays $w[1:n]$ (the weight coefficients) and $x[1:n]$ (the abscissae). For a description of the method see H. Rutishauser, "On a modification of the QD-algorithm with Graeffe-type convergence" [Proceedings of the IFIPS Congress, Munich, 1962].;

begin
  integer k;
  Boolean *test*;
  real m, p;
  array $g[1:n]$;
  procedure *red* (a,f,n); **value** n; **integer** n; **array** a,f;
    comment subprocedure *red* reduces a heptadiagonal matrix a to tridiagonal form as described in the paper loc. cit. Since the bulk of the computing time of the whole method is spent in this subprocedure, it would pay to write it in machine code.;
  begin
    real c; **integer** j,k;
    for $k := 1$ **step** 1 **until** $n-1$ **do**
    begin
      for $j := k$ **step** 1 **until** $n-1$ **do**
      begin
        $c := -f[j] \times a[j,7]/a[j,2]$;
        $a[j,7] := 0$;
        $a[j+1,2] := a[j+1,2] + c \times a[j,5]$;
        $a[j,1] := a[j,1] - c \times f[j] \times a[j,4]$;
        $a[j,6] := a[j,6] - c \times a[j+1,1]$;
        $a[j+1,3] := a[j+1,3] - c \times a[j+1,6]$;
      end j;
      for $j := k$ **step** 1 **until** $n-1$ **do**
      begin
        $c := -f[j] \times a[j,4]/a[j,1]$;
        $a[j,4] := 0$;
        $a[j+1,1] := a[j+1,1] + c \times a[j,6]$;
        $a[j+1,6] := a[j+1,6] + c \times a[j+1,3]$;
        $a[j,5] := a[j,5] - c \times a[j+1,2]$;

$a[j+1,0] := a[j+1,0] - c \times a[j+1,5]$;
      end j;
      for $j := k+1$ **step** 1 **until** $n-1$ **do**
      begin
        $c := -a[j,3]/a[j-1,6]$;
        $a[j,3] := 0$;
        $a[j,6] := a[j,6] + c \times a[j,1]$;
        $a[j-1,5] := a[j-1,5] - c \times f[j] \times f[j] \times a[j,0]$;
        $a[j,2] := a[j,2] - c \times f[j] \times f[j] \times a[j,5]$;
        $a[j,7] := a[j,7] - c \times f[j] \times a[j+1,2]$;
      end j;
      for $j := k+1$ **step** 1 **until** $n-1$ **do**
      begin
        $c := -a[j,0]/a[j-1,5]$;
        $a[j,0] := 0$;
        $a[j+1,2] := a[j+1,2] + c \times f[j] \times a[j,7]$;
        $a[j,5] := a[j,5] + c \times a[j,2]$;
        $a[j,1] := a[j,1] - c \times f[j] \times f[j] \times a[j,6]$;
        $a[j,4] := a[j,4] - c \times f[j] \times a[j+1,1]$;
      end j;
    end k;
  end *red*;
procedure *qdgraeffe* (n,h,g,f); **value** n;
  integer n; **array** h,g,f;
  comment Subprocedure *qdgraeffe* computes for a given finite continued fraction

$$f(z) = \frac{1}{|z} - \frac{q_1}{|1} - \frac{e_1}{|z} - \frac{q_2}{|1} - \cdots - \frac{q_n}{|1}$$

another one, the poles of which are the squares of the poles of $f(z)$. However *qdgraeffe* uses not the coefficients $q_1, \cdots, q_n$ and $e_1, \cdots, e_{n-1}$ of $f(z)$, but the quotients

$$\begin{cases} f_k = q_{k+1}/q_k \\ g_k = e_k/q_{k+1} \end{cases} \qquad (k := 1,2,\cdots,n-1)$$

and the $h_k = \ell n(abs(q_k))$ $(k := 1,2,\cdots,n)$, and the results are delivered in the same form. Procedure *qdgraeffe* can be used independently, but requires subprocedure *red* above;
begin
  integer k; **array** $a[0:n,0:7]$;
  $g[n] := f[n] := 0$;
  for $k := 1$ **step** 1 **until** n **do**
  begin
    $a[k-1,4] := a[k-1,5] := 1$;
    $a[k,1] := a[k,2] := 1 + g[k] \times f[k]$;
    $a[k,6] := a[k,7] := g[k]$;
    $a[k,0] := a[k,3] := 0$;
    comment The array a represents the heptadiagonal matrix Q of the paper loc. cit., but with the modifications needed to avoid the large numbers and with a peculiar arrangement.;
  end k;
  $a[n,5] := 0$;
  $red(a,f,n)$;
  for $k := 1$ **step** 1 **until** n **do**
    $h[k] := 2 \times h[k] + \ell n(abs(a[k,1] \times a[k,2]))$;
    comment A saving might be achieved by economizing the log-computation in the range $.8 \leq x \leq 1.2$;
  for $k := 1$ **step** 1 **until** $n-1$ **do**
  begin

```
          f[k] := f[k] × f[k] × a[k+1,2] × a[k+1,1]/(a[k,1] × a[k,2]);
          g[k] := a[k,5] × a[k,6]/(a[k+1,1] × a[k+1,2])
       end k;
    end qdgraeffe;
L1:  x[1] := q[1] + e[1];
       for k := 2 step 1 until n do
       begin
          g[k−1] := e[k−1] × q[k]/x[k−1];
          x[k] := q[k] + (if k=n then 0 else e[k]) − g[k−1];
          g[k−1] := g[k−1]/x[k];
          w[k−1] := x[k]/x[k−1];
          x[k−1] := ℓn(x[k−1]);
       end k;
       x[n] := ℓn(x[n]);
L2:  p := 1;
L25:  begin
          test := true;
          for k := 1 step 1 until n−1 do
            test := test ∧ abs(g[k] × w[k]) < eps;
          if test then go to L3;
          qdgraeffe (n,x,g,w);
       end;
       p := 2 × p;
       go to L25;
       comment  What follows is a peculiar method to compute
          the w_k from given ratios g_k = w_{k+1}/w_k such that ∑_{k=1}^{n} w_k = 1,
          but the straightforward formulae to do this might well
          produce overflow of exponent.;
L3:  w[1] := m := 0;
       for k := 1 step 1 until n−1 do
       begin
          w[k+1] := w[k] + ln(g[k]);
          if w[k] > m then m := w[k];
       end k;
       for k := 1 step 1 until n do w[k] := exp(w[k]−m);
       m := 0;
       for k := 1 step 1 until n do m := m + w[k];
       for k := 1 step 1 until n do begin w[k] := w[k]/m;
          x[k] := exp(x[k]/p) end;
end  weightcoeff
```

ALGORITHM 126
GAUSS' METHOD
JAY W. COUNTS
University of Missouri, Columbia, Mo.

```
procedure gauss (u,a,y);
real array a,y;  integer u;
comment   This procedure is for solving a system of linear equa-
    tions by successive elimination of the unknowns. The augmented
    matrix is a and u is the number of unknowns. The solution vector
    is y. If the system hasn't any solution or many solutions, this is
    indicated by the go to error where error is a label outside the
    procedure.;
    begin
        integer i,j,k,m,n;
        n := 0;
ck0:  n := n + 1;
        for k := n step 1 until u do if a[k,n] ≠ 0 then go to ck1;
        go to error;
ck1:  if k = n then go to ck2;
        for m := n step 1 until u+1 do
    begin
        temp := a[n,m];  a[n,m] := a[k,m];  a[k,m] := temp
    end;
ck2:  for j := u + 1 step −1 until n do a[n,j] := a[n,j]/a[n,n];
        for i := k + 1 step 1 until u do
        for j := n + 1 step 1 until u + 1 do
        a[i,j] := a[i,j] − a[i,n] × a[n,j];
        if n≠u then go to ck0;
        for i := u step −1 until 1 do
    begin
        y[i] := a[i,u + 1]/a[i,i];
        for k := i − 1 step −1 until 1 do
        a[k,u + 1] := a[k,u + 1] − a[k,i] × y[i]
    end end;
```

ALGORITHM 127
ORTHO
Philip J. Walsh
National Bureau of Standards, Washington, D. C.

procedure $ORTHO(W,Y,Z,n,fn,m,p,r,ai,aui,mui,zei,X,DEV,$
$COF,STD,CV,VCV,gmdt,Q,Q2,E,EP,A,GF,ENF)$;
value n,m,p,r,ai,aui,mui,zei;
real $fn,gmdt$;
array $W,Y,Z,X,DEV,COF,STD,CV,VCV,Q,Q2,E,EP,A,GF,ENF$;
integer n,m,p,r,ai,aui,zei,mui;
switch $at := at1,at2$; **switch** $ze := ze1, ze2$;
switch $au := au1,au2$; **switch** $mu := mu1,mu2, mu3$;
comment $ORTHO$ is a general purpose procedure which is
capable of solving a wide variety of problems. For a detailed
discussion of the applications listed below and other applications, see (1) Philip Davis and Philip Rabinowitz, "A Multiple
Purpose Orthonormalizing Code and Its Uses," *J. ACM 1*
(1954), 183–191, (2) Philip Davis, "Orthonormalizing Codes in
Numerical Analysis," in J. Todd (Ed.), *A Survey of Numerical
Analysis*, Ch. 10 (McGraw-Hill, 1962), (3) Philip Davis and
Philip Rabinowitz, "Advances in Orthonormalizing Computation," in F. L. Alt (Ed.), *Advances in Computers*, Vol. 2, pp. 55–
133 (Academic Press, 1961), (4) Philip J. Walsh and Emilie V.
Haynsworth, General Purpose Orthonormalizing Code, SHARE
Abstr. #850. Applications: (a) orthonormalizing a set of
vectors with respect to a general inner product, (b) least squares
approximation to given functions by polynomial approximations
or any linear combination of powers, rational functions, trans-
cendental functions and special functions, such as those defined
numerically by a set of values, (c) curve fitting of empirical data
in two or more dimensions, (d) finding the best solution in the
l.s.s. to a system of m linear equations in n unknowns $(n \leq m)$,
(e) matrix inversion and solution of linear systems of equations,
(f) expansion of functions in a series of orthogonal functions,
such as a series of Legendre or Chebyshev polynomials.

The following information must be supplied to the procedure.
(We are considering here the approximation feature of the procedure.)

n the number of components per vector (excluding augmentation)

m the number of vectors used in the approximation. For a
polynomial fit of degree t, set $m=t+1$.

p the number of augmented components per vector. A feature
of this procedure is that once the approximating vectors
have been orthonormalized, they may be used in approximating r functions without repeating the orthonormali-
zation procedure on the original approximating vectors.

r the number of functions to be approximated.

ai a switch control concerning the approximating vectors.
With $ai=1$, the procedure selects the first n components
of the first row of $[Z]$, supplied by user. The i powers of
these values are computed and stored into working location $[X]$, $i=0(1)m-1$. This is the usual set up for a polynomial fit. With $ai=2$, the procedure selects the first n
components of the first m rows of $[Z]$ supplied by user and
stores them into working location $[X]$.

aui a switch control concerning augmentation on the approximating vectors. If $p=0$, this switch is ignored. With
$aui=1$, regular augmentation is applied to the vectors in

$[X]$. p zeros are stored after the nth component of the first
m rows of $[X]$. The $(n+i)$th component is replaced by
1.0, $i=1(1)m$. With $aui=2$, special augmentation is applied to the vectors in $[X]$. The p components located after
the nth component of the first m rows of $[Z]$ supplied by
the user augment $[X]$.

zei a switch control concerning augmentation on the functions
to be approximated. If $r=0$, this switch is ignored. With
$zei=1$, regular augmentation is applied to the functions
during the calculation. The n components of the first r
rows of $[Y]$ supplied by user will be augmented by p zeros
when moving $[Y]$ to $[X]$. With $zei=2$, special augmentation is applied. The first n components of the first r rows
of $[Y]$ are the functional values supplied by user. The next
p components of the first r rows of $[Y]$ are special values
also supplied by user.

mui a switch control concerning weights. $[W]$ is an $n \times n$ real,
positive definite, symmetric matrix of weights. It is generally diagonal and often the Identity matrix. $mui=1$
when $[W]=I_n$, the matrix $[W]$ need not be supplied.
$mui=2$ when $[W]$ is diagonal, but not I_n. The procedure is
supplied the n diagonal elements of $[W]$, but stored in the
first row of matrix $[W]$. $mui=3$ when the full weighting
matrix is supplied to the procedure.

The following list of matrix arrays is given to aid the user in
determining the number of components and vectors in the input
and results. $W[1:n,1:n]$, $Y[1:r,1:n+p]$, $Z[1:m,1:n+p]$,
$X[1:m+1,1:n+p]$, $DEV[1:r,1:n]$, $COF[1:r,1:p]$, $STD[1:r]$,
$CV[1:p+1,1:p]$, $VCV[1:r,1:p+1,1:p]$, $Q[1:r,1:m+1]$, $Q2$, E,
$EP[1:r,1:m]$, $A[1:m,1:p]$, $GF[1:m+r]$, $ENF[1:m]$.

The results of the procedure are stored in the following locations. The user must be sufficiently familiar with the theory to
know which results are relevant to his application of the procedure. All vectors are stored row-wise in the matrices listed
below.

X orthonormal vectors
DEV deviations
COF coefficients
STD standard deviations
CV covariance matrix, stored in upper triangular form.
      The $(p+1)$st row contains the square root of the
      diagonal elements of the matrix.
VCV variance-covariance matrices, stored in upper triangu-
      lar form with the $(p+1)$st rows containing the square
      root of the diagonal elements. There are r such
      matrices, the first subscript running over the r values.
$gmdt$ Gram determinant value
Q Fourier coefficients
$Q2$ squared Fourier coefficients
E sum of the squared residuals
EP residuals
A a lower triangular matrix used to calculate the co-
      variance matrix. $CV = A'A$.
GF Gram factors
ENF norms of the approximating vectors;
  begin
integer $npp, npm, m1, n2, m2, r1, rbar, p2, bei, rhi, i18, gai, sii, i,$
  $j, dei, mui, elz1, elz2, k, thi, ali, omi, nii$;
array $PK,XP[1:n+p], QK[1:m+1]$;
real $denom,sum,dk2,dk,fi,ss,ssq$;
switch $be := be1, be2$; **switch** $rh := rh1,rh2$; **switch** $ga :=$

```
            ga1,ga2;
       switch si := si1,si2;  switch de := de1,de2;  switch nu :=
            nu1,nu2;
       switch th := th1,th2,th3;  switch al := al1,al2;
       switch om := om1,om2;
       npp := n+p; npm := n+m; m1 := m-1; n2 := n+1; m2 := m+1;
       r1 := 0;  rbar := r;  p2 := p+1; denom := if n=m then 1.0
       else sqrt(n-m); bei := rhi := i18 := 1;
       if (p≠0) then gai := sii := 2 else gai := sii := 1;
  box1:     go to at[ai];
     at1:     for j := 1 step 1 until n do begin
                X(2,j) := Z(1,j); X[1,j] := 1.0 end;
                for i := 2 step 1 until m1 do begin
                for j := 1 step 1 until n do
                X[i+1,j] := X[i,j] × X[2,j] end; go to box2;
     at2:     for i := 1 step 1 until m do begin
                for j := 1 step 1 until n do
                X[i,j] := Z[i,j] end;
  box2:     if p=0 then go to box3 else go to au[aui];
     au1:     for i := 1 step 1 until m do begin
                for j := n2 step 1 until npp do
                X[i,j] := 0.0;  X[i,n+i] := 1.0 end;  go to box3;
     au2:     for i := 1 step 1 until m do begin
                for j := n2 step 1 until npp do
                X[i,j] := Z[i,j] end;
  box3:     dei := nui := elz1 := elz2 := k := 1;
  box4:     thi := 1;
  box5:     ali := omi := 1;  if p=0 then go to box6 else
                for j := 1 step 1 until p do PK[n+j] := 0.0;
  box6:     go to mu[mui];
     mu1:     for i := 1 step 1 until n do PK[i] := X[k,i];
                go to box7;
     mu2:     for i := 1 step 1 until n do
                PK[i] := X[k,i] × W[1,i];  go to box7;
     mu3:     for i := 1 step 1 until n do begin sum := 0.0;
                for j := 1 step 1 until n do sum := sum + X[k,j] ×
                W[i,j];  PK[i] := sum end;
  box7:     go to om[omi];
     om1:     for i := 1 step 1 until k do begin sum := 0.0;
                for j := 1 step 1 until npp do
                sum := sum + PK[j] × X[i,j];  QK[i] := sum end;
                go to box8;
     om2:     dk2 := 0.0; for i := 1 step 1 until npp do
                dk2 := dk2 + PK[i] × X[k,i];
                dk := sqrt(dk2);
                GF[i18] := dk;  i18 := i18 + 1;
                for i := 1 step 1 until npp do
                X[k,i] := X[k,i]/dk;
                omi := 1;  go to box6;
  box8:     go to de[dei];
     de1:     elz1 := -elz1;  if elz1<0 then go to box8b else
                go to box8a;
  box8a:    for i := 1 step 1 until k-1 do
                QK[i] := -QK[i];  QK[k] := 1.0;
                for i := 1 step 1 until npp do begin
                sum := 0.0;  for j := 1 step 1 until k do
                sum := sum + X[j,i] × QK[j];
                XP[i] := sum end;  go to box9;
  box8b:    ENF[i18] := sqrt (QK[k]);  go to box8a;
     de2:     elz2 := -elz2;  if elz2<0 then go to box8c else
                go to box8a;
  box8c:    for i := 1 step 1 until m do begin
                Q[r1,i] := QK[i];  Q2[r1,i] := QK[i] × QK[i] end;
                Q[r1,m2] := QK[m2];  E[r1,1] := Q[r1,m2]-Q2[r1,1];
                for j := 2 step 1 until m do
                E[r1,j] := E[r1,j-1] - Q2[r1,j];
                fi := 1.0;
                for i := 1 step 1 until m do begin
```
```
       if (fn-fi)>0.0 then begin if E[r1,i]<0.0 then begin
         EP[r1,i] :=  -sqrt(abs(E[r1,i])/(fn-fi));  go to box8d;
       end
       else EP[r1,i] := sqrt(E[r1,i]/(fn-fi));
         go to box8d;  end else E[r1,i] := -1.0;
  box8d:  fi := fi+1.0;  end go to box8a;
  box9:     go to th[thi];
     th1:     for i := 1 step 1 until npp do
                X[k,i] := XP[i];  go to box10;
     th2:     for i := 1 step 1 until n do
                DEV[r1,i] := XP[i];
                for i := 1 step 1 until p do
                COF[r1,i] := -XP[n+i];  thi := 3;  go to th1;
     th3:     go to box11;
  box10:    go to al[ali];
     al1:     omi := ali := 2;  go to box6;
     al2:     if k<m then begin k := k+1;  go to box4;  end
                else go to box12;
  box11:    go to nu[nui];
     nu1:     nui := 2;  go to box14;
     nu2:     ss := dk/denom;  ssq := ss × ss;
                STD[r1] := ss;  go to box14;
  box12:    go to be[bei];
     be1:     for i := 1 step 1 until m do begin
                for j := 1 step 1 until p do
                A[i, j] := X[i, n + j] end;
                gmdt := 1.0;  for i := 1 step 1 until m do
                gmdt := gmdt × (GF[i]/ENF[i]);
                gmdt := gmdt × gmdt;  dei := bei := thi := 2;
                k := k + 1;  go to box13;
     be2:     go to box11;
  box13:    go to ga[gai];
     ga1:     go to box11;
     ga2:     for i := 1 step 1 until p do begin
                for j := i step 1 until p do begin
                sum := 0.0;
                for nii := 1 step 1 until m do
                sum := sum + A[nii, i] × A[nii, j];
                CV[i, j] := sum end end;
                for i := 1 step 1 until p do
                CV[p2, i] := sqrt(CV[i, i]);  gai := 1;  go to box11;
  box14:    go to rh[rhi];
     rh1:     if rbar = 0 then go to final else rbar := rbar -1;
                r1 := r1 + 1;  thi := rhi := 2;  go to ze[zei];
       ze1:     for i := 1 step 1 until n do
                X[m2, i] := Y[r1, i];
                for i := 1 step 1 until p do
                X[m2, n+i] := 0.0;  go to box5;
       ze2:     for i := 1 step 1 until npp do
                X[m2, i] := Y[r1, i];  go to box5;
     rh2:     go to si[sii];
       si1:     go to rh1;
       si2:     for i := 1 step 1 until p do begin
                for j := i step 1 until p do
                VCV[r1, i, j] := ssq × CV[i, j] end;
                for i := 1 step 1 until p do
                VCV[r1, p2, i] := ss × CV[p2, i];  go to rh1;
  final: end ortho
```

CERTIFICATION OF ALGORITHM 127 [F5]
ORTHO [Philip J. Walsh, *Comm. ACM 5* (Oct. 1962)]
Ian Barrodale (Recd. 22 Aug. 1966)
Department of Mathematics, University of Victoria,
  Victoria, B.C., Canada

KEY WORD AND PHRASES: orthogonalization, approximation
CR CATEGORIES: 5.13, 5.17, 5.5

Algorithm 127 contains the following errors.

1. A **begin** must appear between the 6th and 7th lines, i.e. immediately after the integer specification and before the switch declaration. The **begin** following the **comment** and preceding the integer declaration must be removed.

2. In the second integer declaration the identifiers mui, $elz1$, $elz2$ should be nui, $elz1$, $elz2$, respectively.

3. The section of the statement labeled $at1$ that reads $X(2, j) := Z(1, j)$ should read $X[2, j] := Z[1, j]$.

4. Following the statement labeled $box8d$ there should be a semicolon between **end** and **go to** $box8a$.

5. The formal parameter fn is not defined or mentioned in the comment. It appears in the program between the labels $box8c$ and $box8d$. If fn is put equal to n the array EP then contains unbiased estimates of the m standard deviations.

We have not needed the generalized definition of an inner product [1, p. 348] but have often required n (number of components per vector) to be large. We thus replaced the array $W[1:n, 1:n]$ by an array $W[1:n]$ which necessitated the removal of the switch list element $mu3$ from the 8th line, also an alteration to the line before the statement labeled $mu3$ and the removal of the three lines beginning with the statement labelled $mu3$. Consequently that part of the program that appeared in the six lines beginning with the statement labeled $mu2$ and ending with the statement labeled $box7$ then read as follows:

```
mu2:  for i := 1 step 1 until n do
    PK[i] := X[k, i] × W[i];
    box7:  go to om[omi];
```

After the above modifications and corrections had been included the program ran successfully on an English Electric KDF9 computer using both the Whetstone Algol compiler and the Kidsgrove Algol compiler, these codes being proper subsets of Algol 60.

Some of the problems used in testing Algorithm 127 were from approximation theory as applied to boundary value problems of elliptic type. For one such problem linear approximating functions were used in which most of the coefficients of the best approximations are zero. The computed values of the standard deviations sometimes differed by more than 10 percent from both the true values and the unbiased estimates. We also solved the Dirichlet problem described by Davis [1, p. 369]. The set of coefficients obtained for the approximating function agreed only to the third decimal place with those given in [1]. All our calculations were in single-precision floating-point arithmetic.

Rice [2, p. 325] has recently noted that once the Gram-Schmidt orthogonalization method loses orthogonality it produces almost identical vectors. However, Algorithm 127 includes a correcting device which gives a second and better estimate to the true value of an orthonormal vector once the value obtained by Gram-Schmidt is known. Thus although Rice's modifications were included in the program we have not noticed any significant differences in computational behaviour.

References:
1. Davis, P. J. *Orthonormalizing codes in numerical analysis.* In *Survey of Numerical Analysis*, J. Todd (Ed.), McGraw-Hill, New York, 1962, pp. 347–379.
2. Rice, J. R. Experiments on Gram-Schmidt orthogonalization. *Math. Comput. 20* (Apr. 1966), 325–328.

ALGORITHM 128
SUMMATION OF FOURIER SERIES
M. WELLS
University of Leeds, Leeds 2, England*

* Currently with Burroughs Corp., Pasadena, Calif.

procedure *Fourier* (X, r, w, n, A, B);
  value n; **real** X, w, A, B; **integer** r, n;
  comment *Fourier* sums a one-dimensional Fourier series,
    using a recurrence relation described by Watt [*Computer
    J. 1, 4* (1959) 162]. The parameters are the coefficients X, which
    are selected by r, w, the argument and n the total number of
    terms in the series. On exit $A = \sum_{r=0}^{n-1} X_r cos(rw)$ and
    $B = \sum_{r=0}^{n-1} X_r sin(rw)$. *Fourier* is particularly efficient
    where $X_r = 0$ for all $r >$ some r_1 and $X_r \neq 0$ for all $r \leqq r_1$.;
begin real $t, tr, tr1, cosw2$;
  $tr1 := 0$; $cosw2 := 2 \times cos(w)$;
  for $r := n-1$ **step** -1 **until** 0 **do**
  begin if $X \neq 0$ **then go to** *term* **end** search for nonzero term;
  $tr := 0$; **go to** *all zeroes*;
term: $tr := X$; **for** $r := r-1$ **step** -1 **until** 0 **do**
  begin $t := tr \times cosw2 + X - tr1$; $tr1 := tr$; $tr := t$ **end**
    recurrence;
all zeros: $A := tr - tr1 \times cosw2/2$; $B := tr1 \times sin(w)$
end Fourier series

CERTIFICATION OF ALGORITHM 128 [C6]
SUMMATION OF FOURIER SERIES [M. Wells, *Comm.
  ACM 5* (Oct. 1962), 513]
HENRY C. THACHER, JR.* (Recd. 18 Mar. 1964)
Argonne National Lab., Argonne, Ill.
  * Work supported by the U.S. Atomic Energy Commission

The body of *Fourier* was transcribed for the Dartmouth SCALP
translator for the LGP-30 computer. After uniformizing the spelling of *zeros* (lines 5 and 9 in the procedure body), the program
compiled and ran without difficulty.

In the procedure statement for *Fourier*, the actual parameter
corresponding to X should be an expression depending on the
actual parameter corresponding to r.

The SCALP program was tested for the finite series:

$$A = \sum_{r=0}^{n-1} \cos rw = \frac{\sin((n-1)w/2)}{\sin(w/2)} \cos(nw/2) + 1$$

$$B = \sum_{r=0}^{n-1} \sin rw = \frac{\sin((n-1)w/2)}{\sin(w/2)} \sin(nw/2)$$

for $w = 0.1, 0.2, 0.5$ and 1.0, and for $n = 1(1)51$. Although the algorithm appears to be numerically correct, the results showed evi-
dence of serious numerical instability, particularly for small
values of w. For $w = 0.1$, and $n = 51$, the error in A was .00109,
and in B, $-.00231$. Since the largest A for $n < 51$ is 10.5, and the
largest B about 20, the best result obtainable with the 7+ significant digit arithmetic of the SCALP system is about .00001. For
comparison, a program summing the same series using a forward
recurrence based on the addition formulas for the sine and cosine
gave errors of .00012 and $-.00018$. It was, however, only about
half as fast.

ALGORITHM 129
MINIFUN
V. W. WHITLEY
Signal Missile Support Agency, White Sands Missile
Range, N. Mex.

```
procedure MINIFUN (t1, b1, eps, n, ncnt, fmin, xmin, k1,
    GFUN);
  value t1, b1, eps, n, ncnt;  integer n, ncnt, k1;  real fmin;
  real procedure GFUN;  array t1, b1, eps, xmin;
comment MINIFUN is a subroutine to find the minimum of a
```

function of n variables, using the method of steepest descent.
Input is:
1. $t1(i)$, $i = 1,2, \cdots , n$, the upper limits of the search region
2. $b1(i)$, $i = 1,2, \cdots , n$, the lower limits of the search region
3. $eps(i)$, $i = 1,2, \cdots , n$, the convergence criteria. The function must be a minimum in the region $| x(i) - xmin(i) | \leq eps(i)$
4. n, the number of variables (the dimension of the arrays)
5. $ncnt$, the maximum number of iterations. The routine searches for a minimum until $| x(i) - xmin(i) | \leq eps(i)$ for all i, or until $icnt = ncnt$, whichever happens first.

Output is:
1. $fmin$, the minimum value of the function
2. $xmin(i)$, $i = 1, \cdots , n$, the point at which the minimum occurs
3. $k1$, an error code

   If $k1 = 1$, a minimum has been found within the specified number of iterations and the minimum is less than all values of the function at the centers of the planes forming the boundary of the epsilon-cube

   If $k1 = 2$, $\Delta x(i) \leq eps(i)$ but a new minimum has been found

   If $k1 = 3$, $ncnt$ has been exceeded without $\Delta x(i) \leq eps(i)$. In this case, a test is made to see if the current minimum is a minimum in the epsilon-cube.

$MINIFUN$ has been written as a FORTRAN II subroutine and is available from the SMSA Computation Center. It should be noted that the FORTRAN II deck has been tested only on some relatively simple functions of two variables, such as $GFUN$ $(x,y) = \cos(xy)$. The writer does not claim that the algorithm has been thoroughly tested;

```
begin integer j, i, icnt, k;  real w, dmax, alamb, ft;
    array wnew [1:n], xt[1:n], x1b[1:n], xub [1:n],
      delx[1:n], d12x[1:n], xmin[1:n], x[1:n, 1:4], g[1:n, 1:4],
      dxmin[1:n], d2xmn[1:n];
  comment start looking for a minimum at midpoint of region;
  for j := 1 step 1 until n do
    begin wnew[j] := (t1[j] + b1[j])/2;  xt[j] := wnew[j];
      xub[j] := t1[j];  x1b[j] := b1[j];  delx[j] := (xub[j]
        - x1b[j])/5;
      d12x[j] := delx[j]↑2;  xmin[j] := xt[j]
    end;
  fmin := GFUN (xmin);
  for j := 1 step 1 until n do
    begin w := xt[j];  for i := 1 step 1 until 4 do
      begin x[j, i] := x1b[j] + i × delx[j];
        xt[j] := x[j,i];  g[j,i] := GFUN(xt);
      end;
    xt[j] := w;
    dxmin[j] := (g[j,3] - g[j,2])/delx[j];
    d2xmn[j] := (g[j,4] - g[j,3] - g[j,2] + g[j,1])/d12x[j]
```

```
  end;
comment first and second difference quotients have been com-
    puted;
  icnt := 0;  dmax := dxmin[1];  k := 1;
nustep: for j := 2 step 1 until n do
    begin if abs(dmax) < abs(dxmn[j]) then
        begin dmax := dxmin[j];  k := j
        end;
    end;
  alamb := dxmin[k]/d2xmn[k];  w := xt[k] − alamb;
comment a new coordinate has been computed for the variable
```

having the largest first partial derivative. It will be checked to see if the new point still lies within the region and search will continue;

```
  if w < b1[k] then w := b1[k] else if w > t1[k] then w := t1[k];
  xt[k] := w;  ft := GFUN(xt);
  if ft < fmin then go to check else
restart:  if xt[k] < wnew[k] then go to 1bdchk
      else if xt[k] = wnew[k] then go to stnubds
      else if t1[k] > xt[k] then go to nupbds
      else xt[k] := 1.5 × wnew[k];
nupbds:  xub[k] := t1[k];  x1b[k] := 2 × xt[k] − t1[k];  go to
    newde1;
stnubds:  x1b[k] := xt[k] − 0.5 × wnew[k];  xub[k] := xt[k] +
    0.5 × wnew[k];
newde1:  delx[k] := 0.2 ×(xub[k] − x1b[k]);  d12x[k] := delx[k]↑2;
    for i := 1 step 1 until 4 do
    begin x[k,i] := x1b[k] + i × delx[k];  w := xt[k];
        xt[k] := x[k,i];  g[k,i] := GFUN(xt);  xt[k] := w
    end;
    dxmin[k] := (g[k,3] − g[k,2])/delx[k];
    d2xmn[k] := (g[k,4] − g[k,3] − g[k,2] + g[k,1])/d12x[k];
    icnt := icnt + 1;
    if icnt > ncnt then go to outcd else go to nustep;
1bdchk:  if xt[k] ≤ b1[k] then xt[k] := 0.5 × wnew[k]
        else x1b[k] := b1[k];  xub[k] := 2.0 × xt[k] − b1[k];
    go to newde1;
check:  fmin := ft;  xmin[k] := xt[k];
    for j := 1 step 1 until n do if delx[j] > eps[j] then go to restart;
recheck:  for j := 1 step 1 until n do
    begin w := xmin[j];  xmin[j] := w + eps[j];  ft := GFUN
        (xmin);
      if ft < fmin then go to set2;  xmin[j] := w − eps[j];
      ft := GFUN(xmin);  if ft < fmin then go to set2;  xmin[j]
        := w
    end;
  if k1 < 3 then k1 := 1; go to bgend;
set2:  k1 := 2;  go to bgend;
outcd:  k1: =3; go to recheck;
bgend:  end MINIFUN;
```

REMARK ON ALGORITHM 129 MINIFUN
MINIFUN [V. W. Whitley, *Comm. ACM*, Nov. 1962]

E. J. Wasscher
Philips Research Laboratories
N. V. Philips' Gloeilampenfabrieken
Eindhoven-Netherlands

Some errors found in Algorithm 129 *MINIFUN* [*Comm. ACM*, Nov. 1962] are given below.

In addition, the way "steepest descent" is used to compute the minimum of a function of n variables is not entirely satisfactory. The method for computing first derivatives may be improved in two ways:

1. Instead of computing $\frac{f(x+h)-f(x)}{h}$ it is better to take $\frac{f(x+h)-f(x-h)}{2h}$. As $f(x-h)$ has been computed by *MINIFUN* this does not give rise to extra computations.

2. In *MINIFUN* the choice of h seems rather deliberate. Indeed, h is taken as $.2 \times (xub-xlb)$, where xub and xlb are variable bounds of x. In the beginning of the program these bounds are put equal to the fixed bounds bl and ub; afterwards in the iteration process they should tend towards each other, and in the limit they provide the minimum. So especially when a good approximation to the minimum is unknown, bl and ub have to be taken well apart from each other, which means that h is rather large. At the limit, however, h is very small. It is better to take h in such a way that the nominator $f(x+h)-f(x-h)$ attains an appropriate value.

As the method used by *MINIFUN* is the Newton-Raphson method applied to the first derivatives, convergence is not always secured—especially since first and second partial derivatives are estimated with numerical methods.

It should be noted that the test on end of program is not correct. For a further possible decrease of the function one has not to look in the direction of the coordinate axes but in the direction of the steepest descent.

Algol descriptions of some "steepest descent" programs which were written in the symbolic code of the Philips computer Pascal [cf. H. J. Heijn and J. C. Selman, *IRE Trans. EC10* (June 1961), 175–183] are given in Algorithms 203, 204 and 205.

Corrections of *MINIFUN*:

Printing errors: The line below label *nustep* should read:
    begin if $abs(dmax) < abs(dxmin[j])$ **then**
The label 1 *bdchk* should be *lbdchk*

In **comment** *MINIFUN*: $k1 = 2$: a new minimum has not been found.

The **label** nustep should be placed before the statement: $dmax := dxmin[j]$; The declaration of $xmin$ should be removed from the blockhead of the procedure body. The 2-dimensional arrays $x[1:n, 1:4]$ and $g[1:n, 1:4]$ can be replaced by a **real** x and a 1-dimensional **array** $g[1:4]$ respectively.

An improvement could be the insertion of the statement

$$k1 := 1;$$

just before the **label** nustep.

I am having considerable trouble with the obviously important part played by the **array** *wnew*, although it does not change after being set in the first statement of the program. Furthermore it seems to me that *wnew* plays a double rôle: first the component $wnew[k]$ is the value of $xt[k]$ before an iteration on $xt[k]$. But then one should insert another statement after **label** *nustep*: $wnew[k] := xt[k]$; Secondly $wnew[k]$ is to be understood as half the distance between upper and lower bound $t1[k]$ and $b1[k]$, which is only true when $b1[k] = 0$.

Convergence of $delx[j]$ to 0 is only achieved when $xlb[k]$ and $xub[k]$ are tending towards each other. This indicates that $wnew[k]$

should go to 0 too. (See statements after **label** *stnubds*.)

The following modifications could remove these objections (starting with the line above **label** *restart*):

```
          if ft < fmin then go to check else xt[k] := wnew[k];
restart:  if xt[k] < wnew[k] then go to lbdchk;
          if xt[k] = wnew[k] then go to stnubds;
          if xt[k] < t1[k] then go to nupbds;
          xt[k] := 0.5 × (wnew[k] + t1[k]);
nupbds:   xub[k] := t1[k]; xlb[k] := 2 × xt[k] − t1[k]; go to
              newde1;
stnubds:  xlb[k] := xt[k] − 0.5 × (wnew[k] − xlb[k]);
              xub[k] := xt[k] + 0.5 × (wnew[k] − xlb[k]); (etc.)
lbdchk:   if xt[k] = b1[k] then xt[k] := 0.5 × (wnew[k] + b1[k]);
              xlb[k] := b1[k]; xub[k] := 2 × xt[k] − b1[k]; go to
              newde1; (etc.)
```

ALGORITHM 130
PERMUTE
Lt. B. C. Eaves
U.S.A. Signal Center and School, Fort Monmouth, N. J.

procedure *PERMUTE* (A, n, x)
array A; **integer** n, x;
comment Each entry into *PERMUTE* generates the next permutation of the first n elements of A. If A is read as a number $(A[1]A[2] \cdots A[n])$, each generation is larger than the last:
$n := 4, x := 1$

$A[1]$	1	1	1	8	8	8
$A[2]$	1	8	8	1	1	8
$A[3]$	8	1	8	1	8	1
$A[4]$	8	8	1	8	1	1	**end**

Permutations $= \dfrac{4!}{2!2!}$

Identical elements in A reduce the number of permutations. The array should be ordered before the first call on *PERMUTE*. Integer x specifies the first elements whose order should be preserved: $n := 4, x := 3$

$A[1]$	1	1	1	4
$A[2]$	2	2	4	1
$A[3]$	3	4	2	2
$A[4]$	4	3	3	3	**end**

Permutations $= \dfrac{4!}{3!}$

Before the first call on *PERMUTE* for a given array, *first* should be made **true**. If *more* is **true**, then *PERMUTE* was able to give another permutation;

```
begin array B[1:n];  integer f, i, k, m, p;  real r;  own real t;
    if first then t := A[x]; first := false;
    for i := 1 step 1 until n do B[i] := 0;
    for i := n step −1 until 2 do
        begin if A[i] > t∧A[i] > A[i − 1] then go to find;  end;
    more := false; go to exit;
find:  for k := n step −1 until i do
            begin if A[k] > t∧A[k] > A[i −1] then
                begin B[k] := A[k];  m := k;  end;  end;
        for k := n step −1 until i do
        begin if B[k] > 0 ∧B[k] < B[m] then
            begin B[m] := B[k];  f := k;  end;  end;
        r := A[i − 1];  A[i − 1] := B[m];  A[f] := r;
schell:  p := i − 1; m := n − p;
        for m := m/2 − .4 while m > 0 do
            begin k := n − m;
            for f := p + 1 step 1 until k do
                begin i := f;
comp:       if A[i] > A[i + m] then
                begin r := A[i + m];  A[i + m] := A[i];
                A[i] := r; i := i − m;
                if i ≥ p + 1 then go to comp;
                end end end schell;
exit.  end PERMUTE
```

REMARKS ON:

ALGORITHM 87 [G6]
PERMUTATION GENERATOR
[John R. Howell, *Comm. ACM 5* (Apr. 1962), 209]
ALGORITHM 102 [G6]
PERMUTATION IN LEXICOGRAPHICAL ORDER
[G. F. Schrak and M. Shimrat, *Comm. ACM 5* (June (1962), 346]
ALGORITHM 130 [G6]
PERMUTE
[Lt. B. C. Eaves, *Comm. ACM 5* (Nov. 1962), 551]
ALGORITHM 202 [G6]
GENERATION OF PERMUTATIONS IN
LEXICOGRAPHICAL ORDER
[Mok-Kong Shen, *Comm. ACM 6* (Sept. 1963), 517]

R. J. ORD-SMITH (Recd. 11 Nov. 1966, 28 Dec. 1966 and 17 Mar. 1967)
Computing Laboratory, University of Bradford, England

A comparison of the published algorithms which seek to generate successive permutations in lexicographic order shows that Algorithm 202 is the most efficient. Since, however, it is more than twice as slow as transposition Algorithm 115 [H. F. Trotter, Perm, *Comm. ACM 5* (Aug. 1962), 434], there appears to be room for improvement. Theoretically a "best" lexicographic algorithm should be about one and a half times slower than Algorithm 115. See Algorithm 308 [R. J. Ord-Smith, Generation of Permutations in Pseudo-Lexicographic Order, *Comm. ACM 10* (July 1967), 452] which is twice as fast as Algorithm 202.

ALGORITHM 87 is very slow.

ALGORITHM 102 shows a marked improvement.

ALGORITHM 130 does not appear to have been certified before. We find that, certainly for some forms of vector to be permuted, the algorithm can fail. The reason is as follows.

At execution of $A[f] := r$; on line prior to that labeled *schell*, f has not necessarily been assigned a value. f has a value if, and only if, the Boolean expression $B[k] > 0 \land B[k] < B[m]$ is **true** for at least one of the relevant values of k. In particular when matrix A is set up by $A[i] := i$; for each i the Boolean expression above is **false** on the first call.

ALGORITHM 202 is the best and fastest algorithm of the exicographic set so far published.

A collected comparison of these algorithms is given in Table I. t_n is the time for complete generation of $n!$ permutations. Times are scaled relative to t_8 for Algorithm 202, which is set at 100. Tests were made on an ICT 1905 computer. The actual time t_8 for Algorithm 202 on this machine *was* 100 seconds. r_n has the usual definition $r_n = t_n/(n \cdot t_{n-1})$.

TABLE I

Algorithm	t_6	t_7	t_8	r_6	r_7	r_8
87	118	—	—	—	—	—
102	2.1	15.5	135	1.03	1.08	1.1
130	—	—	—	—	—	—
202	1.7	12.4	100	1.00	1.00	1.00

ALGORITHM 131
COEFFICIENT DETERMINATION*
V. H. Smith and M. L. Allen
Georgia Institute of Technology, Atlanta 13, Ga.
* This procedure pertains to research work sponsored in part by NSF Grant G-7361.

procedure DET (n, G, H);
array G, H; **integer** n;
comment Given the first n coefficients of the power series $G(z) = g_1 + g_2 z + g_3 z^2 + \cdots + g_n z^{n-1} + \cdots$, and $H(z) = h_1 + h_2 z + h_3 z^2 + \cdots + h_n z^{n-1} + \cdots$, this procedure determines the coefficients d_i, $i = 1, \cdots, n$, of the power series which is the expansion of the quotient $H(z)/G(z)$. It is assumed that $g_1 \neq 0$. The arrays G and H initially contain the coefficients of $G(z)$ and $H(z)$, respectively. The integer n is the number of known coefficients in the expansion of $G(z)$ and $H(z)$. At the conclusion, H_i contains the coefficient d_i. The procedure may also be useful in calculating residues for certain complex functions. Suppose $F(z) = H(z)/G(z)$ is a complex valued function of a complex variable and that F has a pole of order m at $z = b$, where $H(z) = \sum_{k=1}^{\infty} h_k(z - b)^{k-1}$, $G(z) = \sum_{k=1}^{\infty} g_k(z - b)^{k+m-1}$, and $g_1 \neq 0$, $h_1 \neq 0$. The required residue at $z = b$ is d_m where

$$D(z) = \left[\sum_{k=1}^{\infty} h_k(z - b)^{k-1} \right] \Big/ \left[\sum_{k=1}^{\infty} g_k(z - b)^{k-1} \right]$$

$$= \sum_{j=1}^{\infty} d_j(z - b)^{j-1}.$$

For more on this, one is referred to Einar Hille, "Analytic Function Theory, Vol. I, " Ginn and Co., 1959, pages 242–244;
begin integer i, j, n; **real** *alpha, beta*;
         alpha := $1/G[1]$;
    for $j := 1$ **step** 1 **until** n **do**
    begin *beta* := *alpha* $\times H[j]$;
    for $i := j + 1$ **step** 1 **until** n **do**
    $H[i] := H[i] - (beta \times G[i - j + 1])$ **end**;
    for $j := 1$ **step** 1 **until** n **do**
    $H[j] := H[j] \times alpha$;
end DET

ALGORITHM 132

QUANTUM MECHANICAL INTEGRALS OVER ALL SLATER-TYPE INTEGRALS

J. C. BROWNE

The University of Texas, Austin, Tex.

real procedure: *allslater* $(p,q,pe,qe,np,nq,lp,lq,mp,mq,na,nb)$
internuclear distance: (r);
      real pe,qe,r; **integer** $p,q,np,nq,lp,lq,mp,mq,$ na,nb;

comment The Slater-type orbitals frequently used in quantum mechanical calculations on atoms and molecules are defined as $p = k(np,pe)\ r^{np-1}e^{-(pe)r}\ Y_{1p}^{mp}(\theta, \phi)$, where $k(np,pe)$ is a normalization constant, $Y_1{}^m(\theta,\phi)$ is a spherical harmonic with the phase convention $[Y_1{}^m(\theta,\phi)]^* = (-1)^m Y_1^{-m}(\theta,\phi)$, np is a positive integer, lp is an integer, $lp < np$, mp is an integer, $- lp \leqq mp \leqq lp$; and pe is a real positive constant. Algorithm 110, Y. A. Kruglyak and D. R. Whitman (*Comm. ACM*, July 1962) serves to compute integrals over certain operators of a quite restricted class of Slater-type orbitals, $np \geqq 4$, $lp = 1$, $mp = 0$. The algorithm given here will compute all integrals of the form

$$\int\ p_c(r_c^{n_c})q_c d\tau$$

which can be expressed in terms of the simple $A_n(b)$ and $B_n(a)$ functions. The subscript c denotes either of the two nuclei of a diatomic molecule. These integrals include all those one-electron integrals necessary for a conventional energy calculation on a diatomic molecule. In the arguments of *allslater* p and q are numerical designations for the respective orbitals. p and q are even or odd as they respectively are associated with the "left," a, nucleus or "right," b, nucleus of a diatomic molecule. Global arrays, *fact 1*, of factorials and *binom*, of binomial coefficients are assumed. We first define some procedures utilized by *allslater*. The main program begins at the label *set*;

```
begin real norm, r2, alpha, beta, s, clp, clq, bpci;
  integer nsum, lsum, peven, qeven, podd, godd, limitp, limitq,
  g, h, i, j, n1p, n1q, lmp, lmq, gama, gamb, aidaa, aidab, gam,
  aida, num2; real array avalues [0:21], bvalues[0:21]; real pro-
  cedure c1, bpc, modulus;
  real procedure  c1(l,m,j); value l,m,j, integer l,m,j;
    begin c1 := ((-1)↑j)× fact1[2×(l − j)]/((2↑l) × fact1
    [l-2×j-m]×
    fact1[l − j]× fact1[j])
    end c1;
  real procedure bpc(i, j, k);  value i,j,k; integer i,j,k;
    begin real t; integer m; t := 0;
      for m := 0 step 1 until k do
      begin t := t + ((-1) ↑ (k − m)
        × binom [i, m] × binom [j, k − m]
      end
    end bpc;
  real procedure modulus (i, j);  value i, j;  integer i, j;
    begin modulus := 1 −abs(i ÷ j)× j
    end modulus;
  procedure avector (b, nmax, avalues);  value b, nmax;
    real b;  integer nmax;  real array avalues;
    begin integer m;
      avalues[0] = exp(−b)/b;
      if nmax = 0 then go to exit;
      for m = 1 step 1 until nmax do
      begin avalues[m] = avalues[0] + (m/b) × avalues[m − 1]
    end;
exit:  end avector;
  procedure bvector(a nmax, bvalues); value a, nmax; real a;
    integer nmax; real array bvalues;  real procedure modulus;
  comment  This procedure computes a sequence of values for the
    integral, B_n(a) = ∫_{−1}^{1} x^n e^{−ax}dx, for n = 0 to n = nmax. If a ≧
    alim then B_0(a) is computed and upward recursion is used to
    generate the higher n values. If a < alim then B_{nmax}(a) is com-
    puted by series expansion and downward recursion is used to
    generate the smaller n values. alim is determined within the
    program by a simplification of a result of Gautschi (J. ACM 8,
    21 (1961)). Gautschi has made an analysis of the recursive pro-
    cedures for the B_n(a) which could be taken as a model for workers
    in molecular quantum mechanics;
  begin real fxx, fxy, numerator, denom, sum, factor1, tsum
    factor2, t, aa;  integer m,mn;
    begin if abs(a) ≧ ((nmax+nmax/6+3)/2.3) then
up:     begin fxx := exp(a);
          fxy := 1/fxx;
          bvalues [0] := (fxx-fxy)/a;
          for m := 1 step 1 until nmax do
        begin fxx := − fxx;
            bvalues[m] := (fxx−fxy + m ×
            bvalues[m−1])/a
        end;
        go to exit;
      end up;
down:   begin aa := axa;
          if modulus (nmax, 2)≠0 then
        setodd:  begin numerator := nmax + 2;
            sum := a/numerator;
            factor1 := −2;
            factor2 := 3;
            go to compute;
          end setodd;
        seteven:  begin numerator := nmax + 1;
            sum := 1/numerator;
            factor1 := factor2 := 2;
          end seteven;
        compute: begin denom := numerator + 2;
            t := sum;
            t := (((((t/factor2)×aa)
              /(factor2−1))×numerator)
              /denom;
            tsum := t + sum;
            if (sum−tsum)=0 then
            begin bvalues[nmax] := sum × factor1;
              go to recur;
            end;
            begin factor2 := factor2 + 2;
              numerator := denom;
              sum := tsum;
            go to compute;
          end  compute;
recur:    begin fxx := exp(a);
            fxy := 1/fxx;
            mn := nmax −1;
            if modulus(nmax, 2) ≠ 0 then
            fxx := −fxx;
            for m := mn step −1 until 0 do
            begin fxx = −fxx;
```

$$bvalues[m] := (fxx+fxy + a \times$$
$$bvalues[m+1])/(m+1);$$
$$\textbf{end}$$
$$\textbf{end } recur;$$
$$\textbf{end } down;$$
$$\textbf{end};$$

exit: **end** *bvector*;
set: **begin if** $(mp + mq) \neq 0$ **then**
      begin *allslater* := 0.0; **go to** *exit* **end**;
   set: **begin** *norm* := *sqrt* $(((2\times pe)\uparrow$
       $(2\times np+1) \times (2\times lp+1) \times fact1[lp-mp] \times (2\times qe)\uparrow$
       $(2\times nq+1) \times (2\times lq+1) \times fact1[lq\text{-}mq])/(fact1[2\times$
       $np] \times fact1[lp+mp] \times fact1[2\times nq] \times fact1[lq+mq] \times$
       $4));$
       $nsum := np+nq;$
       $lsum := lp+lq;$
       $r2 := r/2;$
       $norm := norm \times (r2\uparrow(nsum+1+na+nb));$
       $alpha := r2 \times(pe+qe);$
       $beta := r2 \times(((-1)\uparrow p)\times pe + ((-1)\uparrow q) \times qe);$
       $num2 := 2;$
       $avector\ (alpha, nsum, avalues);$
       $bvector\ (beta, nsum, bvalues);$
       $peven := modulus\ (p+1,2);$
       $qeven := modulus\ (q+1,2);$
       $podd := modulus\ (p,2);$
       $qodd := modulus\ (q,2);$
       $limitp := (lp-mp)\div num2;$
       $limitq := (lq-mq)\div num2;$
       $s := 0;$
   end *set*;
sum: **begin for** $g := 0$ **step** 1 **until** *limitp* **do**
      begin $clp := cl(lp,mp,g);$
      for $h := 0$ **step** 1 **until** *limitq* **do**
      begin $clq := cl(lq,mq,h);$
      $n1p := np-lp+2\times g-1;$
      $n1p := nq-lq+2\times h-1;$
      $lmp := lp-mp-2\times g;$
      $lmp := lq-mq-2\times h;$
      $gama := n1p \times peven + n1q \times qeven +1 +na;$
      $gamb := n1p \times podd + n1q \times qodd +1 +nb;$
      $aidaa := lmp \times peven + lmq \times qeven;$
      $aidab := lmp \times podd + lmq \times qodd;$
      $gam = gama + gamb;$
      $aida = aidaa + aidab;$
      for $i := 0$ **step** 1 **until** *gam* **do**
      begin $bpci := bpc(gama, gamb, i);$
      for $j := 0$ **step** 1 **until** *aida* **do**
      begin
      $s := s+ clp \times clq \times bpci \times$
      $bpc(aidaa, aidab, j) \times$
      $avalues[nsum+na+nb-i-j]$
      $\times bvalues[lsum -2 \times (g+h) +i-j];$
      end
      end
      end
      end;
     $allslater := s \times norm;$
   end sum;
exit: **end**;
end *allslater*;

ALGORITHM 133
RANDOM
Peter G. Behrenz
Mathematikmaskinnämnden, Stockholm, Sweden

```
real procedure RANDOM (A, B, X0);
value A, B, X0;
real A, B;
integer X0;
```

comment *RANDOM* generates a rectangular distributed pseudo-random number in the interval $A < B$. $X0$ is an integer starting-value. The first time *RANDOM* is used in a program $X0$ should be a positive odd integer with 11 digits, $X0 < 2^{35} = 34\,359\,738\,368$. The following times *RANDOM* is used, $X0$ should be $X0 = 0$. The mathematical method used is $X_{n+1} = 5\,X_n$ (mod 2^{35}). This sequence has period 2^{33}). *RANDOM* was successfully run on *FACIT EDB* using *FACIT-ALGOL* 1, which is a realization of *ALGOL* 60 for *FACIT EDB*, except for the declarator **own,** which is not included in FACIT-ALGOL 1. To test *RANDOM*, we computed $1/N \sum X_n$ and $1/N \sum X_n^2$ in the interval 0,1 for $N = 500, 1000, 5000$. The starting-value was $X0 = 28\,395\,423\,107$. The results were 0.50625, 0.48632, 0.50304 and 0.34304, 0.31681, 0.33469. Theoretically one expects 0.50000 and 0.33333;

```
begin
integer M35, M36, M37;
own integer X;
if X0 ≠ 0 then begin
X := X0;   M35 := 34 359 738 368;   M36 := 68 719 476 736;
M37 := 137 438 953 472 end;   X := 5 × X;
if X ≥ M37 then X := X − M37;
if X ≥ M36 then X := X − M36;
if X ≥ M35 then X := X − M35;
RANDOM := X/M35 × (B − A) + A end
```

REMARK ON ALGORITHM 133
RANDOM (P. G. Behrenz, *Comm. ACM*, Nov. 1962)
Peter G. Behrenz
Matematikmaskinnämnden, Box 6131, Stockholm 6, Sweden

Replace the declarations in the body of the procedure,
    integer *M35, M36, M37;*   **own integer** *X;*
by:

    own integer *X, M35, M36, M37;*

The sequence of 2^{33} random numbers contains about 15 numbers which are not really random numbers. For details, see R. W. Hamming, *Numerical Methods for Scientists and Engineers*, p. 384 [McGraw-Hill, 1962].

REMARK ON ALGORITHM 133
RANDOM [Peter G. Behrenz, *Comm. ACM 11*, Nov. 1962]
Donald L. Laughlin
Missouri School of Mines and Metallurgy, Rolla, Missouri

Algorithm 133 was translated into Fortran II for the IBM 1620 and run successfully. The starting value was changed to 21 348 759 609 and significant results followed.

For $N = 500$ and 1000, the resulting values were: 0.4990157688, 0.4986269653 and 0.3318717863, 0.3290401482.

CERTIFICATION OF ALGORITHM 133
RANDOM [Peter G. Behrenz, *Comm. ACM*, Nov. 1962]
Jesse H. Poore, Jr.
Louisiana Polytechnic Institute, Ruston, La.

Algorithm 133 was transliterated into Fortran II for the IBM 1620 computer. A monitor program performed the test indicated in Algorithm 133 on the generated numbers.

Results of the test are shown in the following chart. The notation used is identical to that used in the algorithm.

X_0	$\frac{1}{N} \Sigma X_n$	$\frac{1}{N} \Sigma X^2_n$	
	.4986480931	.3280561242	$N = 500$
13543288579	.4840396640	.3141520616	$N = 1000$
	.4996829627	.3321160892	$N = 5000$
	.4971414796	.3297990588	$N = 500$
24376589411	.4997720126	.3326801987	$N = 1000$
	.4986380784	.3319949173	$N = 5000$
	.4962408228	.3339214302	$N = 500$
34359738367	.4974837457	.3335720239	$N = 1000$
	.4929612237	.3253421270	$N = 5000$
	.5313808305	.3691599122	$N = 500$
11324679915	.5167083685	.3498558251	$N = 1000$
	.5043814637	.3383429327	$N = 5000$

ALGORITHM 134

EXPONENTIATION OF SERIES

Henry E. Fettis

Aeronautical Research Laboratories, Wright-Patterson
Air Force Base, Ohio

procedure $SERIESPWR(A, B, P, N)$;

comment This procedure calculates the coefficients $B[i]$ for
the series $(f(x))^P \equiv g(x) \doteq 1 + \sum B[i] \times x \uparrow i, (i = 1, 2, \cdots, N)$
given the coefficients of the series $f(x) = 1 + \sum A[i] \times x \uparrow i.$
P may be any real number;

value A, P, N;

array A, B;

integer N;

begin integer i, k;

real p, s;

  $B[1] := P \times A[1]$;

  for $i := 2$ **step** 1 **until** N **do**

  begin $s := 0$;

    for $k := 1$ **step** 1 **until** $i-1$ **do**

    $S := s + (P \times [i-k] - k) \times B[k] \times A[i-k]$;

    $B[i] := P \times A[i] + (s/i)$

    end for i;

  end $SERIESPWR$

CERTIFICATION OF ALGORITHMS 134 AND 158

EXPONENTIATION OF SERIES [Henry E. Fettis,
Comm. ACM, Oct. 1962 and Mar. 1963]

Henry C. Thacher, Jr.

Reactor Engineering Div., Argonne National Laboratory
Argonne, Ill.

Work supported by the U.S. Atomic Energy Commission.

The bodies of SERIESPWR were transcribed for the Dartmouth Scalp processor for the LGP-30 computer. In addition to
the modifications required by the limitations of this translator,
the following corrections were necessary:

1. Add "**real** P;" to the specifications.
2. Delete "p," from the declarations in the procedure body.
3. (134 only) Replace "S" by "s" and $[i-k]$ by "$(i-k)$" in the
statement $S := s + \cdots$.
4. (158 only) Changes last sentence of comment to "Setting
$P := 0$ gives the coefficients for $ln(f(x))$. In this series, the
constant term is 0, instead of 1 as elsewhere;"
5. (158 only) Add the identifier $P2$ to the declared real variables.
6. (158 only) Make the first statements read:
    "**if** $P = 0$ **then** $P2 := 1$ **else** $P2 := P$;
    $B[1] := P2 \times A[1]$;
7. (158 only) Make the statement of the **for** k loop read

    "$S := S + (P \times (i-k) - k) \times B[k] \times A[i-k]$;"

8. Change the last statement to

    "$B[i] := P2 \times A[i] + S/i$ **end** for i;"

In addition, the following modifications would improve the
efficiency of the program:

1. Remove A from the value list.

2. Omit the statement $B[1] := P \times A[1]$, $(P2 \times A[1]$ in 158
according to correction 6) and change the initial value of i
in the statement following from 2 to 1.

When these changes were made, both procedures produced the
first ten coefficients of the series for $(exp(x)) \uparrow 2.5$ from the first
ten coefficients of the exponential series. The procedures were
also used to generate the binomial coefficients by applying them
to $(1+x)^P$, for $P = 2.0$, and 0.5000000. Algorithm 158 was also
tested with $P := 0$ for $1+x$ and for the series expansions for
$(sin\ x)/x$, $cos\ x$, and $exp\ x$. In all cases, the coefficients agreed
with known values within roundoff.

ALGORITHM 135
CROUT WITH EQUILIBRATION AND ITERATION
WILLIAM MARSHALL MCKEEMAN*
Stanford University, Stanford, Calif.
* This work was supported in part by the Office of Naval Research under contract Nonr 225(37).

```
procedure LINEARSYSTEM (A) order:(n) right-hand sides:(B)
    number of right-hand sides:(m) answers:(X) determinant:(det,
    ex) condition of A:(cnr);
integer n, m, ex;  real det, cnr;  real array A, B, X;
comment, LINEAR SYSTEM uses Crout's method with row
    equilibration, row interchanges and iterative improvement
    for solving the matrix equation AX = B where A is n × n and
    X and B are n × m. As special cases one sees that: for m ≦ 0,
    only the determinant of A is evaluated, for m = 1, the algo-
    rithm solves a system of n equations in n unknowns, for m = n
    and B = the identity matrix, the algorithm inverts A.
```

If the algorithm breaks down for a singular or nearly singular matrix A, exit to a non-local label "singular" is provided. Five auxiliary procedures: *EQUILIBRATE*, *CROUT*, *PRODUCT*, *RESIDUALS* and *SOLVE* are declared with appropriate comments after the end of this procedure. This code is the result of the joint efforts of G. Guthrie, W. McKeeman, Cleve Moler, Margaret Salmon, Alan Shaw and R. Van Wyk. It was written following ideas presented by J. H. Wilkinson as a visiting lecturer in Professor George E. Forsythe's class in Advanced Numerical Analysis at Stanford, 1962;

```
begin  integer array pivot [1:n]; integer i, j, k; real mx;
    real array LU[1:n, 1:n], y, res, mult[1:n];
    comment, remove appropriate factors from the rows of A... ;
    EQUILIBRATE(A, n, mult);
    comment   ... and save the result for the eventual computation
    of residuals during iteration;
    for i := 1 step 1 until n do
        for j := 1 step 1 until n do LU[i,j] := A[i,j];
    comment, decompose the matrix into triangular factors;
    CROUT(LU, n, pivot, det);
    comment, assuming that there was no exit to "singular",
    evaluate the determinant in the form det × (10.0 ↑ ex);
    for i := 1 step 1 until n do y[i] := LU[i,i] × mult[i];
    det := det × PRODUCT(y,1,n,ex);
    comment, now begin to process right-hand sides;
    for k := 1 step 1 until m do
    begin integer i, count, limit;  real normy, kr;
        kr := k;
        comment, scale the right-hand side;
        for i := 1 step 1 until n do res[i] := B[i,k] := B[i,k]/mult[i];
        comment, store the first approximation and its L(1) norm;
        normy := 0;
        SOLVE(LU, n, res, pivot, y);
        for i := 1 step 1 until n do
        begin
            normy := normy + abs(y[i]);
            X[i,k] := y[i]
        end;
        comment, enter the iterating loop. The iteration is termi-
            nated on the integer "limit" which itself is determined on
            the basis of the success of the first iteration and a machine-
            dependent real number designated here by "eps". For
            "eps", the programmer must insert the largest real num-
```

ber such that $eps + 1.0 = 1.0$;

```
        for count := 1, 2 step 1 until limit do
        begin integer i; real t;
            comment, compute the residuals of the solution y;
            RESIDUALS(A,n,B,k,X,res);
            comment ... and find the next increment to the solution;
            SOLVE(LU,n,res,pivot,y);
            comment, set up termination conditions;
            if count = 1 then
            begin real normdy;
                normdy := 0;
                for i := 1 step 1 until n do normdy := normdy+abs(y[i]);
                if normdy = 0 then begin cnr := 1.0; go to enditer end;
                t := normy/normdy;
                comment, The quantity ‖ A ‖·‖ A⁻¹ ‖ (spectral norm)
                    is called the condition number of the matrix A. It is
                    a measure of the difficulty in solving the input equation
                    and appears naturally in error bounds for the solution
                    (see Wilkinson [3]). cnr is a direct measure of the
                    error and experimentally approximates the condition
                    number;
                cnr := ((kr − 1.0) × cnr + 1.0/(eps × t))/kr;
                if t < 2.0 then go to singular;
                limit := ln(eps)/ln(1.0/t);
            end;
            comment, store the new approximation;
            for i := 1 step 1 until n do X[i,k] := X[i,k] := X[i,k]+y[i];
        end iteration;
        enditer:
    end right-hand sides
end LINEAR SYSTEM;
procedure  EQUILIBRATE  (A) order:(n)  multipliers:(mult);
integer n; real array A, mult;
comment, scaling the rows of the matrix A to roughly the same
    maximum magnitude (here, dividing by the largest element)
    allows the procedure CROUT to select effective pivotal elements
    for the Gaussian decomposition of the matrix. The iterating
    procedure will converge to the solution for the equilibrated
    matrix rather than the input matrix. If the matrix is badly
    conditioned then the solution is sensitive to perturbations in
    the input and the scaling division must be done not by the
    largest element but rather by the power of the machine number
    base (2 and 10 for binary and decimal machines, respectively)
    nearest the largest element so as to avoid rounding errors.
    Equilibration is discussed in reference [3] p. 284;
begin integer i;  real mx;
    for i := 1 step 1 until n do
    begin integer j;
        mx := 0.0;  comment, find the largest element;
        for j := 1 step 1 until n do
            if abs(A[i,k]) > mx then mx := abs(A[i,k]);
        if mx = 0.0 then go to singular;
        comment, now store the multiplier and scale the row;
        mult[i] := mx; comment := base ↑ ex for exact scaling;
        if mx ≠ 1.0 then
            for j := 1 step 1 until n do A[i,j] := A[i,j]/mx
    end
end EQUILIBRATE;
```

```
procedure CROUT (A) order:(n) pivots:(pivot) interchanges:(sg).
integer n;  integer array pivot;  real array A;  real sg;
comment, this is Crout's method with row interchanges as
    formulated in reference [1] for transforming the matrix A into
    the triangular decomposition LU with all the L[k,k] = 1.0.
    pivot[k] stores the index of the pivotal row at the k-th stage of
    the elimination for use in the procedure SOLVE;
begin integer i, j, k, imax, p;  real t, quot;
  real procedure IP1 (A) extra term:(t) length:(f);
    integer f; real t; real array A; comment non-local i, j, k;
    comment, IP1 forms a row by column inner product of A,
        namely the sum of A[i,p] × A[p,k] for p := 1, 2, ... , f, and
        then adds the extra term t. If f < 1, the value of IP1 is t.
        This procedure is the inner loop of the algorithm. The pro-
        grammer can expect a substantial advantage from substi-
        tuting a faster and more accurate inner product here;
    begin real sum; integer p;
      sum := t;
      for p := 1 step 1 until f do sum := sum + A[i,p] × A[p,k];
      IP1 := sum
    end IP1;
  sg := 1.0;
  comment, k is the stage of the elimination;
  for k := 1 step 1 until n do
  begin
    t := 0;
    for i := k step 1 until n do
    begin comment, compute L. Note that the first calls on IP1
    are empty;
        A[i,k] := −IP1(A, −A[i,k],k−1);
        if abs(A[i,k]) > t then
            begin t := abs(A[i,k]);  imax := i end
    end;
    if t = 0 then go to singular;
    comment, A[imax,k] is the largest element in the remainder
        of column k. Interchange rows if necessary and record the
        change;
    pivot[k] := imax;
    if imax ≠ k then
    begin
      sg := −sg;
      for j := 1 step 1 until n do
      begin
        t := A[k,j]; A[k,j] := A[imax, j]; A[imax, j] := t
      end
    end;
    comment, compute a column of multipliers;
    quot := 1.0/A[k,k];
    for i := k+1 step 1 until n do A[i,k] := A[i,k] × quot;
    comment, and compute a row of U;
    for j := k+1 step 1 until n do
        A[k,j] := −IP1(A,−A[k,j],k−1)
  end
end CROUT;
real procedure PRODUCT (factors) start:(s) finish:(f)
    exponent:(ex);
integer s,f,ex;  real array factors;
comment, PRODUCT multiplies the numbers stored from index
    s through f inclusive in the array "factors", preventing ex-
    ponent overflow. The answer is normalized so that 1.0 > abs
    (PRODUCT) ≧ 0.1. The exponent appears in ex;
begin integer i;  real p, p1;
  ex := 0;  p := 1.0;
  for i := s step 1 until f do
  begin
      p1 := factors [i];
      if abs(p1) < 0.1 then begin p1 = 10.0 × p1;  ex := ex−1
        end;
```

```
      p := p × p1;
      if p = 0 then begin ex := 0; go to fin end;
  1:  if abs(p) < 0.1 then
        begin p := p × 10.0;  ex : = ex−1; go to 1 end;
  2:  if abs(p)≧1.0 then
        begin p := p/10.0;  ex := ex + 1;  go to 2 end;
  end;
  fin: PRODUCT := p
end PRODUCT;
procedure RESIDUALS (A) order:(n)  right-hand sides:(B)
    column of B:(k) approximate solution:(X) residuals:(res);
    integer n, k; real array A, B, X, res;
comment, RESIDUALS computes b − Ay where b is the kth
    column of the right-hand side matrix B and y is the kth column
    of X;
    real procedure IP2 (A) row: (i) order:(n) approximate
        solution:(X)
    column:(k) extra therm:(t);
    integer i, k, n;  real t real array A, X;
    comment, IP2 forms the inner product of row i of the matrix
        A and column k of the solution matrix X, then adds the
        single term t. It is essential that IP2 be an "accumulating"
        or double precision inner product as discussed in reference
        [3] p. 296. The value of IP2 is the rounded single precision
        result of the double precision arithmetic. The body of the
        procedure is left undefined;
begin integer i;
for i := 1 step 1 until n do
  res[i] := −IP2(A,i,n,X,k,−B[i,k])
end RESIDUALS;
procedure SOLVE (A) order:(n) right-hand side:(b) pivots:
    (pivot) answer:(y);
integer n; integer array pivot; real array A, b, y;
comment, SOLVE processes a right-hand side b and then back-
    solves for the solution y using the LU decomposition provided
    by CROUT;
begin integer k, p;  real t;
  for k := 1 step 1 until n do
  begin
      t := b[pivot[k]];  b[pivot[k]] := b[k];
      for p := 1 step 1 until k−1 do t := t − A[k,p] × b[p];
      b[k] := t
  end ...having modified b by L inverse;
  comment, now the back solution for y;
  for k := n step −1 until 1 do
  begin
      t := b[k];
      for p := k+1 step 1 until n do t := t − A[k,p] × y[p];
      y[k] := t
  end backsolution
end SOLVE
```

REFERENCES

1. GEORGE E. FORSYTHE, Crout with Pivoting. Algorithm 16.
   Comm. ACM 3, 2 (Sept. 1960), 507.
2. DEREK JOHANN ROEK, Simultaneous System of Equations and
   Matrix Inversion Routine. Algorithm 92. *Comm. ACM 5*,
   5 (May 1962), 286.
3. J. H. WILKINSON, Error Analysis of Direct Methods of Matrix
   Inversion, *J. ACM 8*, 3 (July 1961), 281–330.

CERTIFICATION OF ALGORITHM 135
CROUT WITH EQUILIBRATION AND ITERATION
[William Marshall McKeeman,* *Comm. ACM*, Nov. 1962]

WILLIAM MARSHALL MCKEEMAN,
Stanford University, Stanford, Calif.

* This work was supported in part by the Office of Naval Research under contract Nonr 225(37).

A BALGOL translation of the algorithm was tested for accuracy, proper termination and running time on the Burroughs 220. The exact inverse of the Hilbert segment of order 6 can be stored in the 8-decimal-digit floating word of the B220 and was used in the accuracy and termination tests. The Hilbert segment H_6 is very ill-conditioned (for the spectral norm, $\| H_6 \| \cdot \| H_6^{-1} \| = 1.3 \times 10^7$). Hence the number of iterations required should not be taken as typical.

The $[n,n]$ element (mathematically $\frac{1}{11} = .090909 \cdots$) is representative of the behavior of the rest:

	"exact" equilibration (by powers of 10)	equilibration by largest element in row
initial solution	.092587535	.094091506
first iteration	.090877240	.091498265
second iteration	.090909695	.091570311
third iteration	.090909080	.091568310
fourth iteration	.090909091	.091568365
fifth iteration	terminated	.091568364
		terminated

Conclusions: The iterating procedure terminated correctly, or performed one extra iteration in each case. If the equilibration procedure alters the data, the iteration will converge to the solution for the altered matrix. If the matrix is ill-conditioned, as in the case above, the equilibration may cost a great deal more than it gains. As a practical matter, a machine language substitute for EQUILIBRATE which will not cause rounding of the data is probably the best course of action.

The running time is approximately proportional to n^3 as expected. If for a given machine, μ is the floating multiply time in seconds, one can expect that run time will be given by $rt := 1.3 \times \mu \times (n + 7) \uparrow 3$ seconds for a call on LINEARSYSTEM with one right-hand side.

The division of run time between the various phases of the algorithm is as follows:

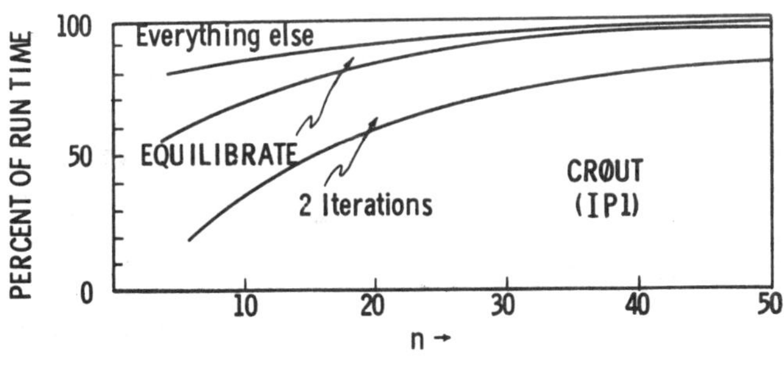

ORDER OF MATRIX

REFERENCE:
1. SAVAGE AND LUKACS, Tables of inverses of finite segment of the Hilbert matrix. In Olga Taussky (Ed.), Contributions to the Solution of Systems of Linear Equations and the Determination of Eigenvalues, pp. 105–108, Nat. Bur. Standards Appl. Math. Series no. 39, U. S. Government Printing Office, Wash., D.C., 1954.

REMARK ON ALGORITHM 135 [F4]
CROUT WITH EQUILIBRATION AND ITERATION
[W. M. McKeeman, *Comm. ACM 5* (Nov. 1962), 555–557, 559]

WILLIAM MARSHALL MCKEEMAN (Recd. 1 Apr. 1964)
Computation Center, Stanford University, Stanford, Calif.

The following corrections to the published algorithm are recommended:

1. Two lines above the bottom line of procedure *SOLVE* one must change

$$y[k] := t \qquad \text{to} \qquad y[k] := t/A[k,k]$$

2. In procedure *EQUILIBRATE*, all occurrences of the subscript k must be changed to j.

3. The statement $cnr := 1.0$ should be added at the start of the body of procedure *LINEARSYSTEM*, so that cnr will have a value the first time it is used.

4. Line 19 from the end of *LINEARSYSTEM* should be changed from

if $normdy = 0$ **then begin** $cnr := 1.0$; **go to** *enditer* **end**;

to read

if $normdy = 0$ **then go to** *enditer*;

This correction makes sure that cnr retains a reasonable value in case $normdy$ should be 0 for some column.

5. The symbol "-" must be removed from the parameter delimiters in the declarations of procedures *LINEARSYSTEM*, *RESIDUALS* and *SOLVE*.

6. Four lines above the bottom line of procedure *LINEARSYSTEM*, delete the first occurrence of $X[i,k] :=$

7. In the third line of the heading of procedure *IP2*, the parameter delimiter

) extra therm:(

should be changed to

) extra term:(

REMARK ON ALGORITHM 135 [F4]
CROUT WITH EQUILIBRATION AND ITERATION
[W. M. McKeeman, *Comm. ACM 5* (Nov. 1962), 553–555, 557; *7* (July 1964), 421]

LOREN P. MEISSNER (Recd. 21 Oct. 1964)
Lawrence Radiation Lab., U. of California, Berkeley.

1. The following error in the published algorithm is noted: The procedure *IP1* forms the sum of $A[i, p] \times A[p, k]$; however, two lines above the bottom line of procedure *CROUT* an attempt is made to use *IP1* to form the sum of $A[k, p] \times A[p, j]$.

A possible way of correcting this is to add a procedure *IP1a* which is identical with *IP1* except that k is written for i and j for k. Since the procedure is used often, making the correction in this way is not unreasonable. A more extensive undertaking would be to modify *CROUT* to use a more general procedure such as *INNERPRODUCT* [1].

2. The following comment is made in view of the reference to this algorithm in a recent Editor's Note [2]: In the use of Algorithm 135 as a determinant evaluator, it may be well to set m, the "number of right-hand sides" to 1 instead of zero and give an arbitrary nonzero right-hand side such as $(1, 0, 0, \cdots)$. This will cause a calculation of the "condition," and possibly an exit to *singular*, to call the user's attention to cases in which the determinant is nonsense.

REFERENCES:
1. FORSYTHE, G. E. Crout with Pivoting. Algorithm 16. *Comm. ACM 3* (Sept. 1960), 507.
2. ROTENBERG, L. J. Remark on Revision of Algorithm 41. *Comm. ACM 7* (Mar. 1964), 144.

ALGORITHM 136
ENLARGEMENT OF A GROUP
M. Wells*
University of Leeds, England
* Currently with Burroughs Corporation, Pasadena, California

```
procedure  Enlarge group (G, n, g, Abelian);
   array G, g;  integer n;  Boolean Abelian;
comment  This procedure combines the element g with the sub-
   group G, of n elements, to form a new group. The Boolean
   Abelian has the value true if the group to which G and g belong
   is Abelian. Two procedures, multiply and equal are assumed
   to be declared: multiply (G[i]) by : (G[j]) to give : (G[k]) will set
   the element G_k equal to the product of the elements G_i and G_j.
   equal (G[i], G[j]) is a Boolean procedure whose value is true
   if, and only if, the elements G_i and G_j are equal. On leaving the
   procedure the enlarged group is in G, and n is equal to the
   number of elements in the new sub-group G. The procedure
   will function correctly if g is included in G on entry. It is prob-
   able that g and the elements of G will be arrays, and the pro-
   cedure body will, in practice, need to be altered considerably.
   The procedure has been used successfully in connection with
   problems of space-group theory;
begin integer i, j, k;
   for i := 1 step 1 until n do
   if equal (G[i], g) then go to not new generator;
   n := n + 1;  G[n] := g;
   for i := n step 1 until n do
   begin for j := 1 step 1 until n do
   begin multiply (G[i], G[j], G[n+1]);
      for k := 1 step 1 until n do
      if equal (G[k], G[n+1]) then go to not new element 1;
      n := n + 1;
not new element 1: if Abelian then go to take next element;
   multiply (G[j], G[i], G[n+1]);
   for k := 1 step 1 until n do
   if equal (G[k], G[n+1]) then go to not new element 2;
   n := n + 1;
not new element 2: take next element:
   end of j-loop;
   end of i-loop;
not new generator: end of group enlargement
```

ALGORITHM 137
NESTING OF FOR STATEMENT I
David M. Dahm & M. Wells*
Burroughs Corp., Pasadena, Calif.
* On leave of absence from the University of Leeds, England.

```
procedure Fors 1 (n, P);
  value n;  integer n;  procedure P;
  comment Fors 1 generates a nest of n for statements with the
    procedure P at their center. Two non-local arrays I and U,
    which give the value of the controlled variable and its upper
    bound for each level are assumed to be declared;
  begin integer j;
    if n = 0 then P
    else for j := 1 step 1 until U[n] do
    begin I[n] :=j;  Fors 1 (n−1,P) end end Fors 1
```

ALGORITHM 138
NESTING OF FOR STATEMENT II
David M. Dahm & M. Wells*
Burroughs Corp., Pasadena, Calif.
  * On leave of absence from the University of Leeds, England. .

```
procedure Fors 2 (P);
  procedure P;
  comment Fors 2 performs the same function as Fors 1, but is
      more economic of storage space. It is expected, however,
      that Fors 1 would be more economic of time. The formal
      parameter n is now replaced by the non-local integer n;
  begin if n = 0 then P
    else for I[n] := 1 step 1 until U[n] do
    begin n := n−1;  Fors 2 (P) end;
    n := n + 1 end Fors 2
```

ALGORITHM 139
SOLUTIONS OF THE DIOPHANTINE EQUATION
J. E. L. PECK
University of Alberta, Calgary, Alberta, Canada

procedure Diophantus (a,b,c); **integer** a,b,c;

comment This procedure seeks the integer solutions of the equation $ax + by = c$, where the integers a,b,c are given. It assumes a non-local integer M, which should be as large as storage will allow, two nonlocal labels *INDETERMINATE* and *NO SOLUTION* and two non-local Boolean variables 'general solution' and 'time permits' which are self explanatory. It also assumes the procedures abs, sign and print;

begin integer n,r,s,d,i; **integer array** $q[1:M]$;

$n := i := 0$; $d := s := \text{abs}(a)$; $r := \text{abs}(b)$;

comment d will become the greatest common divisor of a and b. If $b = 0$ then $d = |a|$. The vector q will retain the successive quotients in the Euclidean algorithm $r_{i-1} = r_i q_i + r_{i+1}$, $i = 1, 2, \cdots, n$, where $0 \leq r_{i+1} < r_i$, $r_0 = |a|$, $r_1 = |b|$, and $r_{n+1} = 0$;

for $i := i + 1$ **while** $r \neq 0$ **do**

begin $n := i$; $d := r$; $q[i] := s \div d$;

$r := s - d \times q[i]$; $s := d$ **end** This records the quotients and the number n of divisions for use below;

if $d = 0$ **then go to if** $c = 0$ **then** *INDETERMINATE*

else *NO SOLUTION*; **comment** The case $d = 0$ occurs when $a^2 + b^2 = 0$. If d now does not divide c then the equation cannot be solved so;

if $(c \div d) \times d \neq c$ **then go to** *NO SOLUTION*;

if $d \neq 1$ **then**

begin $a := a/d$, $b := b/d$; $c := c/d$ **end**, which removes the common factor and reduces the equation to the case where a and b are relatively prime;

begin comment We shall now find u_1 and v_1 in order to express

$1 = au_1 + bv_1$, using the relations $r_n = r_i v_i + r_{i-1} u_i$, $i = n, n-1, \cdots, 1, v_n = 1$, $u_n = 0$, and $r_{i+1} = -r_i q_i + r_{i-1}$, $i = n-1, n-2, \cdots, 1$; **integer** u,v;

if $n = 0$ **then**

begin $v := 0$; $u := 1$ **end**, which takes care of the case $b = 0$

else

begin $v := 1$; $u := 0$;

for $i := n-1$ **step** -1 **until** 1 **do**

begin integer t;

$t := v$; $v := u - v \times q[i]$; $u := t$

end i

end the case $n \neq 0$. It remains now to multiply the equality

$1 = au_1 + bv_1$ through by c;

begin integer $x0, y0$;

$x0 := c \times u \times \text{sign}(a)$; $y0 := c \times v \times \text{sign}(b)$; print $(x0,y0)$;

comment If x_0, y_0 is a particular solution then $x_0 \pm ib$, $y_0 \mp ia$, $i=1,2,\ldots$ gives the general solution. Therefore;

if general solution **then**

begin $u := b$; $v := a$;

A: print$(x0 + u, y0 - v)$; print$(x0-u, y0 + v)$;

$u := u + b$; $v := v + a$;

if time permits **then go to** A

end general solution and

end solution.

end u,v

end Diophantus.

CERTIFICATION OF ALGORITHM 139 [A1]
SOLUTIONS OF THE DIOPHANTINE EQUATION
[J.E.L. Peck, *Comm. ACM* 5 (Nov. 1962), 556]
HENRY J. BOWLDEN (Recd. 30 Sept. 1964 and 5 Nov. 1964)
Westinghouse Electric Corp., R&D Ctr., Pittsburgh, Pa.

Algorithm 139 was transcribed into Burroughs Extended ALGOL after the following typographical error was corrected: On the line following "**if** $d \neq 1$ **then**" replace "$a := a/d$," by "$a := a/d$;".

The cases shown in the table were tried, with the results shown in columns 4 and 5. These solutions are correct, but perhaps not too useful. Of course, a definition of "useful" in this context would be rather subjective; in any case, the user can always obtain an arbitrary solution "useful" for his purpose. We have chosen to regard a small value of x as a criterion for usefulness, and obtain this by inserting, just before "*print* $(x0, y0)$", the statements

$c := x0 \div b$; $x0 := x0 - c \times b$; $y0 := y0 + c \times a$;

The following remarks have to do with matters of programming taste rather than accuracy.

(a) A **value** part of form **value** a, b, c; should be inserted to avoid side effects.

(b) The results should be passed back to the calling program for use by the caller. This requires the addition of two call-by-name parameters $(x0, y0)$, and the removal of the declaration **integer** $x0, y0$;. The provisions for printing the results should be omitted.

(c) The procedure contains a deliberate possibility of an infinite loop. This is unacceptable on most operating systems and should be omitted.

(d) The provision of an array (q) "as large as storage will allow" is rather indefinite. The algorithm as given provides no test to prevent exceeding this arbitrary size. The number of partial quotients in the Euclidean algorithm may be shown to be no more than five times the number of decimal digits in the (largest of the) coefficients a, b, c, so a size of five times the number of digits in the largest integer to be considered is sufficient.

The algorithm, modified as suggested above, gives the results in columns 6 and 7 of the table below. The execution time on the B-5000 was approximately 40 milliseconds.

			original		modified	
a	b	c	$x0$	$y0$	$x0$	$y0$
1000	23	1046	−2092	91002	−22	1002
0	0	0	indeterminate			
57	−103	47009	2209423	1222234	73	−416
10	12	578	−289	289	−1	49
10	12	97	no solution			

ALGORITHM 140
MATRIX INVERSION
P. Z. INGERMAN
University of Pennsylvania, Philadelphia, Penn.

```
procedure invert (a) of order:(n) with tolerance:(eps) and
    error exit:(oops);
value n, eps; array a; integer n; real eps; label oops;
comment  This procedure inverts a matrix by using elementary
    row operations. Although the method is not particularly good
    for ill-conditioned matrices, the simplicity of the algorithm
    and the fact that the inversion occurs in place make it useful
    on occasion;
begin integer i;
    for i := 1 step 1 until n do
    begin integer j, k;  real q;
        q := a[i,i];
        if abs(q)≤abs(eps) then go to oops;
        a[i,i] := 1;
        if q≠1 then for k := 1 step 1 until n do a[i,k] := a[i,k]/q;
        for j := 1 step 1 until n do
            if i≠j then
                begin q := a[j,i];  a[j,i] := 0;
                    for k := 1 step 1 until n do
                        a[j,k] := a[j,k]−q×a[i,k] end end end
```

CERTIFICATION OF ALGORITHM 140
MATRIX INVERSION [P. Z. Ingerman, *Comm. ACM*,
Nov. 1962]
RICHARD GEORGE*
Argonne National Laboratory, Argonne, Ill.

* Work supported by the United States Atomic Energy Commission.

Algorithm 140 was tested on the LGP-30, using SCALP, a load-and-go compiler from the Dartmouth College Computation Center, and it was shown to be syntactically correct.

It is indeed a simple procedure. It is so simple because the author has eliminated the very necessary search for largest elements and the row interchanges. As a result, this procedure will fail to invert many non-singular matrices. To be invertable by this procedure, a matrix must be such that all of its leading diagonal submatrices will have non-zero determinants.

One would do well to avoid this algorithm and use one (such as 120) which employs the pivoting process.

ALGORITHM 141
PATH MATRIX
P. Z. INGERMAN
University of Pennsylvania, Philadelphia, Penn.

```
procedure  find path (a, n);
value n;  Boolean array a;  integer n;
comment  This procedure is merely an Algol implementation
  of the method of Warshall (JACM 9(1962), 11–12). Some ad-
  vantage is taken of the characteristics of the problem to in-
  crease the efficiency;
begin integer i, j, k;
  for j := 1 step 1 until n do
  for i := 1 step 1 until n do
  if a[i,j] ∧ i≠j then
  for k := 1 step 1 until n do
  a[i,k] := a[i,k]∨a[j,k] end findpath
```

ALGORITHM 142
TRIANGULAR REGRESSION
W. L. HAFLEY AND J. S. LEWIS
Aluminum Company of America, Pittsburgh, Penn.

```
procedure trireg (n, nob, dep, pmax);
  real pmax;  integer n, nob, dep;
  comment trireg is a multiple regression procedure which
    develops and inverts only the upper triangular portion of a
    correlation matrix of order n. The i,jth (i ≦ j) matrix element
    is r(ci+j) where the c's are cram numbers (ref. Algorithm 67,
    J. Caffey, Comm. ACM 4, July 1961). dep < n dependent
    variables are regressed simultaneously. Read (u) is an input
    procedure for single elements. The input consists of nob ob-
    servations on n variables. The first dep variables are con-
    sidered dependent and the remaining n − dep are considered
    independent variables. Independent variables are dropped
    when the pivotal element exceeds pmax during the inversion.
    Total variable storage is 14 + 3n + n(n+1)/2;
begin integer i1, i2, i3, c1, c2, c3, df;  integer array c[1:n];
  real d, p, a; real array r[1:n(n+1)/2], v[1:n], m[1:n];
initial:  df := 0; for i1 := 1 step 1 until n do m[i1] := 0;
          for i1 := 1 step 1 until n(n+1)/2 do r[i1] := 0;
input:    for i1 := 1 step 1 until nob do
          begin for i2 := 1 step 1 until n do Read (v[i2]);
            c1 := 0;  for i2 := 1 step 1 until n do
              begin d := v[i2];  m[i2] := m[i2] + d;
                for i3 := i2 step 1 until n do
                  begin c1 := c1 + 1;  r[c1] := r[c1] + v[i3] × d end
              end i2;
          end i1;
correlation:  c1 := 1;  a := 1/nob;  for i1 := 1 step 1 until n do
              begin v[i1] := 1/sqrt(r[c1] − (m[i1]↑2)×a);
                r[c1] := 1;  c1 := c1 + n − i1
              end i1;
              c1 := 1;  for i1 := 1 step 1 until n do
              begin d := a × m[i1];  p := v[i1];  c1 := c1 + 1;
                m[i1] := d;
                for i2 := i1 + 1 step 1 until n do
                begin r[c1] := (r[c1]−d×m[i2]) × v[i2] × p;
                end i2;
              end i1;
              comment variable i may be dropped from the
                regression by setting vi = 0 and df equal to the
                number of variables dropped;
cram:  i1 := − n;  i2 := n + 1;  for i3 := 1 step 1 until n do
          begin i1 := i1 + i2 − i3;  c[i3] := i1
          end i1;
inversion:  for i1 := dep + 1 step 1 until n do
            begin c1 := c[i1];  if v[i1] ≠ 0 then
              begin p := 1/r[c1+i1];  if p > pmax then
              begin df := df + 1;  go to YY end else
                begin r[c1+i1] := p;  for i2 := 1 step 1
                    until i1 − 1 do
                  begin c2 := c[i2];  a := p × r[c2+i1];
                    for i3 := i2 step 1 until n while i3 ≠ i1 do
                    begin if i3 < i1 then
                      begin c3 := c[i3];  d := r[c3+i1] end
                      else d := −r[c1+i3];
                      r[c2+i3] := r[c2+i3] + d × a
                    end i3;
                  end i2;
                for i2 := i1 + 1 step 1 until n do
                begin a := p × r[c1+i2];  c2 := c[i2];
                  for i3 := i2 step 1 until n do
                  r[c2+i3] := r[c2+i3] − a × r[c1+i3];
                end i2;
            ZZ: for i2 := 1 step 1 until i1 − 1 do
                begin c2 := c[i2+i1];  r[c2] := − p × r[c2]
                end i2;
                for i2 := c1 + i1 + 1 step 1 until n + c1 do
                r[i2] := p × r[i2]
              end
            end else
            YY: begin p := 0;  r[c1+i1] := 0;  go to ZZ end
coeff:  d := 1/(nob−n+dep−l+df);  for i1 := 1 step 1 until
          dep do
        if v[i1] ≠ 0 then
        begin a := 0;  p := 1/v[i1];  c1 := c[i1];  for i2 := dep
          +1 step 1 until n do
          begin if r[i2] ≠ 0 then
            begin r[c1+i2] := −r[c1+i2] × v[i2] × p;  a :=
              a + r[c1+i2] × m[i2]
            end
          end i2;
          v[i1] := (2−r[c1+i1]) × d/(v[i1]↑2)
          comment: v[1:dep] now contains the mean square
            deviations from regressions for the dependent vari-
            ables. The coefficients of determination R² may be
            obtained as r[c1+i1] − 1;
          r[c1+i1] := m[i1] − a else
          begin c1 := c[i1];  for i2 := c1 + i1 step 1 until
            c1 + n do r[i2] := 0
          end
        end
  comment The r-array now contains the constants and coeffi-
    cients of regression, and the inverse of the correlation matrix of
    the independent variables that have been kept. The following
    example will help to locate the information in the r array.
  Example:  n = 6  dep = 3
```

$$
\begin{array}{ccc|ccc}
r_1 & r_2 & r_3 & r_4 & r_5 & r_6 \\
r_7 & r_8 & r_9 & r_{10} & r_{11} \\
r_{12} & r_{13} & r_{14} & r_{15}
\end{array}
\qquad
\begin{array}{c}
b_{01} \\ b_{02} \\ b_{03}
\end{array}
\qquad
\begin{array}{ccc}
b_{11} & b_{21} & b_{31} \\
b_{12} & b_{22} & b_{32} \\
b_{13} & b_{23} & b_{33} \\ \hline
r^{11} & r^{12} & r^{13} \\
 & r^{22} & r^{23} \\
 & & r^{33}
\end{array}
$$

Below the main block (left) are $r_{16}\ r_{17}\ r_{18}$; $r_{19}\ r_{20}$; r_{21}.

The variances and covariances of the regression coefficients for
the jth dependent variable can be determined by—

$$\text{Var}\,(b_{ij}) = r^{ii} \times v_j \times v_i^2$$
$$\text{Covar}\,(b_{ij}b_{kj}) = r^{ik} \times v_j \times v_i \times v_k;$$

```
end trireg
```

ALGORITHM 143
TREESORT 1
Arthur F. Kaupe, Jr.
Westinghouse Electric Corp., Pittsburgh, Penn.

```
procedure  TREESORT 1 (UNSORTED, n, SORTED, k);
  value n, k;
integer n, k;  array UNSORTED, SORTED;
comment  TREESORT 1 is a revision of TREESORT (AL-
  GORITHM 113) which requires neither the "packed" array m
  nor the machine procedures pack, left half, right half, and mini-
  mum. The identifier infinity is used as nonlocal real variable
  with value greater than any element of UNSORTED;
begin integer i, j;  array m1 [1:2×n−1];
  integer array m2 [1:2×n−1];
procedure  minimum;  if m1 [2×i] ≦ m1[2×i+1] then
  begin m1[i] := m1[2×i];  m2[i] := m2[2×i] end else
  begin m1[i] := m1[2×i+1];  m2[i] := m2[2×i+1] end mini-
    mum;
for i := n step 1 until 2 × n − 1 do begin m1[i] := UNSORTED
  [i−n+1];  m2[i] := i end
for i := n − 1 step −1 until 1 do minimum;
for j := 1 step 1 until k do
  begin SORTED [j] := m1[1];  i := m2[1];  m1[i] := infinity;
    for i := i ÷ 2 while i > 0 do minimum end
end TREESORT 1
```

ALGORITHM 144
TREESORT 2
Arthur F. Kaupe, Jr.
Westinghouse Electric Corp., Pittsburgh, Penn.

```
procedure  TREESORT 2 (UNSORTED, n, SORTED, k, ordered);
  value n, k;
integer n, k;  array UNSORTED, SORTED;  Boolean proce-
  dure ordered;
comment  TREESORT 2 is a generalized version of TREESORT
  1. The Boolean procedure ordered is to have two real argu-
  ments. The array SORTED will have the property that ordered
  (SORTED[i], SORTED[j]) is true when j > i if ordered is a
  linear order relation;
begin integer i, j;  array m1 [1:2×n−1];  integer array m2
  [1:2×n−1];
procedure minimum;  if ordered (m1[2×i], m1[2×i+1]) then
  begin m1[i] := m1[2×i];  m2[i] := m2[2×i] end else
  begin m1[i] := m1[2×i+1];  m2[i] := m2[2×i+1] end mini-
    mum;
for i := n step 1 until 2 × n − 1 do begin m1[i] := UNSORTED
  [i−n+1];  m2[i] := i end
for i := n − 1 step −1 until 1 do minimum;
for j := 1 step 1 until k do
  begin SORTED[j] := m1[1];  i := m2[1];  m1[i] := infinity;
    for i := i ÷ 2 while i > 0 do minimum end
end TREESORT 2
```

ALGORITHM 145

ADAPTIVE NUMERICAL INTEGRATION BY SIMPSON'S RULE

William Marshall McKeeman*

Stanford University, Stanford, Calif.

* This work was supported in part by the Office of Naval Research under contract Non4 225(37).

```
real procedure  Integral (F) limits: (a, b) tolerance: (eps);
real procedure  F;  real a, b, eps;
begin  comment Integral will numerically approximate the
    integral of the function F between the limits a and b by the
    application of a modified Simpson's rule. Although eps is a
    measure of the relative error of the result, the actual error
    may be very much larger (e.g. whenever the answer is small
    because a positive area cancelled a negative area). The pro-
    cedure attempts to minimize the number of function evalua-
    tions by using small subdivisions of the interval only where
    required for the given tolerance;
  integer level;
  real procedure Simpson (F, a, da, Fa, Fm, Fb, absarea,est,eps);
  real procedure F;  real a, da, Fa, Fm, Fb, absarea, est, eps;
  begin comment Recursive Simpson's rule;
    real dx, x1, x2, est1, est2, est3, F1, F2, F3, F4, sum;
    dx := da/3.0;  x1 := a + dx;  x2 := x1 + dx;
    F1 := 4.0 × F(a+dx/2.0);  F2 := F(x1);
    F3 := F(x2);  F4 := 4.0 × F(a+2.5×dx);
    est1 := (Fa+F1+F2) × dx/6.0;
    est2 := (F2+Fm+F3) × dx/6.0;
    est3 := (F3+F4+Fb) × dx/6.0;
    absarea := absarea-abs(est) + abs(est1) + abs(est2) + abs(est3);
    sum := est1 + est2 + est3;
    level := level + 1;
    Simpson := if (abs(est-sum) ≤ eps × absarea ∧ est ≠ 1.0) ∨
        level ≥ 7 then sum
      else Simpson (F, a, dx, Fa, F1, F2, absarea, est1, eps/3.0)
        + Simpson (F, x1, dx, F2, Fm, F3, absarea, est2, eps/3.0)
        + Simpson (F, x2, dx, F3, F4, Fb, absarea, est3, eps/3.0);
    level := level −1;
  end Simpson;
  level := 1;
  Integral := Simpson (F, b-a, F(a), 4.0 × F((a+b)/2.0), F(b),
        1.0, 1.0, eps)
end Integral 13
```

CERTIFICATION OF ALGORITHM 145

ADAPTIVE NUMERICAL INTEGRATION BY SIMPSON'S RULE [W. McKeeman. Comm. ACM, Dec. 1962]

Wm. M. McKeeman

Stanford University, Stanford, Calif.

Suggested changes in the code:

1. Replace all occurrences of eps/3.0 by eps/1.7.
2. Replace level ≥ 7 by level ≥ 20.
3. The second parameter a in the final call of Simpson was omitted; insert it.

With the above cnanges, a BALGOL translation of Integral nas been tested successfully on a large number of functions. An example of its behavior is given below:

Machine: Burroughs 220, 8 decimal digit floating-point mantissa.

$f(x) = 1.0/sqrt(abs(x))$; which has a pole at the origin.

$a = -9.0$; $b = 1000.0$; correct answer = 206.0;

eps	computer answer	relative error
0.1	200.22251	0.028
0.01	206.00226	0.0000107
0.001	206.00092	0.0000045
0.0001	205.99985	0.0000007

If the recursion was allowed to go thirty levels deep we found:

0.0001	206.00005	0.0000002

The graph below shows the adaptive clustering of the points of evaluation around the pole of the function (taken from the first example above).

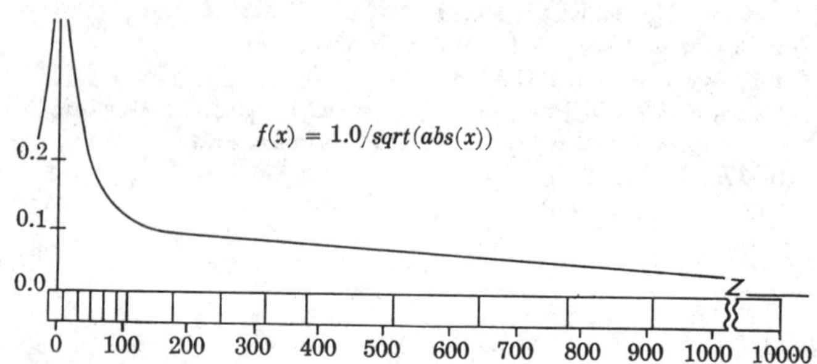

Each vertical line represents a point of evaluation for the function during the execution of the call:

integral$(f, -9.0, 10000.0, 0.1)$;

REMARK ON ALGORITHM 145 [D1]

ADAPTIVE NUMERICAL INTEGRATION BY SIMPSON'S RULE [William Marshall McKeeman, Comm. ACM 6, (Dec. 1962), 604]

M. C. Pike (Recd. 5 Oct. 1964 and 23 Nov. 1964)

Statistical Research Unit of the British Medical Research Council, University College Hospital Medical School, London, United Kingdom

This procedure was tested on the ICT Atlas computer and found satisfactory after the following three modifications were made:

(1) add "**real** absarea;" on the line following "**integer** level;",
(2) add "absarea := 1.0;" on the line following "level := 1;",
(3) substitute

"Integral := Simpson (F, a, b−a, F(a), 4.0×F((a+b)/2.0),
        F(b), absarea, 1.0, eps)"

for
>>*"Integral := Simpson (F, b−a, F(a), 4.0×F((a+υ)/2.0), F(b),*
>>   *1.0, 1.0, eps)"*.

These corrections are necessary since *absarea* appears on the left-hand side of an assignment statement, namely, in line 10 of the **real procedure** *Simpson*, and yet when *Simpson* is called in the third to last line of the **real procedure** *Integral* the actual parameter for *absarea* is given as 1.0.

The author wishes to thank the referee for helpful suggestions.

ALGORITHM 146
MULTIPLE INTEGRATION
WILLIAM MARSHALL MCKEEMAN*
Stanford University, Stanford, Calif.
    * This work was supported in part by the Office of Naval Research under contract Non4 225(37).

```
real procedure  MultipleIntegral (F) limits: (a, b) order: (n)
  tolerance: (eps);
real procedure  F;  real array a, b;  real eps;  integer n;
begin  comment  F is a function of n variables which are stored
    in an internal array x. MultipleIntegral approximates the
    multiple integral of F between the n pairs of limits stored in
    the parameter arrays a and b. For a mesh of k steps on each
    axis, the number of function evaluations required for an
    integral of nth order is approximately k↑n. One consequence
    is that the practical limit on n is quite small. Another is that
    any inefficiency in the (undefined) procedure Integral will
    reflect itself to the nth power in MultipleIntegral. The adap-
    tive procedure Integral is recommended;
  real array x[1:n+1];  integer axis;
  real procedure Integral (F) limits: (a, b) tolerance: (eps);
  real procedure F;  real a, b, eps;
  begin  comment  The body of procedure Integral is left
      undefined. For it one may substitute any procedure of the
      same name that evaluates the integral of a function of a single
      variable between the real limits a and b;
  end Integral;
  real procedure MI(y);  real y;
  begin comment Recursive multiple integration;
    x[axis] := y;
    axis := axis −1;
    MI := if axis = 0 then F(x) else
      Integral (MI, a[axis], b[axis], eps/n);
    axis := axis +1;
  end MI;
  axis := n + 1;
  MultipleIntegral := MI(0)
end MultipleIntegral
```


CERTIFICATION OF ALGORITHM 146
MULTIPLE INTEGRATION [W. M. McKeeman,
    Comm. ACM 5 (Dec. 1962), 604]
NIKLAUS WIRTH (Recd. 6 Jan. 1964)
Computer Science Div., Stanford U., Stanford, Calif.

Algorithm 146 was translated into a generalized ALGOL [cf. N. Wirth, A generalization of ALGOL, *Comm. ACM 6* (Sept. 1963), 547–554] and successfully run on the Stanford IBM 7090 computer. Algorithm 60, Romberg Integration [*Comm. ACM 4* (June 1961), 255; *5* (Mar. 1962), 168; *5* (May 1962), 281] was used for the real procedure *Integral*.

The main disadvantage of Algorithm 146 is that the bounds of the domain of integration must be constant, i.e. the domain must always have the form of a rectangular hyperbox.

ALGORITHM 147
PSIF
D. Amit
Ministry of Defense, Israel

real procedure $psif(x, a, tan, ln)$ exit: $(errexit)$;
value x, a; **label** $errexit$; **real procedure** tan, ln;

comment Computes the logarithmic derivative of the factorial function defined by:

$$\Psi(x) = \frac{(x!)'}{x!} = \frac{\Gamma'(x+1)}{\Gamma(x+1)}.$$

We make use of the expansion: (1) $\Psi(x) = lnx + 1/2x - 1/12x^2 + 1/120x^4 - 1/252x^6 + \epsilon$, (2) $\epsilon < 1/240x^2$ and of the recursion relation, (3) $\Psi(x) = \Psi(x+n) - (1/(x+1) + \ldots + 1/(x+n))$. For $x < -1$ we use: (4) $\Psi(-x) = \pi tan\ \pi(x+0.5) + \Psi(x-1)$. The value of x is increased up to a. Then Ψ is calculated by (3) and (1). The error is then less than $1/240a^8$;

begin real psi, pei; $psi := 0$;
if $x > -1 \wedge x \neq 0$ **then go to** pos;
if $x = 0$ **then begin** $psi := -0.5772156649$; **go to** $exit$ **end**;
begin integer $x1$; $x1 := x$;
if $x = x1$ **then go to** $errexit$ **end**
comment psi is infinite;
$pei := 3.141592654$; $x := -x - 1$;
$psi := pei \times tan(pei \times (x+0.5))$;
$pos:$ **if** $x \geq a$ **then go to** $large$;
$x := x + 1$; $psi := psi - 1/x$; **go to** pos;
$large:$ **begin real** y; $y := 1/x$;
$psi := psi + ln(x) + y/2 - y \uparrow 2/12 + y \uparrow 4/120 - y \uparrow 6/252$;
$exit:$ $psif := psi$;
end $psif$

CERTIFICATION OF ALGORITHM 147
PSIF [D. Amit, *Comm. ACM.*, Dec. 62]
Henry C. Thacher, Jr.*
Reactor Eng. Div., Argonne National Lab., Argonne, Ill.

* Work supported by the U. S. Atomic Energy Commission.

The following minor errors were noted in this algorithm:
a. (3) in the comment should read $\epsilon < 1/240\ x^8$.
b. The function tan is not a standard ALGOL function. It should be declared, or replaced by $sin(\ )/cos(\ )$.
c. The block labelled *large* should be closed by inserting **end** immediately after 252.

The efficiency of the program would be improved by the following modifications:
a. Let the statement

if $x = 0$ **then begin** ... **end**;

be the first statement of the procedure body.
b. Delete the condition $x \neq 0$ from the if clause,

if $x > -1 \wedge x \neq 0$ **then** ...

c. Delete the declaration of pei, and the assignment of the value of 3.141592654 to pei in the statement

$psi := pei \times sin(pei \times (x+0.5))/cos(pei \times (x+0.5))$;
replace pei by the value 3.141592654.
d. Replace the block labelled large by:
$large:$ **begin real** y; $x := 1/x$; $y := x \times x$;
$psi := psi - ln(x) + x/2 - ((y/252 - 0.008333333333) \times y + 0.08333333333) \times y$ **end**;

With these changes, the body of the procedure was translated and run on the LGP-30 computer using the Dartmouth SCALP processor. The program was used to tabulate $psif(x)$ for $x = -1$ (0.5)0(0.005)1.250. With $a = 3.0$ the results agreed with tabulated values to within 3 in the 6th decimal place. This is considered satisfactory, since one decimal place is lost in applying the recurrence. Running time, including output on the Flexowriter and computation of new values of the independent variable, averaged about 30 seconds per value.

It should be observed that $psif(x)$ is $\Psi(x+1)$ as tabulated, for example, by Jahnke-Emde-Losch.

CERTIFICATION OF ALGORITHM 147 [S14]
PSIF [D. Amit, *Comm. ACM* **5** (Dec. 1962), 605]
Ronald G. Parsons* (Recd. 7 Dec. 1966 and 5 Aug. 1969)
Stanford Linear Accelerator Center, Stanford University, Stanford, CA 94305

* Present address: Department of Physics, The University of Texas, Austin, TX 78712. Work supported by the US Atomic Energy Commission.

KEY WORDS AND PHRASES: gamma function, logarithmic derivative, factorial function, psi function
CR CATEGORIES: 5.12

The following errors were noted in this algorithm in addition to those noted by Thacher [2].
a. (4) in the comment should read "For $-x < -1$ we use: (4) $\Psi(-x) = \Psi(x-1) + \pi \cot (\pi x)$".
b. At the end of the first comment add: "Note that $psif(x) \equiv \Psi(x)$ is $\psi(x+1)$ as defined, for example, by Jahnke-Emde-Lösch" (see [1]).
c. The statement in the algorithm before the label pos should read: $psi := pei \times cos(pei \times x)/sin(pei \times x)$; These errors caused the procedure to give incorrect results for $psif(x, a)$ for $x < -1$.
d. The arguments tan and ln should be deleted from the parameter list and **real procedure** tan, ln; should be deleted from the specification part of the procedure heading.

With these changes and those of Thacher, the procedure was translated into Burroughs B5500 extended ALGOL and run on the Stanford B5500. $psif(x, a)$ was tabulated for $x = -2.9(0.1)5.0$ with $a = 8.0$ The results agreed with tabulated values to within $1/(240a^8)$.

REFERENCES:
1. Jahnke-Emde-Lösch. *Tables of Higher Functions* (6th Ed.). McGraw-Hill, New York, 1960.
2. Thacher, H. C., Jr. Certification of Algorithm 147. *Comm. ACM 6* (Apr. 1963), 168.

ALGORITHM 148
TERM OF MAGIC SQUARE
D. M. Collison
Elliott Brothers (London) Ltd., Borehamwood, Herts.

```
integer procedure magicterm (x, y, n);  value x, y, n;  integer
    x, y, n;
comment  for the magic square s[1:n, 1:n], magicterm generates
    the element s[x, y], where n > 2 and n is odd. De la Loubère's
    method is used;
begin integer b, c;
    b := y − x + (n−1) ÷ 2;  c := y + y − x;
    if b ≧ n then b := b − n else if b < 0 then b := b + n;
    if c > n then c := c − n else if c ≦ 0 then c := c + n;
    magicterm := b × n + c
end magicterm
```


CERTIFICATION OF ALGORITHM 148
TERM OF MAGIC SQUARE [D. M. Collinson, *Comm.
    ACM*, Dec. 1962]
J. N. R. Barnecut
University of Alberta, Calgary; Calgary, Alberta, Canada

MAGICTERM was translated into Fortran for the IBM 1620.
The procedure was tested for terms of squares up to order 13.
Correct results were obtained. For determination of complete
squares operating time was not significantly different from Algorithm 118.

CERTIFICATION OF ALGORITHM 148
TERM OF MAGIC SQUARE [D. M. Collison, *Comm.
    ACM*, Dec. 62]
Dmitri Thoro
San Jose State College, San Jose, Calif.

This algorithm was translated into Fortran and Forgo for the
IBM 1620. No changes in the program were necessary. The elements of magic squares of odd orders up to 15 were generated
satisfactorily.

ALGORITHM 149
COMPLETE ELLIPTIC INTEGRAL
J. N. MERNER
Burroughs Corp., Pasadena, Calif.

comment The following two procedures, along with a test program were compiled and run by Peter Naur on the DISADEC computer. Compilation time for the 9 pass compiler was less than 10 seconds. The elliptic integral of the form

$$\int_0^{\pi/2} \frac{dt}{\sqrt{a^2 \cos^2 t + b^2 \sin^2 t}}$$

is evaluated by replacing a and b by their arithmetic and geometric means, respectively. *ELIP 2* is a nonrecursive procedure to accomplish the same thing;

real procedure *ELIP 1* (a, b); **value** a, b; **real** a, b;
  ELIP 1 $:=$ **if** $abs(a-b) <_{10} -8 \times a$
    then $3.14159265/2/a$
    else *ELIP 1* $((a+b)/2, sqrt (a \times b))$;
real procedure *ELIP 2* (a, b); **value** a, b; **real** a, b;
  begin real C;
  $L:$ $C := (a+b)/2$; $b := sqrt (a \times b)$; $a := c$;
  if $abs(a-b) <_{10} -8 \times a$ **then** *ELIP 2* $:= 3.14159265/2/a$
  else go to L **end**

CERTIFICATION OF ALGORITHM 55
COMPLETE ELLIPTIC INTEGRAL OF THE FIRST
    KIND [John R. Herndon, *Comm. ACM*, Apr. 1961]
    and
CERTIFICATION OF ALGORITHM 149
COMPLETE ELLIPTIC INTEGRAL [J. N. Merner,
  Comm. ACM, Dec. 1962]
HENRY C. THACHER, JR.*
Reactor Eng. Div., Argonne National Laboratory,
Argonne, Ill.

  * Work supported by the U.S. Atomic Energy Commission.

The bodies of Algorithm 55 and of the second procedure of Algorithm 149 were tested on the LGP-30 computer using SCALP, the Dartmouth "LOAD-AND-GO" translator for a substantial subset of ALGOL 60. The floating-point arithmetic for this translator carries 7+ significant digits.

In addition to modifications required because of the limitations of the SCALP subset, the following need correction:
  In Algorithm 55:
    1. The constant 0.054555509 should be 0.054544409.
    2. The function *log* should be *ln*.
  In procedure ELIP 2 of Algorithm 149, the statement $a := c$ should be $a := C$.

The parameters of Algorithm 149 are related to the complete elliptic integral of the first kind by: $K = a \times ELIP(a, b)$ where the parameter $m = k^2 = 1 - b/a$.

The maximum approximation error in Algorithm 55 is given by Hastings as about $0.6_{10}-6$. In addition there is the possibility of serious cancellation error in forming the complementary parameter $t = 1 - k \times k$. For k near 1, errors as great as 4 significant digits were sustained. In these regions, the complementary parameter itself is a far more satisfactory parameter.

The accuracy obtainable with Algorithm 149 is limited only by the arithmetic accuracy and the amount of effort which it is desired to expend. Six-figure accuracy was obtained with 5 applications of the arithmetic-geometric mean for $a = 1000$, $b = 2$, and with one application for $a = 500$, $b = 500$.

Neither algorithm is satisfactory for $k = 1$. The behavior for Algorithm 55 will be governed by the error exit from the logarithm procedure. Under these circumstances, Algorithm 149 goes into an endless loop. Algorithm 149 may also go into an endless loop of the terminating constant ($_{10}-8$ in the published algorithm) is too small for the arithmetic being used. For the SCALP arithmetic it was found necessary to increase this tolerance to $5.0_{10}-7$. The resulting values of the elliptic integrals were, however, accurate to within 2 in the 7th significant digit (6th decimal).

The relative efficiency of the two algorithms will depend strongly on the efficiency of the square-root and logarithm subroutines. With most systems, Algorithm 55 will provide sufficient accuracy, and will be more efficient. If a square-root operation or a highly efficient square-root subroutine is available, Algorithm 149 may well be the better method.

ACM Transactions on Mathematical Software, Vol. 4, No. 1, March 1978, Page 95.

REMARK ON ALGORITHM 149

Complete Elliptic Integral [S21]
[J.N. Merner, *Comm. ACM 5*, 12 (Dec. 1962), 605]

Ove Skovgaard [Recd 18 October 1976 and 14 February 1977]
Institute of Hydrodynamics and Hydraulic Engineering, Technical University of Denmark, DK-2800 Lyngby, Denmark

The text following the colon at the end of the fifth paragraph in [4] should read as follows: $K = a \times ELIP1(a,b)$ or $K = a \times ELIP2(a,b)$, where $m = k^2 = 1 - (b/a)^2$.

  A better procedure is given in [1, p. 86, *procedure cel*1]. This procedure can for some computers be made slightly more efficient by eliminating the last assignment statement $m := m/2$ in the loop, replacing the second assignment statement $m := kc + m$ with $m := (kc + m) \times .5$ and replacing the last assignment statement $cell := pi/m$ with $cell := (pi/2)/m$. Note the variable m should not be confused with the parameter $m = k^2$.

A more efficient, but less portable procedure is defined in [2] and implemented, for example, in the FUNPACK package [3].

REFERENCES

1. BULIRSCH, R. Numerical calculation of elliptic integrals and elliptic functions. *Numer. Math. 7* (1965), 78–90. Prepublication for the planned volume of *Special Functions of the Handbook for Automatic Computation*, Springer, Berlin.
2. CODY, W.J. Complete elliptic integrals. In Hart, J.F., et al., *Computer Approximations*. Wiley, New York, 1968, pp. 150–154 and pp. 335–339.
3. CODY, W.J. The FUNPACK package of special function subroutines. *ACM Trans. Mathematical Software 1*, 1 (March 1975), 13–25.
4. THACHER, H.C., JR. Certification of Algorithm 149: Complete elliptic integral. *Comm. ACM 6*, 4 (April 1963), 166–167.

ALGORITHM 150
SYMINV2
H. Rutishauser
Eidg. Technische Hochschule, Zurich, Switzerland

```
procedure syminv2(a,n) result: (a) exit: (fail); value n; in-
    teger n; array a; label fail;
    comment  syminv2 obtains inverse of a symmetric matrix a of
        order n by a method which is similar to that given by Busing
        and Levy [Comm. ACM 5 (1962), 446] but requires no inter-
        changes of rows and columns nor storage space for an ad-
        ditional matrix Q, yet is numerically equivalent. The pro-
        cedure requires the upper triangular part of a to be given and
        overwrites it by the upper triangular part of the inverse which
        is again denoted by a. All pivots are chosen on the diagonal,
        and if all further diagonal elements which are eligible as
        pivots vanish (this is impossible for a positive definite matrix
        a) then exit through fail occurs;
    begin
        real bigajj;
        integer i, j, k;
        real array p, q[1:n];
        Boolean array r[1:n];
        for i := 1 step 1 until n do r[i] := true;
grand loop:
        for i :=1 step 1 until n do
        begin
search for pivot:
            bigajj := 0;
            for j := 1 step 1 until n do
            begin
                if r[j] ∧ abs(a[j,j]) > bigajj then
                begin
                    bigajj := abs(a[j,j]);
                    k := j
                end;
            end;
            if bigajj = 0 then go to fail;
preparation of elimination step i:
            r[k] := false;
            q[k] := 1/a[k,k];
            p[k] := 1;
            a[k,k] := 0;
            for j := 1 step 1 until k−1 do
            begin
                p[j] := a[j,k];
                q[j] := (if r[j] then −a[j,k] else a[j,k]) × q[k];
                a[j,k] := 0
            end;
            for j := k+1 step 1 until n do
            begin
                p[j] := if r[j] then a[k,j] else −a[k,j];
                q[j] := −a[k,j] × q[k];
                a[k,j] := 0
            end;
elimination proper:
            for j := 1 step 1 until n do
                for k := j step 1 until n do
                    a[j,k] := a[j,k] + p[j] × q[k]
        end grand loop
end syminv2
```

REMARK ON ALGORITHM 150

SYMINV2 [H. Rutishauser, Comm. ACM, Feb. 1963]
Arthur Evans, Jr.
Carnegie Institute of Technology, Pittsburgh, Pennsyl-
vania

The identifier "a" appears twice in the procedure heading as
a formal parameter. It is not clear that this situation has any
meaning in Algol. Indeed, it is not at all obvious how one might
translate the procedure. If the actual parameters corresponding
to the two formal parameters with the same identifier are different
there is no way for the translator (or for the reader) to distinguish
which 'a' is to be used. Further, it would take a detailed examination of the published algorithm to determine how this situation
might be corrected. It is certainly not clear that it would be safe
merely to delete one occurrence of the formal parameter 'a', since
the operation of the algorithm might require that two separate
matrices be available.

REMARK ON ALGORITHM 150

SYMINV2 [H. Rutishauser, Comm. ACM, Feb. 1963]
H. Rutishauser
Eidg. Technische Hochschule, Zurich, Switzerland

procedure *syminv* 2 (a, n) result : (a) exit : (fail); ··· indicates that the value of parameter "a" is changed by the computing process (the matrix a is changed into its inverse, whereby
the given matrix is destroyed). In any procedure call, the two
actual parameters corresponding to the two a's must be identical,
otherwise the action of the procedure will be undefined (by virtue
of the substitution rule). The user may also change the procedure
heading into *syminv* 2 (a, n) exit : (fail); ··· without changing
the effect of the procedure.

Editor's Note: The ALCOR group has adopted the rule that
if the value of a parameter is changed by the execution of the
procedure, then the parameter should be listed twice. Although
the Algol 60 Report does not forbid listing a formal parameter
twice, it would appear that a compiler which thus restricts the
language could not accept some of the examples given in the
Algol 60 Report.

CERTIFICATION OF ALGORITHM 150
SYMINV2 [H. Rutishauser, *Comm. ACM 6* (Feb. 1963), 67]

PETER NAUR (Recd 27 Sept. 63)
Regnecentralen, Copenhagen, Denmark

Since the translator refuses to run programs with more than one occurrence of the same identifier in a formal parameter list, the second *a* was taken out when this procedure was run with the GIER ALGOL system [cf. also the discussion in *Comm. ACM 6* (July 1963), 390]. Otherwise it ran smoothly. For testing the accuracy, segments of the Hilbert matrix were inverted and the results multiplied by the original segment and compared with the unit matrix. The largest deviation in any element was found to be:

Order	Max. deviation from elements of the unit matrix	Order	Max. deviation from elements of the unit matrix
2	$-1.49_{10}-8$	6	$-7.32_{10}-3$
3	$-2.38_{10}-7$	7	$-3.59_{10}-1$
4	$-1.53_{10}-5$	8	$-2.95_{10}\,1$
5	$-3.36_{10}-4$	9	$-1.25_{10}\,1$

These figures may be compared directly with the ones related to Algorithms 120, *INVERSION II*, and *gjr* [*Comm. ACM 6* (Aug. 1963), 445]. A comparison shows that all three algorithms yield about the same accuracy, with *syminv2* being the best in most cases, however. This is not too surprising since the knowledge that the matrix is symmetric ought to simplify the calculation considerably.

The lengths of the three procedures after translation are as follows:

	Number of GIER words
syminv2	216
INVERSION II	279
gjr	302

Execution times for *syminv2* in GIER ALGOL are:

Order	Time (sec)
5	1
10	3.5
15	10.5
20	23

This is about half the time of execution of *INVERSION II* or *gjr*.

ALGORITHM 151

LOCATION OF A VECTOR IN A LEXICO-
   GRAPHICALLY ORDERED LIST

Henry F. Walter
United States Steel Corp., Applied Research Laboratory,
  Monroeville, Penn.

```
integer procedure LOCATE (min, n, c, v, combinatorial);
value v;  integer min, n, c;  integer array v;
integer procedure combinatorial;
comment  This procedure locates the position, LOCATE, of a
   given vector in a list of vectors without searching the list. The
   list consists of all the combinations of n consecutive digits taken
   d at a time. Min is the smallest of the n integers. Each vector
   (combination) is written in ascending order from left to right,
   as, for example, 378 and the vectors are listed lexicographically,
   by which is meant, that, considered as d digit numbers, the
   vectors are listed in ascending order. For example, with min = 1,
   d = 3, and n = 6, the vectors in order are 123, 124, 125, 126, 134,
   135, ..., 456. Given the vector, v = 356, the procedure locates
   this vector as the 19th in the list;
begin integer i, r, max, part, whole;
   r := 1;  v [0] := min − 1;  max := min − 1+n;
   for i := 0 step 1 until c−1 do
   begin part := c − i − 1;
     ask:  if v[i+1] − v[i] > 1 then
       begin whole := max − v[i] − 1;
         r := r + combinatorial (whole, part);
         v[i] := v[i] + 1;
         go to ask
       end;
   end;
   locate := r
end;
```

ALGORITHM 152
NEXCOM
JOHN HOPLEY
Peat, Marwick, Mitchell & Co., London, England

```
procedure nexcom (char, N, setcomplete, nullvector);
array char;  integer N;
label setcomplete, nullvector;
comment  char is a column vector containing N elements each of
    which is either 1 or 0. Nexcom transforms char into another
    vector containing the same number of 1's and 0's, but in a differ-
    ent sequence. Starting with char in the state of having 1 in each of
    the element positions 1, ..., r and zeros elsewhere then repeated
    application of nexcom generates all nCr patterns of char. The
    procedure terminates if the presented vector char has 1 in each
    of the positions N, N−1, ... N−r+1 and zeros elsewhere. Termi-
    nation is indicated by exit through the formal label 'setcomplete'.
    If char is the null vector then procedure exists through the
    formal label 'nullvector';
begin integer n, p, m;
comment  find the first 1 in char;
for n := 1 step 1 until N do if
char [n] = 1 then go to A;
go to nullvector;
comment  how many adjacent 1's;
A:  p := 0;
for m := n + 1 step 1 until N do
if char [m] = 1 then p := p + 1 else go to B;
comment  Have all combinations been generated;
B:  if p + n = N then go to setcomplete;
comment  Set up next combination; char[n+p+1] := 1;
for m := n + p step − 1 until n do char [m] := 0;
for m := 1 step 1 until p do char [m] := 1;
end nexcom;
```

ALGORITHM 153
GOMORY
F. L. Bauer
Johannes Gutenberg-Universität, Mainz, Germany

```
procedure Gomory (a, m, n) result: (a) exit: (no solution);
  value m, n;
  integer m, n;
  integer array a;
  label no solution;
comment  Gomory algorithm for all-integer programming. The
  objective of this procedure is to determine the integer solution
  of a linear programming problem with integer coefficients only.
  The tableau-matrix a consists of m + 1 rows and n columns.
  The top row of a is the objective row, the last column represents
  the right-hand sides. The tableau-columns, with the exception
  of the last column, have to be lexicographically positive. The
  algorithm is finished if all entries in the last column, except the
  top most entry, are nonnegative. Then the top most entry of
  the last column represents the value of the objective function.
  The other entries of the last column define the coordinates of
  the optimal solution. There are always the same variables con-
  nected with the same rows. The exit no solution is used if a row
  is found which has a negative entry in the last column, but
  otherwise only nonnegative entries;
begin   integer i, k, j, l, r;
        real lambda;
        integer array t[1:n−1], c[1:n];
1:      for i := 1 step 1 until m do if a[i,n] < 0 then
        begin r := i;  go to 2 end;
        go to end;
2:      for k := 1 step 1 until n−1 do if a[r,k] < 0 then
        go to 4;
        go to no solution;
4:      l := k;
        for j := k+1 step 1 until n−1 do if a[r,j] < 0 then
          begin i := 0;
            3: if a[i,j] < a[i,l] then l := j else
               if a[i,j] = a[i,l] then
               begin i := i+1;  go to 3 end
          end;
        for j := 1 step 1 until n−1 do if a[r,j] < 0 then
          begin if a[0,l] ≠ 0 then t[j] := entier(a[0,j]/a[0,l])
            else t[j] := 1
          end;
        lambda := abs(a[r,1]/t[1]);
        for j := 2 step 1 until n−1 do if a[r,j] < 0 then
          begin if abs(a[r,j]/t[j]) > lambda then
          lambda := abs(a[r,j]/t[j]) end;
        for j := 1 step 1 until n do if j≠l then
          begin c[j] := entier(a[r,j]/lambda);
            if c[j] ≠ 0 then
            for i := 0 step 1 until m do a[i,j] := a[i,j] + c[j] ×
            a[i,l]
          end;
        go to 1;
end:  end;
```

CERTIFICATION OF ALGORITHM 153
GOMORY [F. L. Bauer, *Comm. ACM 6*, Feb. 1963]
B. Lefkowitz and D. A. D'Esopo*
Stanford Research Institute, Menlo Park, California
  * Work supported by Office of Naval Research.

GOMORY was hand-coded in Balgol for the Burroughs 220
and in Fortran for the CDC 1604. The following corrections should
be made:
   The statement
```
lambda := abs(a[r,1]/t[1]);
```
should read
```
lambda := abs(a[r,l]/t[l]);
```
The statement
```
for j := 2 step 1 until n−1 do if a[r,j] < 0 then
```
should read
```
for j := 1 step 1 until n−1 do if a[r,j] < 0 then
```
The following changes to Bauer's program were made to increase its efficiency and reduce storage requirements.
   Change the statement
```
begin integer i, k, j, 1, r;
```
to read
```
begin integer i, k, j, 1, r, c, t;
```
Change the statement
```
real lambda;
```
to read
```
real lambda, lambd;
```
Delete the statement
```
integer array t[1: n−1], c[1: n];
```
Before the statement
```
for j := 1 step 1 until n−1 do if a[r,j] < 0 then
```
insert the statement
```
lambda := 1.0;
```
Change the statement
```
begin if a[0,l] ≠ 0 then t[j] := entier(a[0,j]/a[0,l])
```
to read
```
begin if a[0,l] ≠ 0 then t := entier(a[0,j]/a[0,l])
```
Change the statement
```
else t[j] := 1
```
to read
```
else t := 1
```
After the statement
```
else t[j] := 1
```
insert the statements
```
lambd := −a[r,j]/t;
lambda := if lambda < lambd then lambd else lambda;
```
Delete the statements starting with
```
lambda := abs(a[r,1]/t[1]);
```
up to and including
```
lambda := abs(a[r,j]/t[j]) end;
```
Change the statement
```
begin c[j] := entier (a[r, j]/lambda);
```
to read
```
begin c := entier(a[r,j]/lambda);
```
Change the statement
```
if c[j] ≠ 0 then
```
to read
```
if c ≠ 0 then
```
Change the statement
```
for i := 0 step 1 until m do a[i,j] := a[i,j] + c[j] ×
```
to read

$$\textbf{for } i := 0 \textbf{ step } 1 \textbf{ until } m \textbf{ do } a[i,j] := a[i,j] + c \times$$

The "tie-breaking" procedure embodied in the three statements beginning at

$$3: \textbf{if } a[i,j] < a[i,l] \textbf{ then } l := j \textbf{ else}$$

will fail if the two columns being compared are identical. Although this cannot happen on the first iteration, it may occur later. To test for this condition change the two statements beginning with

$$\textbf{begin } i := i + 1; \quad \textbf{go to } 3 \textbf{ end}$$

to read

$$\textbf{begin } i := i + 1; \quad \textbf{if } i > m \textbf{ then go to } 31 \textbf{ else go to } 3 \textbf{ end};$$
$$31: \textbf{end};$$

The revised algorithm yielded satisfactory answers on a ten equation-seven variable problem in 159 iterations and a 35-equation 14-variable problem in 447 iterations.

The following comments may be helpful for preparing a problem for GOMORY. The problem constraints must be stated in the form:

$$\sum_j a_{ij} x_j + s_i = b_i$$

where the s_i are slack variables. The columns representing these slack variables need not appear in the initial tableau-matrix a.

Since the only variables in the solution that will necessarily be non-negative are the s_i, any non-negativity constraints on the other variables must be among the above equations (e.g. the constraint $x_j \geqq 0$ is represented by $-x_j + s_k = 0$).

The size of the integers in the b vector substantially affects the number of iterations.

The requirement that all but the last tableau-columns be lexicographically positive means that the first nonzero element in these columns must be positive.

EDITOR's NOTE: Prof. Bauer wishes to indicate that for the Algorithm 153, GOMORY, credit is due to Ch. Witzgall, who wrote the draft.

ALGORITHM 154
COMBINATION IN LEXICOGRAPHICAL ORDER
CHARLES J. MIFSUD
Armour Research Foundation, ECAC Annapolis, Md.

```
procedure COMB1 (n,r,I); integer n, r; integer array I;
comment The distinct combinations of the first n integers
    taken r at a time are generated in I in lexicographical order
    starting with an initial combination of the r integers 1, 2, ⋯ ,
    r. Each call of the procedure, after the first, must have in I
    the previous generated combination. The Boolean variable
    first is nonlocal to COMB1 and must be true before the first call.
    Thereafter first remains false until all combinations have been
    generated. When calling COMB1 with I containing n − r +·1,
    n − r + 2, ⋯ , n, I is left unchanged and first is set true;
begin integer s, j;
    if first then begin for j := 1 step 1 until r do
        I[j] := j;
        first := false;  go to EXIT end;
    begin if I[r] < n then begin I[r] := I[r] + 1;  go to EXIT
            end;
        for j := r step −1 until 2 do
            if I[j−1] < n − r + j − 1 then
                begin I[j−1] := I[j−1] + 1;
                    for s := j step 1 until r do
                        I[s] := I[j−1] + s − (j−1);  go to EXIT end end;
    first := true;
EXIT : end
```

CERTIFICATION OF ALGORITHM 154
COMBINATION IN LEXICOGRAPHICAL ORDER
[Charles J. Mifsund, *Comm. ACM*, Mar. 1963]
K. M. BOSWORTH
I.C.T. Ltd., Hayes, Middlesex, England

This procedure was tested

$$\textbf{for } r := 1 \textbf{ step } 1 \textbf{ until } n \text{ with } n = 6$$

with correct results.

COLLECTED ALGORITHMS FROM CACM

ALGORITHM 155
COMBINATION IN ANY ORDER
CHARLES J. MIFSUD
Armour Research Foundation, ECAC Annapolis, Md.

```
procedure  COMB2 (m,M,n,r,s,S,TOTAL);  integer array m,
        M, S;  integer n, r, s,TOTAL;
comment  Each call of COMB2 generates a distinct combina-
    tion S, (if possible) of the n integer values of J taken r (r>1)
    at a time if J consists of m[1] integers each equal to M[1], and
    m[2] integers each equal to M[2], and so on, there being s integers
    available. TOTAL must be set to zero before the first call of
    COMB2 and thereafter TOTAL is increased by one after each
    new combination is generated. To speed up the machine opera-
    tion arrange the s integers in M such that m[1] ≧ m[2] ≧ ··· ≧
    m[s];
begin integer i, j, t, p;  own integer array J[1:n], I[1:r];  own
Boolean first;
    if TOTAL = 0 then begin
      t := 1;  p := 0;
      for j := 1 step 1 until s do
        begin p := p + m[j];
          for i := t step 1 until p do
            begin J[i] := M[j];
              t := t + 1 end end;
                first := true end;
  1:  COMB1 (n,r,I);
      if first then go to EXIT;
        if I[1] = 1 then go to 2 else go to 3;
  2:  for j := 2 step 1 until r do
      if  (J[I[j]]=J[I[j]−1]) ∧ (I[j]>I[j−1]+1) then go to 1;
        go to 4;
  3:  if J[I[1]] = J[I[1]−1] then go to 1 else go to 2;
  4:  for j := 1 step 1 until r do
        S[j] := J[I[j]];
        TOTAL := TOTAL + 1;
EXIT:  end
```

CERTIFICATION OF ALGORITHM 155
COMBINATION IN ANY ORDER [Charles J. Mifsud,
Comm. ACM, Mar. 1963]
K. M. BOSWORTH
I.C.T. Ltd., Hayes, Middlesex, England

This procedure was tested using

$$m[1] = 4 \quad m[2] = 3 \quad m[3] = 2 \quad m[4] = 2$$

$$M[1] = 4 \quad M[2] = 7 \quad M[3] = 9 \quad M[4] = 16$$

and for $r := 1$ **step 1 until** s

It is correctly generated for $r = 1$ the four combinations 4, 7, 9, 16, as well as the ten combinations for $r = 2$, the eighteen combinations for $r = 3$, and the twenty-six combinations for $r = 4$.

Changes made due to compiler limitations were (i) systematic changes of upper case letters where there was conflict due to having only one case of letters, (ii) transfer of **own** declared variables to non-local variables, and (iii) integer labels to identifiers.

ALGORITHM 156
ALGEBRA OF SETS
CHARLES J. MIFSUD
Armour Research Foundation, ECAC Annapolis, Md.

```
procedure INOUT (A,n,SUM);  real array A;  integer n;
          real SUM;
comment  SUM = ∑₁ Aᵢ − ∑₂ AᵢAⱼ + ∑₃ AᵢAⱼAₖ − ···±
  A₁A₂ ··· Aₙ is formed where the symbols ∑₁ , ∑₂ , ∑₃ , ··· ,
  ∑ₙ₋₁ stand for summation of the possible combinations of the
  numbers A₁ , A₂ , ··· , Aₙ taken one, two, three, ··· , (n−1)
  at a time;
  begin real j, part, T;  integer i, r;  integer array I[1:n];
  Boolean first;
    r := SUM := 0;  j := − 1;
  B: first := true;  r := r + 1;  part := 0;
  A:  COMB1 (n,r,I);
      if first then begin j := −1 × j;  part := j × part;
        SUM := SUM + part;
        if r < n then go to B else go to EXIT end;
      T := 1;
      for i := 1 step 1 until r do
        T := A[I[i]] × T;
      part := part + T;  go to A;
EXIT:  end
```

CERTIFICATION OF ALGORITHM 156
ALGEBRA OF SETS [Charles J. Mifsud, *Comm. ACM*,
Mar. 1963]
K. M. BOSWORTH
I.C.T. Ltd., Hayes, Middlesex, England

One correction required in this procedure is the systematic
change of label A to avoid conflict with the formal parameter
array A.

The procedure was then tested for $n = 9$ and $Ai = i$, $i = 1$,
$\cdots$, n, producing the correct answer $SUM = 1$.

Two other tests with arbitrary values of Ai and $n = 4$ were also
correct.

ALGORITHM 157
FOURIER SERIES APPROXIMATION
CHARLES J. MIFSUD
Armour Research Foundation, ECAC Annapolis, Md.

procedure *FOURIER* (N,f,a,b); **real array** f, a, b; **integer** N;
comment *Fourier* determines $2N+1$ constants a_p $(p=0,1,\cdots,N)$, b_p $(p=1,2,\cdots,N)$ in such a way that the equations $f_n = 1/2a_o + \sum_{p-1}^{N}(a_p \cos 2\pi np/(2N+1) + b_p \sin 2\pi np/(2N+1))$ are satisfied, where the f_n are given numbers. The f_n may be thought of as the $2N+1$ values of a function $f(x)$ at the points $x_n = 2\pi n/(2N+1)$. The method used to generate a_p, b_p was formulated by G. Goertzel in "Mathematical Methods for Digital Computers" (John Wiley and Sons, Inc., 1960);

```
begin real array S, C[1:2], u[0:2];  real TEMP, pi;
  integer p, i;
  pi := 3.14159265;  C[2] := 1;  S[2] := 0;
  C[1] := cos(2×pi/(2×N+1));
  S[1] := sin(2×pi/(2×N+1));
  for p := 0 step 1 until N do
    begin u[1] := u[2] := 0;
      for i := 2 × N step −1 until 1 do
        begin u[0] := f[i] + 2 × C[2] × u[1] − u[2];
          u[2] := u[1];  u[1] := u[0]end;
    a[p] := 2/(2×N+1) × (f[0]+u[1]×C[2]−u[2]);
    b[p] := 2/(2×N+1) × u[1] × S[2];
    TEMP := C[1] × C[2] − S[1] × S[2];
    S[2] := C[1] × S[2] + S[1] × C[2];
    C[2] := TEMP end end
```

REMARK ON ALGORITHM 157
FOURIER SERIES APPROXIMATION [C. J. Mifsud,
Comm ACM, Mar. 1963]
RICHARD GEORGE*
Argonne National Laboratory, Argonne, Ill.

This algorithm was written in FAP language for the 32-K IBM 704. It was tested on a sawtooth curve, and the sawtooth was recreated by summing the expansion up through the $2N + 1$ constants, with excellent results.

* Work supported by the United States Atomic Energy Commission.
The arrays S, C and u are never referenced with a variable subscript. For a saving of time, I suggest that simple variables be used instead.

By declaring one additional real variable, one can bring the phrase

$$2/(2 \times N + 1)$$

outside of the **for** loops, because N does not change through the procedure. This results in a saving of $4N+2$ mult-ops.

REMARK ON ALGORITHM 157
FOURIER SERIES APPROXIMATION [Charles J. Mifsud, *Comm. ACM*, Mar. 1963]
GEORGE R. SCHUBERT*
University of Dayton, Dayton, Ohio

* Undergraduate research project, Computer Science Program, Univ. of Dayton.

Algorithm 157 has been modified to fit $2N$ data points and has run successfully on the Burroughs 220 using BALGOL. With the modifications, $2N$ constants a_p $(p=0, 1, \cdots, N)$ and b_p $(p=1, 2, \cdots, N-1)$ are determined such that the equation $f_n = a_0/2 + \sum_{p-1}^{N-1} (a_p \cos \pi np/N + b_p \sin \pi np/N) + a_N/2 \cos \pi n$ is satisfied.

In the modified procedure, the second and third lines after the integer declaration should read:

```
C[1] := cos (pi/N);
S[1] := sin (pi/N);
```

The second **for** statement should read:

```
for i := 2 × N −1 step −1 until 1 do
```

The lines containing the a and b coefficients should read:

```
a[p] := (f[0]+u[1]×C[2]−u[2])/N;
b[p] := (u[1]×S[2])/N;
```

REFERENCE: R. W. Hamming, *Numerical Methods for Scientists and Engineers*, pp. 68–73 (McGraw-Hill, 1962).

ALGORITHM 158 (ALGORITHM 134, REVISED)
EXPONENTIATION OF SERIES

HENRY E. FETTIS
Aeronautical Research Laboratories, Wright-Patterson Air Force Base, Ohio

procedure *SERIESPWR* (A,B,P,N); **value** A, P, N;
    array A, B; **integer** N;
comment This procedure calculates the first N coefficients $B[i]$ of the series $g(x) = f(x) \uparrow P$ given the first N coefficients of the series

$$f(x) = 1 + \sum A[i] \times x \uparrow i \quad (i=1,2,-,-,N).$$

P may be any real number. Setting $P := 0$ gives the coefficients for $LN(g(x))$;
    begin integer i, k;
    real P, S;
      if $P = 0$ **then** $B[1] = A[1]$;
  else $B[1] := P \times A[1]$;
    for $i := 2$ **step** 1 **until** N **do**
    begin $S := 0$;
      for $k := 1$ **step** 1 **until** $i - 1$ **do**
      $S := S + (P \times (N-k) - k) \times B[k] \times A[N-k]$;
      $B[i] := P \times A[i] + (S/i)$
      end for i;
end *SERIESPWR*

CERTIFICATION OF ALGORITHM 158
EXPONENTIATION OF SERIES [H. E. Fettis, *Comm. ACM*, Mar. 1963]

J. DENNIS LAWRENCE
Lawrence Radiation Laboratory, Livermore, Calif.

This procedure was translated into FORTRAN and run on the Remington-Rand LARC Computer. Three changes are necessary.
  (1) The last line of the comment should read
        for the natural logarithm of $f(x)$;
  (2) The third line from the end should read

$$S := S + (P \times (i-k) - k) \times B[k] \times A[i-k];$$

(This line was given correctly in algorithm 134.)
  (3) The second line from the end apparently should read

$$B[i] := A[i] := (S/i);$$

for the case $P = 0$ only. Probably the best way to incorporate this is by making two changes:
  (a) Change the **if** clause to read
  if $P = 0$ **then** $R := 1$ **else** $R := P$; $B[1] := R \times A[1]$;
  (b) Change the second line from the end to read

$$B[i] := R \times A[i] + (S/i);$$

A large number of examples were run quite successfully; the following give representative samples.
  (1) $(1+2x+3x^2+0.5x^3)^2 = 1+4x+10x^2+13x^3+11x^4+3x^5+0.25x^6$ (using $A[4] := A[5] := A[6] := 0$).
  (2) Setting $P := 1$ gives $B[i] := A[i]$.

(3) Let $f(x) = e^x = 1 + \sum_{i=1}^{n} \frac{1}{i!} x^i$ and let $P = ln2 = .693147181$.

Then $g(x) = 2^x = 1 + \sum_{i=1}^{n} \frac{(ln2)^i}{i!} x^i$. (See Table 1.)

(4) Let $f(x) = e^x$ and $P = -1$. Then $g(x) = e^{-x}$. For $P = 0$, apparently the constant term of $g(x)$ should be zero instead of one.

TABLE 1

	$A[i]$	$B[i]$
1	1.000000000	0.693147181
2	0.500000000	0.240226507
3	0.166666667	0.055504109
4	0.041666667	0.009618129
5	0.008333333	0.001333356
6	0.001388889	0.000154035
7	0.000198413	0.000015253
8	0.000024802	0.000001322
9	0.000002756	0.000000102
10	0.000000276	0.000000007

(5) Let $f(x) = e^x$ and $P = 0$. Then $g(x) = x$.

(6) Let $f(x) = \sum_{i=0}^{n} x^n$ and $P = 0$. Then $g(x) = ln(1-x^n) - ln(1-x) = \sum_{i=1}^{n} \frac{1}{i} x^i$. (See Table 2.)

TABLE 2

	$A[i]$	$B[i]$
1	1.0	1.000000000
2	1.0	0.500000000
3	1.0	0.333333340
4	1.0	0.250000000
5	1.0	0.200000000
6	1.0	0.166666670
7	1.0	0.142857140
8	1.0	0.125000000
9	1.0	0.111111110
10	1.0	0.100000000
11	1.0	0.090909100
12	1.0	0.083333330
13	1.0	0.076923080
14	1.0	0.071428580
15	1.0	0.066666660

CERTIFICATION OF ALGORITHMS 134 AND 158
EXPONENTIATION OF SERIES [Henry E. Fettis,
  Comm. ACM, Oct. 1962 and Mar. 1963]
Henry C. Thacher, Jr.
Reactor Engineering Div., Argonne National Laboratory
  Argonne, Ill.
Work supported by the U.S. Atomic Energy Commission.

The bodies of SERIESPWR were transcribed for the Dartmouth Scalp processor for the LGP-30 computer. In addition to the modifications required by the limitations of this translator, the following corrections were necessary:

1. Add "**real** P;" to the specifications.
2. Delete "p," from the declarations in the procedure body.
3. (134 only) Replace "S" by "s" and $[i-k]$ by "$(i-k)$" in the statement $S := s + \cdots$.
4. (158 only) Changes last sentence of comment to "Setting $P := 0$ gives the coefficients for $ln(f(x))$. In this series, the constant term is 0, instead of 1 as elsewhere;"
5. (158 only) Add the identifier $P2$ to the declared real variables.
6. (158 only) Make the first statements read:
    "**if** $P = 0$ **then** $P2 := 1$ **else** $P2 := P$;
      $B[1] := P2 \times A[1]$;
7. (158 only) Make the statement of the **for** k loop read

    "$S := S+(P\times(i-k)-k) \times B[k] \times A[i-k]$;"

8. Change the last statement to

    "$B[i] := P2 \times A[i] + S/i$ **end** for i;

In addition, the following modifications would improve the efficiency of the program:

1. Remove A from the value list.
2. Omit the statement $B[1] := P \times A[1]$; ($P2\times A[1]$ in 158 according to correction 6) and change the initial value of i in the statement following from 2 to 1.

When these changes were made, both procedures produced the first ten coefficients of the series for $(exp(x)) \uparrow 2.5$ from the first ten coefficients of the exponential series. The procedures were also used to generate the binomial coefficients by applying them to $(1+x)^P$, for $P = 2.0$, and 0.5000000. Algorithm 158 was also tested with $P := 0$ for $1+x$ and for the series expansions for $(sin\ x)/x$, $cos\ x$, and $exp\ x$. In all cases, the coefficients agreed with known values within roundoff.

ALGORITHM 159
DETERMINANT
DAVID W. DIGBY
Oregon State University, Corvallis, Ore.

```
real procedure  Determinant (X,n);
value n;  integer n;  array X;
comment  Determinant calculates the determinant of the n-by-
   n square matrix X, using the combinatorial definition of the
   determinant. This algorithm is intended as an example of a
   recursive procedure which is somewhat less trivial than Factorial
   (Algorithm 33);
begin real D;  integer i; Boolean array B[1:n];
  procedure Thread (P,e,i);
    value P, e, i;  real P;  integer e, i;
    if i > n then D := D + P × (−1) ↑ e else if P ≠ 0 then
      begin integer j, f;
        f := 0;
        for j := n step −1 until 1 do
          if B[j] then f := f + 1 else
            begin
              B[j] := true;
              Thread (P×X[i,j],e+f,i+1);
              B[j] := false;
            end of loop;
      end of Thread;
  for i := 1 step 1 until n do
    B[i] := false;
  D := 0;
  Thread (1,0,1);
  Determinant := D;
end Determinant;
```


CERTIFICATION OF ALGORITHM 159
DETERMINANT [David W. Digby, *Comm. ACM*,
   March 1963]
ARNOLD LAPIDUS
Courant Institute of Mathematical Sciences, New York
   University, New York, N. Y.

Algorithm 159 was translated into FORTRAN II for the IBM
7090 as part of a test of FORTRAN subroutines designed to facilitate
the implementation of recursive procedures. As expected, the
numerical results were poor. For the Hilbert matrices $H_n = (a_{ij})$,
$a_{ij} = 1/(i+j-1)$, results were as follows:

n	Det H_n (true)	Det H_n (computed by Algorithm 159)
2	8.333 333 3 (− 2)	8.333 333 2 (− 2)
3	4.629 629 6 (− 4)	4.629 623 1 (− 4)
4	1.653 439 2 (− 7)	1.651 933 4 (− 7)
5	3.749 295 1 (−12)	−2.910 383 0 (−11)

Determinants of order 4 and 6 with integer elements were also
evaluated. The algorithm gave full accuracy for these.

ALGORITHM 160
COMBINATORIAL OF M THINGS
TAKEN N AT A TIME
M. L. WOLFSON AND H. V. WRIGHT
United States Steel Corp., Monroeville, Penn.

```
integer procedure combination (m, n);
  value n;  integer m, n;
  comment  calculates the number of combinations of m things
    taken n at a time. If n is less than half of m, then the program
    calculates the combinations of m things taken m − n at a time
    which is the exact equivalent of m things taken n at a time;
  begin integer p, r, i;
    p := m − n;
    if n < p then begin p := n;  n := m − p end;
    if p = 0 then begin r := 1;  go to exit end;
    r := n + 1;
    for i := 2 step 1 until p do r := (r × (n+i))/i;
  exit:  combination := r
  end combination
```

CERTIFICATION OF ALGORITHM 160
COMBINATORIAL OF M THINGS TAKEN N AT
A TIME [M. L. Wolfson and H. V. Wright, *Comm. ACM*,
Apr. 1963]
DMITRI THORO
San Jose State College, San Jose, Calif.

Algorithm 160 was translated into FORTRAN II and FORGO for
the IBM 1620. Correct results were obtained for values of m up to
20.

CERTIFICATION OF ALGORITHM 160
COMBINATORIAL OF M THINGS TAKEN N AT
  A TIME [M. L. Wolfson and H. V. Wright, *Comm.
  ACM*, April 1963]
ROBERT F. BLAKELY
Indiana Geological Survey, Bloomington, Ind.

Algorithm 160 was translated into ALGO, a compiler for the
Control Data Corp. G-15 computer (formerly the Bendix G-15).
With the restriction that $m \geqq n \geqq 0$, correct results were obtained for all integer values of m and n, where $0 \leqq m \leqq 10$. Several
other values were tested and all results were correct.

ALGORITHM 161
COMBINATORIAL OF M THINGS
TAKEN ONE AT A TIME, TWO AT A TIME,
UP TO N AT A TIME
H. V. WRIGHT AND M. L. WOLFSON
United States Steel Corp., Monroeville, Penn.

```
procedure combination vector (m, n, v);
integer m, n;  integer array v;
comment   calculates all combinations of m things taken from 1
    to n at a time. The result is a vector, v, within which the first
    element is the combination of m things taken 1 at a time, the
    second element is the combinations of m things taken 2 at a time,
    the third element taken 3 at a time, ···, and the nth element
    taken n at a time.
begin integer i;
    v[1] := m;
    for i := 2 step 1 until n do
        v[i] := (v[i−1] × (m−i+1))/i;
end combination vector
```

CERTIFICATION OF ALGORITHM 161
COMBINATORIAL OF M THINGS TAKEN ONE AT
A TIME, TWO AT A TIME, UP TO N AT A TIME
[H. V. Wright and M. L. Wolfson, *Comm. ACM*, Apr.
1963]
DMITRI THORO
San Jose State College, San Jose, Calif.

Algorithm 161 was translated into FORTRAN II and FORGO for
the IBM 1620. Correct results were obtained for values of m up
to 20.

CERTIFICATION OF ALGORITHM 161
COMBINATORIAL OF M THINGS TAKEN ONE AT
    A TIME, TWO AT A TIME, UP TO N AT A TIME
    [H. V. Wright and M. L. Wolfson, *Comm. ACM*,
    Apr. 1963]
DAVID H. COLLINS
Indiana Geological Survey, Bloomington, Ind.

Algorithm 161 was translated into ALGO, a compiler for the
Control Data Corp. G-15 computer (formerly the Bendix G-15).

With the restriction that $m \geq n \geq 1$, correct results were obtained for all integer values of m and n, where $1 \leq m = n \leq 15$.
Several other values were tested (including cases where $m \neq n$)
and all results were correct.

ALGORITHM 162
XYMOVE PLOTTING
FRED G. STOCKTON
Shell Development Co., Emeryville, Calif.

```
procedure xymove (XZ, YZ, XN, YN);  value XZ, YZ, XN, YN;
integer XZ, YZ, XN, YN;
```

comment *xymove* computes the code string required to move the pen of a digital incremental X,Y-plotter from an initial point (XZ, YZ) to a terminal point (XN, YN) by the "best" approximation to the straight line between the points. The permitted elemental pen movement is to an adjacent point in a plane Cartesian point lattice, diagonal moves permitted. The eight permitted pen movements are coded

$$1 = +Y, \quad 2 = +X+Y, \quad 3 = +X, \quad 4 = +X-Y,$$
$$5 = -Y, \quad 6 = -X-Y, \quad 7 = -X, \quad 8 = -X+Y.$$

The approximation is "best" in the sense that each point traversed is at least as near the true line as the other candidate point for the same move.

xymove does not use multiplication or division.;

begin integer $A, B, D, E, F, T, I, move;$

comment *code* (J) is a procedure which returns a value of *code* according to the following table:

J	1	2	3	4	5	6	7	8
code	1	2	3	2	3	4	5	4
J	9	10	11	12	13	14	15	16
code	5	6	7	6	7	8	1	8

plot (*move*) is a procedure which sends *move* to the plotter as a plotter command.;

```
    if XZ = XN ∧ YZ = YN then go to return;
    A := XN − XZ;  B := YN − YZ;  D := A + B;  T :=
    B − A;  I := 0;
    if B ≧ 0 then I := 2;
    if D ≧ then I := I + 2;
    if T ≧ 0 then I := I + 2;
    if A ≧ 0 then I := 8 − I else I := I + 10;
    A := abs(A);  B := abs(B);  F := A + B;  D := B − A;
    if D ≧ 0 then begin T := A;  D := −D end else T := B;
    E := 0;
repeat:  A := D + E;  B := T + E + A;
        if B ≧ 0 then begin E := A;  move := code(I);
        F := F − 2 end
        else begin E := E + T;  F := F − 1;
        move := code(I−1) end;
        plot(move);
        if F > 0 then go to repeat;
    return:
end
```

CERTIFICATION OF ALGORITHM 162
XYMOVE PLOTTING [Fred G. Stockton, *Comm. ACM*, Apr. 1963]
WILLIAM E. FLETCHER
Bolt, Beranek and Newman Inc., Los Angeles, Calif.

The line in the body of the procedure which read:

$$\textbf{if } D \geqq \textbf{ then } I := I + 2;$$

was corrected to read:

$$\textbf{if } D \geqq 0 \textbf{ then } I := I + 2;$$

With this one change the body of the procedure was transliterated into DECAL-BBN and successfully run on a PDP-1 computer utilizing the cathode ray tube output to display the path of a simulated digital incremental plotter.

REMARK ON ALGORITHM 162 [J6]
XYMOVE PLOTTING [F. G. Stockton, *Comm. ACM 6* (Apr. 1963), 161; *6* (Aug. 1963), 450]
D. K. CAVIN (Recd. 10 Feb. 1964)
Oak Ridge National Laboratory, Oak Ridge, Tenn.

The following modifications were made to Algorithm 162 to decrease the average execution time. The last nine lines of Algorithm 162 are replaced by the following:

```
        move := code(I−1);  I := code(I);
repeat:  A := D + E;  B := T + E + A;
        if B ≧ 0 then begin E := A; F := F − 2; plot(I) end
            else begin E := E + T; F := F − 1; plot(move) end;
        if F > 0 then go to repeat;
    return:
end
```

It is obvious that on any movement containing more than two elemental pen movements the use of the code procedure in the loop is redundant, since no more than two of the eight permitted pen movements are necessary for the approximation of any line. Therefore moving the call of the code procedure outside of the basic loop reduces the execution time whenever the X, Y movement requires more than two elemental pen movements. The procedures were coded in CODAP, the assembly language for the CDC 1604-A, and this modified version was approximately 40 percent faster in the loop than the original version. The timing comparisons used numbers in the range −2000 to 2000 with heavy emphasis on the subrange −150 to 150. The typographical error noted in the certification (*Comm. ACM*, August 1963) was corrected in both codes.

[A referee verifies that Algorithm 162 does indeed run, as changed.—G.E.F.]

ALGORITHM 163
MODIFIED HANKEL FUNCTION
Henry E. Fettis
Aeronautical Research Laboratories, Wright-Patterson
Air Force Base, Ohio

procedure *EXPK* (P, X, E); **real** P, X, E;

comment this procedure calculates the modified Hankel Function $e^x K_p(x)$ to within a given accuracy E from the integral representation:

$$e^x K_p(x) = \int_0^\infty e^{x(1-\cosh t)} \cosh (pt)dt;$$

```
begin real F, G, H, R, S, T, U, Y, Z, ZP;
  R := 0.0;
H := 1.0;
  iteration: begin
      G := R;
      T := .5 × H;
      S := 0;
      Z := exp(T);
      U := Z × Z;
        integration: begin
          Y := X × (1−.5×(Z+1/Z));
          if P = 0 then ZP := 1
          else ZP := Z ↑ P;
          F := .5 × exp(Y) × (ZP+1/ZP);
          S := S + F;
          Z := Z × U;
          end;
          if F ≧ E then go to integration
          else R := H × S;
          H := .5 × H;
        end;
      if abs (R−G) ≧ E then go to iteration
      else EXPK: = R
  end EXPK
```

CERTIFICATION OF ALGORITHM 163
MODIFIED HANKEL FUNCTION [Henry E. Fettis,
Comm. ACM, Apr. 1963]
Henry C. Thacher, Jr.*
Argonne National Laboratory, Argonne, Ill.

Since this algorithm is a function declaration, the procedure declaration should be:

real procedure *EXPK(D, X, E)*; $\cdots$

Otherwise, no syntactical errors were noticed.

The body of the procedure was translated and run on the LGP-30 computer, using the Dartmouth Scalp system. Results for $E = 0.0001$, $X = 0.1(0.1)1.0$, $P = 0$, 0.3333333, 0.6666667 and 1.000000 agreed with values tabulated in Jahnke-Emde-Losch to the 3-4D given in the tables, except for errors discovered in the table of $2/\pi K_{2/3}(x)$.

With $X = 0$, the program ended in floating-point overflow. The algorithm itself, or the call of the procedure, should include a test to insure that the variable is greater than *eps*, where *eps* is chosen to prevent exceeding machine capacity.

The algorithm was found to be excessively slow. Times on the LGP-30 were of the order of 6 minutes. A considerable saving in time could be realized by improving the quadrature formula, currently the simple midpoint formula, repeated completely for each iteration. A more effective method would be a modified Romberg algorithm. A procedure based on the latter approach is being developed in this division.

* Work supported by the U. S. Atomic Energy Commission.

ALGORITHM 164
ORTHOGONAL POLYNOMIAL LEAST SQUARES
SURFACE FIT
R. E. CLARK, R. N. KUBIK, L. P. PHILLIPS
The Babcock & Wilcox Co., Atomic Energy Div.,
Lynchburg, Va.

procedure surfacefit $(x, u, y, w, z, nmax, mmax, imax, jmax)$
result: $(beta, phi, zcomp, minsqd, minsqdcomp, sumdifcomp, maxdifcomp)$;
real array $x, u, y, w, z, phi, beta, zcomp$;
integer $nmax, mmax, imax, jmax$;
real $minsqd, minsqdcomp, sumdifcomp, maxdifcomp$;
comment this is a transliteration of an operating program written in Burroughs ALGOL for the B-220. It fits, in the least squares sense, a polynomial function of two independent variables to values of a dependent variable specified at points on a rectangular grid in the plane of the independent variables. The use of orthogonal polynomials leads to a particularly simple system of linear equations rather than the ill-conditioned system which arises from the usual normal equations. It also provides a measure of the improvements resulting from each new term included which further leads, in this algorithm, to an automatic selection of a "best" degree polynomial function as determined by Gauss' criterion. The initial normalization of the variables results in significant reduction of round off errors in many cases. This algorithm is developed more fully in BAW-182. For a very similar approach to' this and related problems see Cadwell, J. H., "Least Squares Surface Fitting Program", *The Computer J. 3* (1961), 266 and Cadwell, J. H., and Williams, D. E., "Some Orthogonal Methods of Curve and Surface Fitting," *The Computer J. 4* (1961), 260. A further reference is Gauss, C. F., "Theoria Combinationis Observationum Erroribus Minimis Obnoxial," *Gauss Werke 4* (Gottingen 1873), 3–93. $x[i]$ and $y[j]$ are the independent variables, $z[i, j]$ is the dependent variable. $u[i]$ and $w[j]$ represent the weights corresponding to $x[i]$ and $y[j]$, respectively. $nmax$ is one more than the maximum degree of x to be considered. $mmax$ is one more than the maximum degree of y to be considered. $imax$ is the number of x's, and $jmax$ is the number of y's. $beta[n, m]$ is a measure of the improvement resulting from the inclusion of the $x^n y^m$th term. $phi[n, m]$ is the polynomial coefficient for the $x^n y^m$th term. Note the degree of the resulting polynomial may be less than the maximum degree specified as a result of the application of Gauss' criterion. $zcomp$ is the computed dependent variable.

$$minsqd = \left(\frac{\sum_{i,j} u[i] \cdot w[j] \cdot z[i,j]^2 - \sum_{n,m} beta[n,m]}{imax \cdot jmax}\right)^{1/2}$$

$$minsqdcomp = \left(\frac{\sum_{i,j} u[i] \cdot w[j](z[i,j] - zcomp[i,j])^2}{imax \cdot jmax}\right)^{1/2}$$

$$sumdifcomp = \frac{\sum_{i,j} |z[i,j] - zcomp[i,j]|}{imax \cdot jmax}$$

$$maxdifcomp = max\,|z(i,j) - zcomp((i,j)|$$

$minsqd$ and $minsqdcomp$ are equal if computation is exact. In practice they will not be equal due to the imprecise nature of calculation. A wide discrepancy indicates excessive errors in calculation;

```
begin
real array  a, b, denpa[1:nmax], c, d, denqa[1:mmax],
    alpha[1:nmax, 1:mmax], p[1:nmax, 1:imax], q[1:mmax, 1:jmax],
    pc[1:nmax, 1:nmax], qc[1:mmax, 1:mmax];
integer  n, m, i, j, s, t, r;
real  sumx, sumy, sumz, meanx, meany, meanz, numa, dena, denb,
    numc, denc, dend, alph, sumzsq, gausscrit, trialgausscrit, betasum,
    rescomp, poly;
comment  normalization of variables;
sumx := sumy := sumz := 0.0;
for i := 1 step 1 until imax do
    sumx := sumx + x[i];
meanx := sumx/imax;
for i := 1 step 1 until imax do
    x[i] := x[i] - meanx;
for j := 1 step 1 until jmax do
    sumy := sumy + y[j];
meany := sumy/jmax;
for j := 1 step 1 until jmax do
    y[j] := y[j] - meany;
for i := 1 step 1 until imax do begin
    for j := 1 step 1 until jmax do
        sumz := sumz + z[i,j] end;
meanz := sumz/(imax × jmax);
for i := 1 step 1 until imax do begin
    for j := 1 step 1 until jmax do
        z[i, j] := z[i, j] - meanz end;
comment  evaluate orthogonal polynomials;
numa := dena := 0.0;
for i := 1 step 1 until imax do begin
    p[1, i] := 1.0;
    numa := numa + u[i] × x[i];
    dena := dena + u[i] end;
a[2] := numa/dena;
for i := 1 step 1 until imax do
    p[2, i] := x[i] - a[2];
for n := 3 step 1 until nmax do begin
    numa := dena := denb := 0.0;
    for i := 1 step 1 until imax do begin
        numa := numa + u[i] × x[i] × p[n-1] ↑ 2;
        dena := dena + u[i] × p[n-1, i] ↑ 2;
        denb := denb + u[i] × p[n-2, i] ↑ 2 end;
    a[n] := numa/dena;  b[n] := dena/denb;
    for i := 1 step 1 until imax do
        p[n, i] := (x[i]-a[n]) × p[n-1, i] - b[n] × p[n-2, i] end;
numc := denc := 0.0;
for j := 1 step 1 until jmax do begin
    q[1, j] := 1.0;
    numc := numc + w[j] × y[j];
    denc := denc + w[j] end;
c[2] := numc/denc;
for j := 1 step 1 until jmax do
    q[2, j] := y[j] - c[2];
for m := 3 stcp 1 until mmax do begin
    numc := denc := dend := 0.0;
    for j := 1 step 1 until jmax do begin
        numc := numc + w[j] × y[j] × q[m-1, j] ↑ 2;
        denc := denc + w[j] × q[m-1, j] ↑ 2;
        dend := dend + w[j] × q[m-2, j] ↑ 2 end;
    c[m] := numc/denc;  d[m] := denc/dend;
    for j := 1 step 1 until jmax do
```

```
    q[m, j] := (y[j]−c[m]) × q[m−1, j] − d[m] × q[m−2, j] end;
comment  evaluate contribution of each orthogonal polynomial
  to the minimization of the residuals;
for n := 1 step 1 until nmax do begin
  denpa[n] := 0.0;
    for i := 1 step 1 until imax do
    denpa[n] := denpa[n] + u[i] × p[n, i] ↑ 2 end;
    for m := 1 step 1 until mmax do begin
    denqa[m] := 0.0;
      for j := 1 step 1 until jmax do
      denqa[m] := denqa[m] + w[j] × q[m, j] ↑ 2 end;
    for n := 1 step 1 until nmax do begin
      for m := 1 step 1 until mmax do begin
      alph := 0.0;
        for i := 1 step 1 until imax do begin
          for j := 1 step 1 until jmax do
          alph := alph + u[i] × w[j] × z[i, j] × p[n, i] × q[m, j]
            end;
      alpha[n, m] := alph/(denpa[n]×denqa[m]);
      beta[n, m] := alpha[n, m] × alph; end  end;
comment  application of Gauss' criterion to determine the de-
  gree polynomial which yields the closest fit to the given data.
  Gauss' criterion is, strictly speaking, applicable only to cases
  where the weights u[i] and w[j] are unity;
sumzsq := 0.0;
for i := 1 step 1 until imax do begin
  for j := 1 step 1 until jmax do
  sumzsq := sumzsq + u[i] × w[j] × z[i, j] ↑ 2 end;
s := t := 1;
for n := 1 step 1 until nmax do begin
  betasum := 0.0;
  for m := 1 step 1 until mmax do begin
    for r := 1 step 1 until n do
    betasum := betasum + beta[r, m];
    if betasum > sumzsq then trialgausscrit := 0.0
    else
    trialgausscrit := (sumzsq−betasum)/(imax×jmax−n×m);
    if n = 1 ∧ m = 1 then gausscrit := trialgausscrit;
    if gausscrit = trialgausscrit then begin
      if n × m < s × t then begin
        s := n;
        t := m end end;
    if gausscrit > trialgausscrit then begin
      gausscrit := trialgausscrit;
      s := n;
      t := m end end end;
nmax := s;
mmax := t;
minsqd := (gausscrit× (imax×jmax−nmax×mmax)/(imax×jmax))
  ↑ ½;
comment  evaluation of orthogonal polynomial coefficients;
for n := 1 step 1 until nmax do begin
  pc[n, m] := 1.0;
  for s := 1 step 1 until n − 1 do begin
    pc[n, s] := −a[n] × pc[n−1, s];
    if s ≠ 1 then pc[n, s] := pc[n, s] + pc[n−1, s−1];
    if s ≠ n − 1 then pc[n, s] := pc[n, s] − b[n] × pc[n−2, s]
      end end;
  for m := 1 step 1 until mmax do begin
  qc[m, m] := 1.0;
  for t := 1 step 1 until m − 1 do begin
    qc[m, t] := −c[m] × qc[m−1, t];
    if t ≠ then qc[m, t] := qc[m, t] + qc[m−1, t−1];
    if t ≠ m − 1 then qc[m, t] := qc[m, t] − d[m] × qc[m−2, t]
      end end;
comment  evaluation of approximating polynomial coefficients;
for s := 1 step 1 until nmax do begin
  for t := 1 step 1 until mmax do begin
  phi[s, t] := 0.0;
```

```
    for n := s step 1 until nmax do begin
      for m := t step 1 until mmax do
      phi[s, t] := phi[s, t] + alpha[n, m] × pc[n, s] × qc[m, t]
        end end end;
comment  evaluation of dependent variables using the approxi-
  mating  polynomial;
minsqdcomp := sumdifcomp := maxidifcomp := 0.0;
for i := 1 step 1 until imax do begin
  for j := 1 step 1 until jmax do begin
  zcomp[i, j] := 0.0;
    for s := nmax step − 1 until 1 do begin
    poly := phi[s, mmax];
      for t := mmax − 1 step 1 until 1 do
      poly := poly × y[j] + phi[s, t];
    zcomp[i, j] := zcomp[i, j] × x[i] + poly end;
  rescomp := z[i, j] − zcomp[i, j];
  zcomp[i, j] := zcomp[i, j] + meanz;
  minsqdcomp := minsqdcomp + u[i] × w[j] × rescomp ↑ 2;
  sumdifcomp := sumdifcomp + abs(rescomp);
  if abs (rescomp) > maxdifcomp then
  maxdifcomp := abs(rescomp) end end;
minsqdcomp := (minsqdcomp/(imax × jmax) ) ↑ ½;
sumdifcomp := sumdifcomp/(imax × jmax);
end surfacefit
```

CERTIFICATION OF ALGORITHM 164
ORTHOGONAL POLYNOMIAL LEAST SQUARES SURFACE FIT [R. E. Clark, R. N. Kubik, L. P. Phillips, *Comm. ACM*, April 1963]

C. V. BITTERLI
Johns Hopkins Univ. Applied Physics Lab., Silver Spring, Md.

The SURFACEFIT algorithm was translated into FORTRAN and successfully run on an IBM 7094. It was necessary to make the following corrections:

(a) 12th line after

  comment evaluate orthogonal polynomials;

should read

$$numa := numa + u[i] \times x[i] \times p[n−1,i] \uparrow 2;$$

(b) 2nd line after

comment evaluation of orthogonal polynomial coefficients;

should read

$$pc[n,n] := 1.0;$$

(c) 12th line after

comment evaluation of orthogonal polynomial coefficients;

should read

$$\textbf{if } t \neq 1 \textbf{ then } qc[m,t] := qc[m,t] + qc[m−1,t−1];$$

(d) 8th line after

comment evaluation of dependent variables using the approximating polynominal

should read

$$\textbf{for } t := mmax −1 \textbf{ step } −1 \textbf{ until } 1 \textbf{ do}$$

The following function was used to generate data for checking this algorithm:

$$z = 1 − x + y − xy + x^2 − y^2$$

for $x = 0, 1, 2, 3, 4$

and $y = 0, 1, 2, 3, 4$

The resulting polynomial was:

$$z = x - 5y - xy + x^2 - y^2$$

which is correct for the normalized variables.

It should be pointed out in the **comment** for this procedure that the resulting polynomial is in the normalized variables and not the original variables.

ALGORITHM 165
COMPLETE ELLIPTIC INTEGRALS
HENRY C. THACHER, JR.*
Reactor Eng. Div., Argonne National Lab., Argonne, Ill.

* Work supported by the U.S. Atomic Energy Commission.

```
procedure KANDE(ml, K, E, tol, alarm);
value m1, tol;
real m1, K, E, tol;
label alarm;
```

comment this procedure computes the complete elliptic integrals $K(m1) = \int_0^{\pi/2} (1 - (1 - m1) \sin^2\nu)^{-1/2} \, d\nu$ and $E(m1) = \int_0^{\pi/2} (1 - (1 - m1) \sin^2\nu)^{1/2} \, d\nu$ by the arithmetic-geometric-mean process. The accuracy is limited only by the accuracy of the arithmetic.

Except for the provision of tests for pathological values of the parameter, the calculation of K is only a slight modification of the second procedure of Algorithm No. 149 (*Comm. ACM. 5* (Dec. 1962), 605). These integrals may also be approximated to limited (6D) accuracy by Algorithms 55 and 56 (*Comm. ACM. 4* (Apr. 1961), 180). Unless the square-root is exceptionally fast, the latter algorithms are probably more efficient for 6D-accuracy.

The complementary parameter, $m1$, is chosen as the independent variable, rather than the parameter, m, the modulus, k or the modular angle α, because of the possibility of serious loss of significance in generating $m1$ from the other possible independent variables when $m1$ is small and $dK/dm1$ is very large. These variables are related by $m1 = 1 - m = 1 - k^2 = \cos^2\alpha$.

The formal parameter, *tol*, determines the relative accuracy of the result. To prevent entering a nonterminating loop, *tol* should not be less than twice the relative error in the square root routine. If $m1 \leq 0$ or if $m1 > 1$, the procedure exits to alarm. $K(0) = \infty$ while $E(0) = 1.00000000$.

The body of this procedure has been tested using the Dartmouth SCALP processor for the LGP-30. With $tol = 5_{10} - 7$, results agreed with tabulated values to within 3 in the seventh significant digit;

```
begin real a, b, c, sum, temp;
  integer fact;
  if m 1 > 1 V m1 ≤ 0 then go to alarm;
  a := fact := 1;
  b := sqrt(m1);
  temp := 1 − m1;
  sum := 0;
iter:  sum := sum + temp;
  c := (a − b)/2;
  fact := fact + fact;
  temp := (a + b)/2;
  b := sqrt (a × b);
  a := temp;
  temp := fact × c × c;
  if abs(c) ≧ tol × a V temp > tol × sum then go to iter;
  sum := sum + temp;
  K := 3.141592654/(a + b);
  comment  pi must be given to the full accuracy desired;
  E := K × (1 − sum/2)
end
```

CERTIFICATION OF ALGORITHM 165 [S21]
COMPLETE ELLIPTIC INTEGRALS [Henry C. Thacher Jr., *Comm. ACM 6* (Apr. 1963), 163]

I. FARKAS (Recd. 1 Aug. 1968)
Dept. of Computer Science, University of Toronto, Toronto, Ontario, Canada

KEY WORDS AND PHRASES: special functions, complete elliptic integral of the first kind, complete elliptic integral of the second kind
CR CATEGORIES: 5.12

One misprint and one semantic error were found in Algorithm 165:

1. The procedure heading

    procedure *KANDE* (*ml,K,E,tol,alarm*);

should read

    procedure *KANDE* (m1,K,E,tol,alarm);

2. The second statement in the procedure body

$$a := fact: = 1;$$

should read

$$fact: = 1; a: = 1;$$

because *fact* and *a* are of different types.

Algorithm 165 was translated into FORTRAN IV on an IBM 7094-II, whose single-precision mantissa has 27 significant bits (about 8 significant decimal digits). Because our *SQRT* program has a relative accuracy of $.75_{10} - 8$, *tol* was chosen $3_{10} - 8$. K and E were generated for m1 = (.01(.01)1.0) (to 27 bits) and the results obtained were compared with tables in [1]. For m1 = .01 E differed by two units in the last place; for all other values of m1, the maximum absolute error was one unit in the last place. The time taken to activate *KANDE* for the above 100 values of m1 was 0.1 sec.

REFERENCE:

1. ABRAMOWITZ, M., AND STEGUN, I. A. (Eds.) *Handbook of Mathematical Functions*. NBS Appl. Math. Ser. 55, US Govt. Printing Off., Washington, D. C., 1964.

ALGORITHM 166
MONTECARLO
R. D. RODMAN
Burroughs Corp., Pasadena, Calif.

procedure *montecarlo* (*n*, *a*, *row*, *tol*, *mxm*, *inv*, *test*, *count*);
  value *n*, *row*, *tol*, *mxm*; **integer** *n*, *row*, *mxm*, *count*;
  real *tol*; **real array** *a*, *inv*, *test*;
  comment this procedure will compute a single row of the
    inverse of a given matrix using a monte carlo technique.
    n is the size of the matrix, array *a* is the matrix, *row* indicates
    which inverse row is to be computed, *tol* is a tolerance factor
    and thus a criterion for terminating the process, *mxm* is 1000
    times the maximum number of random walks to be taken,
    after which the process is terminated, array *inv* contains the
    inverse row, array *test* contains the innerproduct of *inv* with
    the *row*th column of *a*, *count* is the number of random walks
    executed upon termination. **real procedure** *RANDOM* must
    be declared in the blockhead of **procedure** *MONTE CARLO*
    and generates a single random number between 0 and 1. If
    a is the matrix to be inverted, the absolute value of the largest
    eigenvalue of the matrix $I - a$ (I is the unit matrix) must be
    less than one to assure convergence. This procedure is easily
    adapted to finding a single unknown from a set of simultaneous
    linear equations;
  begin integer *i*, *j*, *k*, *nwk*, *lastwalk*, *walk*; **real** *res*, *p*, *g*;
  real array *sum*[1:*n*], *v*[1:*n*, 1:*n*];
start: *p* := $(n-1)/n \times n$;
      for *i* := 1 **step** 1 **until** *n* **do for** *j* := 1 **step** 1 **until** *n* **do**
      v[*i*,*j*] := **if** $i \neq j$ **then** $-a[i,j]/p$ **else** $(1-a[i,j])/p$;
      nwk := 1000;
      count := *res* := 0;
      for *k* := 1 **step** 1 **until** *n* **do** *test* [*k*] := *sum* [*k*] := 0;
start1: *lastwalk* := *row*; *g* := 1;
start2: *walk* := $(RANDOM/p) + 1$;
      if *walk* > *n* **then go to** *stop*;
      g := *v*[*lastwalk*,*walk*] $\times$ *g*; *lastwalk* := *walk*;
      go to *start2*;
stop: *count* := *count* + 1; *sum*[*lastwalk*] := *sum*[*lastwalk*] + *g*;
      if *count* < *nwk* **then go to** *start1*;
      for *k* := 1 **step** 1 **until** *n* **do** *inv*[*k*] := $n \times sum[k]/count$;
      for *i* := 1 **step** 1 **until** *n* **do for** *k* := 1 **step** 1 **until** *n* **do**
      test[*i*] := *inv*[*k*] $\times$ *a*[*k*, *i*] + *test*[*i*];
      for *i* := 1 **step** 1 **until** *row*−1, *row*+1 **step** 1 **until** *n* **do**
      res := *abs*(*test*[*i*]) + *res*; *res* := *abs*(*test*[*row*]−1) + *res*;
      if *res* < *tol* **then go to** *exit*;
      if *count* $\geq$ 1000 $\times$ *mxm* **then go to** *exit*;
      nwk := *nwk* + 1000; *res* := 0;
      for *k* := 1 **step** 1 **until** *n* **do** *test* [*k*] := 0;
      go to *start1*;
exit: **end** of monte carlo inversion procedure

REMARK ON ALGORITHM 166
MONTECARLO INVERSE [R. D. Rodman, *Comm. ACM*, Apr. 1963]
R. D. RODMAN
Burroughs Corp., Pasadena, Calif.

  The algorithm contained two errors:
  (1) The line which reads
    start: *p* := $(n-1)/n \times n$;
should read
    start: *p* := $(n-1)/n \uparrow 2$;
  (2) The line which reads
    start2: *walk* := (**random**/*p*) + 1;
should read
    start2: *walk*: = **entier** ((**random**/*p*) + 1);
  After making the preceding corrections, procedure montecarlo
was transliterated into EXTENDED ALGOL and run successfully
on the Burroughs B-5000. Convergence occurred in all cases where
the matrix satisfied the conditions set down in the comment statement of the algorithm. It was found that convergence was quickest
and the routine most practical for matrices with eigenvalues small
relative to one.

ALGORITHM 167
CALCULATION OF CONFLUENT DIVIDED
DIFFERENCES
W. Kahan and I. Farkas
Institute of Computer Science, University of Toronto,
Canada

real procedure $DVDFC(n, X, V, B, W)$; **integer** n;
real array X, V, B, W;
comment $DVDFC$ ca.culates the forward divided difference
$\triangle f(X_1, X_2, \cdots X_n)$. n is an integer which takes the values
$n = 1, 2, 3, \cdots$ in turn. X is a real array of dimension at least
n in which $X[i] = X_i$ for $i = 1, 2, \cdots, n$. The values X_i need
not be distinct nor in any special order, but once the array X
is chosen it will fix the interpretation of the arrays B and V.
If $X[1], X[2], \cdots, X[n]$ are in monotonic order, then the effect
of roundoff upon any nth divided difference is no more than
would be caused by perturbing each $f(X[i])$ by n units at most
in its last significant place. But if the X's are not in monotonic order, the error can be catastrophic if some of the divided
differences are relatively large. V is a real array of dimension
at least n containing the values of the function $f(X)$ and perhaps its derivatives at the point X_i. $V[i] = f^m(X_i)/m!$ and
$m = m_i$ for $i = 1, 2, 3, \cdots, n$. m_i is the number of times that
the **value** of X_i has previously appeared in the array X. B is
a real array of dimension at least n containing backward divided
differences. Before a reference to $DVDFC$ is executed one should
have $B[i] = \triangle f(X_i, X_{i+1}, \cdots, X_{n-1})$ for $i = 1, 2, \cdots, n-1$.
After that reference to $DVDFC$ is executed one will find $B[i] =
\triangle f(X_i, X_{i+1}, \cdots, X_{n-1}, X_n)$ for $i = 1, 2, \cdots, n-1, n$. When
$n = 1$ the initial state of B is irrelevant. W is a real array of
dimension $(2 + \bar{m})$ at least, where $\bar{m}$ is the maximum value of
m_i for $i = 1, 2, \cdots, n$. W is used for work space;

```
    begin real DENOM;  integer i, j, NK, NIN;
    if n = 1 then go to L1;
    NK := 1;
    for i := 1 step 1 until n do
    begin
    if X[i] = X[n] then begin NK := NK + 1;
        W[NK] := V[i] end
    end i;
    for i := n step −1 until 2 do
    begin W[1] := B[i − 1];  B[i] := W[2];
    NIN := if n − i + 2 < NK then n − i + 2 else NK;
    for j := NIN step −1 until 2 do
    begin
    DENOM := X[n] − X[i + j − 3];
    if DENOM ≠ 0 then go to L2;
    W[j] := W[j + 1];
    if NK − j − 1 ≠ 0 then go Cont;
    NK := NK − 1;
    go to Cont;
L2:   W[j] := (W[j] − W[j − 1])/DENOM;
Cont:  end j
    end i;
    B[1] := W[2];
    go to L3;
L1:  B[1] := V[1];
L3:  DVDFC := B[1]
    end DVDFC
```

The following program segment is an example of how $DVDFC$ can
be used to construct a table of forward or backward differences.
for $n := 1$ **step** 1 **until** N **do**
begin
$X[n] := \cdots$; $V[n] := \cdots$; $F[n] := DVDFC(n, X, V, B, W)$
end;
The array F can be used in $FNEWT(z, N, X, F, R, D, E)$ or the
array B in $BNEWT(z, N, X, B, P, D, E)$. See algorithms "Newton interpolation with forward (backward) divided differences."
$DVDFC$ has been written as a Fortran II function and is available from I.C.S., University of Toronto;

CERTIFICATION OF ALGORITHM 167
   CALCULATION OF CONFLUENT DIVIDED DIF-
   FERENCES [W. Kahan and I. Farkas, *Comm.
   ACM*, Apr. 1963]
CERTIFICATION OF ALGORITHM 168
   NEWTON INTERPOLATION WITH BACKWARD
   DIVIDED DIFFERENCES [W. Kahan and I.
   Farkas, *Comm. ACM*, Apr. 1963]
CERTIFICATION OF ALGORITHM 169
   NEWTON INTERPOLATION WITH FORWARD
   DIVIDED DIFFERENCES [W. Kahan and I.
   Farkas, *Comm. ACM*, Apr. 1963.]
Henry C. Thacher, Jr.*
Argonne National Laboratory, Argonne, Ill.

The bodies of these procedures were tested on the LGP-30
computer using the Dartmouth Scalp compiler. Compilation and
execution revealed no syntactical or mathematical errors.

It is to be noted that, although with Algorithm 169, reducing
the value of N from that used to generate F leads to an interpolation polynomial based on fewer points, this is not true for Algorithm 168. This flexibility could be supplied by adding an
additional formal parameter, *deg*, say, to the procedure, and by
making the **for** statement read:

$$\text{"for } i := N - deg \text{ step 1 until } N \text{ do} \cdots \text{"}$$

The logic of the error estimate in Algorithms 168 and 169 is not
entirely clear. However, it appears that the estimate can be adjusted for different precision of arithmetic by adjusting the constant $3_{10}-8$ appropriately. For the Scalp arithmetic, this constant
was changed to $1_{10}-7$.

The algorithms were tested on the examples given by Milne-
Thomson [*The Calculus of Finite Differences*, p. 4, Macmillan,
1951] and by Milne [*Numerical Calculus*, p. 204, Princeton, 1949].
In both examples, Algorithm 167 reproduced the divided difference table, and both Algorithms 168 and 169 reproduced the input
values. As a check of the calculation of confluent divided differences, values of the exponential function of its first two derivatives at $x = 5.0$ and 6.0 were used. The difference table shown in
Table A was obtained.

* Work supported by the U. S. Atomic Energy Commission.

TABLE A

n	$X[n]$	$V[n]$	$B[n]$	$B[n-1]$	$B[n-2]$	$B[n-3]$	$B[n-4]$	$B[n-5]$
1	5.0	148.4132	148.4132					
2	5.0	148.4132	148.4132	148.4132				
3	6.0	403.4288	403.4287	255.0155	106.6023			
4	6.0	403.4288	403.4287	403.4287	148.4132	41.81091		
5	5.0	74.20658	148.4132	255.0155	148.4132	41.81091	9.415191	
6	6.0	201.7144	403.4287	255.0155	148.4132	53.30115	11.49023	2.075043

The forward differences lie along the top diagonal.

Use of these results with B_{NEWT} and with F_{NEWT} gave the following results, for $N = 6$.

z	B_{NEWT}			F_{NEWT}		
	P	D	E	R	D	E
5.000000	148.4132	148.4132	$.4567298 \times 10^{-4}$	148.4132	148.4132	$.7420658 \times 10^{-5}$
5.500000	244.6973	244.6924	$.4173722 \times 10^{-4}$	244.6973	244.6924	$.3078276 \times 10^{-4}$
6.000000	403.4287	403.4287	$.2017143 \times 10^{-4}$	403.4287	403.4287	$.7441404 \times 10^{-4}$

ALGORITHM 168
NEWTON INTERPOLATION WITH
BACKWARD DIVIDED DIFFERENCES
W. Kahan and I. Farkas
Institute of Computer Science, University of Toronto,
Canada

procedure $BNEWT(z, N, X, B, P, D, E)$; **value** z, N;
real z, P, D, E; **integer** N; **real array** X, B;
comment X is a real array of dimension at least N in which
$X[i] = X_i$ for $i = 1, 2, 3, \cdots, N$. The values X_i need not be
distinct nor in any special order, but once the array X is chosen
it will fix the interpretation of the array B. B is a real array of
dimension at least N and contains the backward divided differences $B[i] = \triangle f(X_i, X_{i+1}, \cdots, X_N)$ $i = 1, 2, \cdots, N$. If two
or more of the values X_i are equal then some of the B's must
be confluent divided differences, see algorithm: "Calculation of
confluent divided differences." P is the value of the following
polynomial in z of degree $N-1$ at most, $B(N) + (z-X_N)\cdot$
$\{B(N-1) + (z-X_{N-1})\{B(N-2) + \cdots + (z-X_2)B(1)\} \cdots \}\}$.
This polynomial is an interpolation polynomial which would,
but for rounding errors, match values of the function $f(x)$ and
any of its derivatives that $DVDFC$ might have been given. D
is the value of the derivative of P. E is the maximum error in
P caused by roundoff during the execution of $BNEWT$. The
error estimate is based upon the assumption that the result of
each floating point arithmetic operation is truncated to 27 significant binary digits as is the case in FORTRAN programs on
the 7090. $BNEWT$ has been written as a FORTRAN II subroutine
and is available from I.C.S., University of Toronto;
begin real $z1$; **integer** i;
$P := D := E := 0$;
for $i := 1$ **step** 1 **until** N **do**
begin
$z1 := z - X[i]$;
$D := P + z1 \times D$;
$P := B[i] + z1 \times P$;
$E := abs(P) + E \times abs(z1)$
end;
$E := (1.5 \times E - abs(P)) \times 3_{10} - 8$
end $BNEWT$

CERTIFICATION OF ALGORITHM 167
    CALCULATION OF CONFLUENT DIVIDED DIF-
        FERENCES [W. Kahan and I. Farkas, *Comm.
    ACM*, Apr. 1963]
CERTIFICATION OF ALGORITHM 168
    NEWTON INTERPOLATION WITH BACKWARD
        DIVIDED DIFFERENCES [W. Kahan and I.
    Farkas, *Comm. ACM*, Apr. 1963]
CERTIFICATION OF ALGORITHM 169
    NEWTON INTERPOLATION WITH FORWARD
        DIVIDED DIFFERENCES [W. Kahan and I.
    Farkas, *Comm. ACM*, Apr. 1963.]
Henry C. Thacher, Jr.*
Argonne National Laboratory, Argonne, Ill.

The bodies of these procedures were tested on the LGP-30
computer using the Dartmouth SCALP compiler. Compilation and
execution revealed no syntactical or mathematical errors.

It is to be noted that, although with Algorithm 169, reducing
the value of N from that used to generate F leads to an interpolation polynomial based on fewer points, this is not true for Algorithm 168. This flexibility could be supplied by adding an
additional formal parameter, *deg*, say, to the procedure, and by
making the **for** statement read:

"**for** $i := N - deg$ **step** 1 **until** N **do** $\cdots$ "

The logic of the error estimate in Algorithms 168 and 169 is not
entirely clear. However, it appears that the estimate can be adjusted for different precision of arithmetic by adjusting the constant $3_{10}-8$ appropriately. For the SCALP arithmetic, this constant
was changed to $1_{10}-7$.

The algorithms were tested on the examples given by Milne-
Thomson [*The Calculus of Finite Differences*, p. 4, Macmillan,
1951] and by Milne [*Numerical Calculus*, p. 204, Princeton, 1949].
In both examples, Algorithm 167 reproduced the divided difference table, and both Algorithms 168 and 169 reproduced the input
values. As a check of the calculation of confluent divided differences, values of the exponential function of its first two derivatives at $x = 5.0$ and 6.0 were used. The difference table shown in
Table A was obtained.

TABLE A

n	$X[n]$	$V[n]$	$B[n]$	$B[n-1]$	$B[n-2]$	$B[n-3]$	$B[n-4]$	$B[n-5]$
1	5.0	148.4132	148.4132					
2	5.0	148.4132	148.4132	148.4132				
3	6.0	403.4288	403.4287	255.0155	106.6023			
4	6.0	403.4288	403.4287	403.4287	148.4132	41.81091		
5	5.0	74.20658	148.4132	255.0155	148.4132	41.81091	9.415191	
6	6.0	201.7144	403.4287	255.0155	148.4132	53.30115	11.49023	2.075043

The forward differences lie along the top diagonal.
Use of these results with B_{NEWT} and with F_{NEWT} gave the following results, for $N = 6$.

z	B_{NEWT}			F_{NEWT}		
	P	D	E	R	D	E
5.000000	148.4132	148.4132	$.4567298 \times 10^{-4}$	148.4132	148.4132	$.7420658 \times 10^{-5}$
5.500000	244.6973	244.6924	$.4173722 \times 10^{-4}$	244.6973	244.6924	$.3078276 \times 10^{-4}$
6.000000	403.4287	403.4287	$.2017143 \times 10^{-4}$	403.4287	403.4287	$.7441404 \times 10^{-4}$

ALGORITHM 169
NEWTON INTERPOLATION WITH
FORWARD DIVIDED DIFFERENCES
W. Kahan and I. Farkas
Institute of Computer Science, University of Toronto,
Canada

procedure $FNEWT(z, N, X, F, R, D, E)$; **value** z, N;
  real z, R, D, E; **integer** N; **real array** X, F;
comment X is a real array of dimension at least N in which
  $X[i] = X_i$ for $i = 1, 2, \cdots, N$. The values X_i need not be distinct nor in any special order, but once the array X is chosen
  it will fix the interpretation of the array F. F is a real array
  of dimension at least N and contains the forward divided
  differences $F[i] = \Delta f(X_1, X_2, \cdots, X_i)$ $i = 1, 2, \cdots, N$. If
  two or more of the values X_i are equal then some of the F's
  must be confluent divided differences, see algorithm: "Calculation of confluent divided differences." R is the value of the following polynomial in z of degree $N-1$ at most, $F(1) + (z-X_1) \cdot$
  $\{F(2) + (z-X_2)\{F(3) + \cdots + (z-X_{N-1})F(N)\} \cdots \}\}$. This
  polynomial is an interpolation polynomial which would, but
  for rounding errors, match values of the function $f(x)$ and any
  of its derivatives that $DVDFC$ might have been given. D is the
  value of the derivative of R. E is the maximum error in R
  caused by roundoff during the execution of $FNEWT$. The
  error estimate is based upon the assumption that the result of
  each floating-point arithmetic operation is truncated to 27
  significant binary digits as is the case in Fortran programs
  on the 7090. $FNEWT$ has been written as a Fortran II subroutine and is available from I.C.S., University of Toronto;
begin real $z1$; **integer** i;
$R := D := E := 0$;
for $i := N$ **step** -1 **until** 1 **do**
begin
$z1 := z - X[i]$;
$D := R + z1 \times D$;
$R := F[i] + z1 \times R$;
$E := abs(R) + abs(z1) \times E$
end;
$E := (1.5 \times E - abs(R)) \times 3_{10} - 8$
end $FNEWT$

CERTIFICATION OF ALGORITHM 167
  CALCULATION OF CONFLUENT DIVIDED DIF-
    FERENCES [W. Kahan and I. Farkas, *Comm.
    ACM*, Apr. 1963]
CERTIFICATION OF ALGORITHM 168
  NEWTON INTERPOLATION WITH BACKWARD
    DIVIDED DIFFERENCES [W. Kahan and I.
    Farkas, *Comm. ACM*, Apr. 1963]
CERTIFICATION OF ALGORITHM 169
  NEWTON INTERPOLATION WITH FORWARD
    DIVIDED DIFFERENCES [W. Kahan and I.
    Farkas, *Comm. ACM*, Apr. 1963.]
Henry C. Thacher, Jr.*
Argonne National Laboratory, Argonne, Ill.

The bodies of these procedures were tested on the LGP-30
computer using the Dartmouth Scalp compiler. Compilation and
execution revealed no syntactical or mathematical errors.
  It is to be noted that, although with Algorithm 169, reducing
the value of N from that used to generate F leads to an interpolation polynomial based on fewer points, this is not true for Algorithm 168. This flexibility could be supplied by adding an
additional formal parameter, *deg*, say, to the procedure, and by
making the **for** statement read:

"for $i := N - deg$ step 1 until N do $\cdots$ "

  The logic of the error estimate in Algorithms 168 and 169 is not
entirely clear. However, it appears that the estimate can be adjusted for different precision of arithmetic by adjusting the constant $3_{10}-8$ appropriately. For the Scalp arithmetic, this constant
was changed to $1_{10}-7$.
  The algorithms were tested on the examples given by Milne-
Thomson [*The Calculus of Finite Differences*, p. 4, Macmillan,
1951] and by Milne [*Numerical Calculus*, p. 204, Princeton, 1949].
In both examples, Algorithm 167 reproduced the divided difference table, and both Algorithms 168 and 169 reproduced the input
values. As a check of the calculation of confluent divided differences, values of the exponential function of its first two derivatives at $x = 5.0$ and 6.0 were used. The difference table shown in
Table A was obtained.

TABLE A

n	$X[n]$	$V[n]$	$B[n]$	$B[n-1]$	$B[n-2]$	$B[n-3]$	$B[n-4]$	$B[n-5]$
1	5.0	148.4132	148.4132					
2	5.0	148.4132	148.4132	148.4132				
3	6.0	403.4288	403.4287	255.0155	106.6023			
4	6.0	403.4288	403.4287	403.4287	148.4132	41.81091		
5	5.0	74.20658	148.4132	255.0155	148.4132	41.81091	9.415191	
6	6.0	201.7144	403.4287	255.0155	148.4132	53.30115	11.49023	2.075043

The forward differences lie along the top diagonal.
Use of these results with B_{NEWT} and with F_{NEWT} gave the following results, for $N = 6$.

z	B_{NEWT}			F_{NEWT}		
	P	D	E	R	D	E
5.000000	148.4132	148.4132	$.4567298 \times 10^{-4}$	148.4132	148.4132	$.7420658 \times 10^{-5}$
5.500000	244.6973	244.6924	$.4173722 \times 10^{-4}$	244.6973	244.6924	$.3078276 \times 10^{-4}$
6.000000	403.4287	403.4287	$.2017143 \times 10^{-4}$	403.4287	403.4287	$.7441404 \times 10^{-4}$

ALGORITHM 170
REDUCTION OF A MATRIX CONTAINING
POLYNOMIAL ELEMENTS
Paul E. Hennion
Giannini Controls Corp., Astromechanics Res. Div.,
Berwyn, Penn.

real procedure *POLYMATRIX* (*A, NCOL, N, COE, NP1*);
  value *A, NCOL, N*; **real array** *A*; **integer** *NCOL, N*;
comment this procedure will expand a general determinant,
  where each of the elements are polynomials in the Laplace complex variable. This program is useful for the investigation of
  dynamic stability problems when using the transfer function
  approach. The process is one of triangularization of a polynomial matrix with real coefficients whereupon multiplication
  of the diagonal elements the determinant polynomial is formed.
  The polynomial matrix as defined herein is a matrix whose
  elements are polynomials of the form $\sum_{i=0}^{N} a_i x^i$. When such a
  matrix is triangularized, all elements below the main diagonal
  are nulled. Then upon expanding, the nonvanishing terms are
  those formed by the product of these diagonal elements. Hence
  stability criteria may be checked by evaluating the roots of the
  characteristic equation thus formed using some suitable root
  extracting routine.

  Consider the polynomial matrix with quadratic elements
  ($N = 2$). In this case the three-dimensional input matrix A is
  size $A[1:NCOL, 1:NCOL, 1:M]$, where $NCOL$ is the order of
  the matrix and $M = N \times NCOL + 1$. Here the first subscript
  of A refers to the row, the second to the column, and the third
  to the polynomial coefficient. Therefore, prior to entry the constant term of a general polynomial element is contained in
  $A[i, j, 1]$, the linear term is contained in $A[i, j, 2]$, and the
  quadratic term in $A[i, j, 3]$. Upon completion of the routine, the
  coefficients of the determinant polynomial are contained in
  $COE[1:M]$. The constant coefficient being in $COE[1]$, the linear
  coefficient in $COE[2]$, the quadratic coefficient in $COE[3]$, etc.
  The variable $NP1$ will specify the number of coefficients of the
  determinant polynomial. In general $NP1 \neq M$ since some terms
  may vanish during the expansion.

  If the polynomials comprising the matrix elements are not
  all of equal degree, set N prior to entry equal to the degree of
  the highest ordered polynomial;
begin real *sa, sb*; **integer** *i, j, k, j1, j2, j3, j4, j5, j6, j7, j8, j9, j10,*
  j11, NP1, M; **array** *C1*[1:M], *C2*[1:M], *COE*[1:M];
  integer array *MAT* [1:NCOL, 1:NCOL];
start: $M := N \times NCOL + 1$; **for** $i := 1$ **step** 1 **until** $NCOL$ **do**
  begin for $j := 1$ **step** 1 **until** $NCOL$ **do begin** $MAT[i,j] := 0$;
  for $k := 1$ **step** 1 **until** M **do begin**
  if $A[i,j,k] \neq 0$ **then** $MAT[i,j] := k$; **end end end**; $j1 := 1$:
$L0$: $j9 := 0$; **for** $i := j1$ **step** 1 **until** $NCOL$ **do begin**
  if $MAT[i,j1] < 0$ **then go to** *exit*;
  else if $MAT[i, j1] = 0$ **then go to** $L1$
  else $j9 := j9 + 1$; $j3 := i$;
$L1$: **end**; **if** $(j9 - 1) < 0$ **then go to** *exit*
  else if $(j9 - 1) > 0$ **then go to** $L2$
  else if $(j3 - j1) < 0$ **then go to** *exit*
  else if $(j3 - j1) = 0$ **then go to** $L12$
  else for $j := j1$ **step** 1 **until** $NCOL$ **do**
  begin $j2 := MAX(MAT[j3,j], MAT[j1,j])$; $j4 := MAT[j3,j]$;
  $MAT[j3,j] := MAT[j1, j]$; $MAT[j1,j] := j4$;

for $k := 1$ **step** 1 **until** $j2$ **do**
  begin $sa := A[j3, j,k]$; $A[j3,j,k] := A[j1, j, k]$;
  $A[j1,j,k] := -sa$; **end end**; **go to** $L12$;
$L2$: $j3 := j1 + 1$; **for** $i := j3$ **step** 1 **until** $NCOL$ **do begin**
$L3$: **if** $(MAT[i,j1]) < 0$ **then go to** *exit*
  else if $[MAT[i,j1]] = 0$ **then go to** $L11$
  else if $(MAT[j1,j1]) < 0$ **then go to** *exit*
  else if $(MAT[j1,j1]) = 0$ **then go to** $L4$
  else if $(MAT[i,j1] - MAT[j1,j1]) \geqq 0$ **then go to** $L5$ **else**
$L4$: **for** $j := j1$ **step** 1 **until** $NCOL$ **do begin**
  $j2 := MAX(MAT[j1,j], MAT[i, j])$; $j4 := MAT[j1,j]$;
  $MAT[j1,j] := MAT[i,j]$; $MAT[i,j] := j4$;
  for $k := 1$ **step** 1 **until** $j2$ **do begin** $sa := A[i,j,k]$;
  $A[i,j,k] := A[j1,j,k]$; $A[j1,j,k] := -sa$;
  end end; **go to** $L3$;
  comment Interchange row i with $j1$;
$L5$: $j7 := MAT[i,j1]$; $j5 := MAT[j1,j1]$; $j6 := j7 - j5$;
  $sb := A[i,j1,j7]/A[j1,j1,j5]$;
  if $(abs(sb) - 4) < 0$ **then go to** $L6$
  else if $(j6) < 0$ **then go to** *exit*
  else if $(j6) = 0$ **then go to** $L4$ **else**
$L6$: **for** $j := j1$ **step** 1 **until** $NCOL$ **do begin** $j5 := MAT[j1, j]$;
  for $k := 1$ **step** 1 **until** $j5$ **do begin** $j7 := k + j6$;
  if $(j7 - M) > 0$ **then go to** $L10$ **else**
$L7$: **if** $(abs(A[i,j,j7] - sb \times A[j1,j,k]) - 2_{10} - 8) \leqq 0$ **then go to** $L8$
  else $A[i,j,j7] := A[i,j,j7] - sb \times A[j1,j,k]$;
  go to $L9$;
$L8$: $A[i,j,j7] := 0$;
$L9$: **end end**;
$L10$: **for** $j := j1$ **step** 1 **until** $NCOL$ **do begin**
  $j7 := MAX(MAT[i,j], MAT[j1,j] + j6)$; $MAT[i,j] := 0$;
  for $k := 1$ **step** 1 **until** M **do begin if** $(A[i,j,k]) \neq 0$ **then**
  $MAT[i,j] := k$ **end end**;

$L11$: **end**; **go to** $L0$;
$L12$: $j1 := j1 + 1$; **if** $(j1 - NCOL) < 0$ **then go to** $L0$ **else**
  for $j := 1$ **step** 1 **until** $NCOL$ **do begin**
  $j2 := MAT[j,j]$;
  for $k := 1$ **step** 1 **until** $j2$ **do** $C1[k] := A[j,j,k]$;
$L13$: **if** $(j-1) < 0$ **then go to** *exit*
  else if $(j-1) = 0$ **then go to** $L14$
  else for $k := 1$ **step** 1 **until** $NP1$ **do** $C2[k] := COE[k]$;
  for $k := 1$ **step** 1 **until** M **do** $COE[k] := 0$;
  if $(j2) < 0$ **then go to** *exit*
  else if $(j2) = 0$ **then go to** $L15$
  else for $k := 1$ **step** 1 **until** $j2$ **do begin**
  for $j10 := 1$ **step** 1 **until** $NP1$ **do begin**
  $j11 := k + j10 - 1$;
  $COE[j11] := COE[j11] + C1[k] \times C2[j10]$;
  end end; $NP1 := j11$; **go to** $L15$;
$L14$: **for** $k := 1$ **step** 1 **until** $j2$ **do** $COE[k] := C1[k]$;
  $NP1 := j2$;
$L15$: **end**;
exit: **end** *POLYMATRIX*

REMARK ON ALGORITHM 170
REDUCTION OF A MATRIX CONTAINING POLYNOMIAL ELEMENTS [P. E. Hennion, *Comm. ACM*, Apr. 1963]

P. E. HENNION
Giannini Controls Corp., Berwyn, Penn.

Four typographical errors were found upon reviewing the procedure. The following corrections should be made:
(1) The increment for the **for** statement of line start:, should be 1.
(2) The colon at the end of the third line after line start:, should be replaced by a semicolon.
(3) The semicolon at the end of the first line after line LO:, may be removed.
(4) The last statement of the first column should read:

$$MAT[i,j] := k; \textbf{ end end};$$

CERTIFICATION OF ALGORITHM 170 [F3]
REDUCTION OF A MATRIX CONTAINING POLYNOMIAL ELEMENTS [P. E. Hennion, *Comm. ACM* *6* (April 1963), 165; *6* (Aug. 1963), 450]

KAREN B. PRIEBE (Recd. 18 Dec. 1963 and 18 Feb. 1964)
Woodward Governor Co., Rockford, Ill.

Algorithm 170 was translated into FAST for the NCR 315 and gave satisfactory results with the following corrections:
1. **real procedure** ... **integer** $NCOL, N$; should be replaced by

    procedure $POLYMATRIX$ ($A, NCOL, N, COE, NP1$);
      value $NCOL, N$; **real array** A, COE;
      integer $NCOL, N, NP1$;

2. At the end of the first comment add:
The global integer procedure MAX is assumed and furnishes the maximum of two integers.
3. **integer** $i, j, k, \ldots COE[1:M]$;
should be replaced by
  integer $i, j, k, j1, j2, j3, j4, j5, j6, j7, j8, j9, j10, j11, M$;
  array $C1, C2[1:N{\times}NCOL+1]$;

4. Immediately after *start*: the statement

$$NP1 := N + 1;$$

should be added, and the third line after *start*: i.e.,

    for $k := 1$ **step** 1 **until** M **do begin**

should be replaced by

    for $k := 1$ **step** 1 **until** $NP1$ **do begin**

5. The third line after $L10$: i.e.,

    for $k := 1$ **step** 1 **until** M **do** ...

should be replaced by

    for $k := 1$ **step** 1 **until** $j7$ **do** ...

The last two changes simply shorten both of the indicated FOR statements.

[EDITOR'S NOTE. In addition to the above corrections, we have two comments on the Remark on Algorithm 170 by Hennion, *loc. cit.*, p. 450:
First, the semicolon at the end of the first line after $L0$ MUST be removed.
Second, correction (4) is irrelevant.

The referee confirms that a transcription into Burroughs Extended ALGOL of the program as corrected by Mrs. Priebe runs on the B5000.—G.E.F.]

Note. There is no algorithm for the number 171.
Inadvertently this number was never assigned.

Note. There is no algorithm for the number 172.
Inadvertently this number was never assigned.

ALGORITHM 173
ASSIGN
Otomar Hájek
Research Institute of Mathematical Machines, Prague,
  Czechoslovakia

procedure *assign* (*a*) the value of : (*b*) with dimension : (*dim*)
  indices : (*ind*) bounds : (*low, up*) tracer : (*j*);
value *dim;*   **integer** *dim, ind, low, up, j*;
comment This procedure uses Jensen's device (cf. Algol
  Report, procedure Innerproduct) twice: the *a*, *b* may depend on
  ind and also *ind, low, up* may depend on *j*;
begin
  $j := dim$;
  for *ind* := *low* **step** 1 **until** *up* **do**
    if $dim > 1$
    then
      begin
        assign $(a, b, dim-1, ind, low, up, j)$;
        $j := dim$
      end
    else $a := b$
end assign;
comment The obvious use of *"assign"* is in assigning the value
  of one array to another. The point here is that one procedure
  declaration serves for all the dimensions used. In fact, the
  dimension may even be a variable: thus a procedure essentially
  identical with *"assign"* was used by the author in implementing
  the recursive own process in an Algol compiler.

  However, in addition to this, *"assign"* can have further
  functions, as illustrated below. The activation *assign* $(a,$ (**if**
  $i=1$ **then false else** $a) \lor b_{i,i}$, 1, i, 1, n, j) will calculate the
  join-trace of a Boolean 2-dimensional array *b*.
  assign $(a_{i_1.i_2}$, (**if** $i_3=1$ **then** 0 **else** $a_{i_1.i_2}$) $+ b_{i_1.i_3} \times c_{i_3,i_2}$,
  3, i_j, 1, **if** $j = 1$ **then** n **else if** $j = 2$ **then** m **else** p, j)
  will assign to *a* the matrix product of *b*, *c*. It may be noticed that,
  more generally, *"assign"* will perform all the tensor operations,
  e.g. tensor multiplication, alternation, etc.

CERTIFICATION OF ALGORITHM 173
ASSIGN [Otomar Hájek, *Comm. ACM*, June 1963]
R. S. Scowen
English Electric Co. Ltd., Whetstone, Leicester, England

  Algorithm 173 (ASSIGN) has been tested successfully using
the Deuce Algol 60 compiler. The only changes necessary were
the addition of specifications for the formal parameters *a*, *b*
(Deuce Algol 60 compiler requires specifications for all formal
parameters).
  The author's example, *assign* $(a[i[1], i[2]]$, (**if** $i[3]=1$ **then** 0.0
else $a[i[1], i[2]]) + b[i[1], i[3]] \times c[i[3], i[2]]$, 3, $i[j]$, 1, **if** $j = 1$
then n **else if** $j = 2$ **then** m **else** p, j);
did form the matrix product $B \times C$ and store it in A.
  The algorithm was also used to read a matrix into the computer
using the procedure call
  assign $(b[i[1], i[2]]$, *read*, 2, $i[j]$, 1,
  if $j = 1$ **then** n **else** p, j);
(*read* is a real procedure which takes the value given by the next
number on the input tape).

These examples took about three times as long to run as the
simpler equivalent statements
  for $i := 1$ **step** 1 **until** n **do**
  for $j := 1$ **step** 1 **until** m **do**
  begin
    $a[i, j] := 0.0$;
    for $k := 1$ **step** 1 **until** p **do**
    $a[i, j] := a[i, j] + b[i, k] \times c[k, j]$
  end;
and
  for $j := 1$ **step** 1 **until** p **do**
  for $i := 1$ **step** 1 **until** n **do**
  $b[i, j] := read$;

CERTIFICATION OF ALGORITHM 173
ASSIGN [O. Hájek, *Comm. ACM*, July 1963]
Z. Filsak and L. Vrchovecká
Research Institute of Mathematical Machines, Prague,
  and Computing Center Kancelářské stroje, Prague

  The algorithm was modified for input to the Elliott-Algol
system as follows. In Elliott-Algol, name-called parameters in
recursive procedures are prescribed. Luckily, the only parameter
which varies during the recursive call in the body of *Assign* is
called by value (it is the parameter *dim* which determines depth
of recursion). The body of *Assign* was replaced by (i) a procedure
declaration *Ass(dim)*, whose body is that of the original *Assign*,
but with the recursive call of *Assign* replaced by that of *Ass*,
and (ii) a single statement, the activation of *Ass(dim)*.
  The resulting procedure was tested (on the National-Elliott
803 in the Computing Center), on a rather large set of examples,
including those described in the text following Algorithm 173.
It was found that in the last example, matrix multiplication,
indices i_1 and i_3 should be interchanged throughout.
  No changes of the algorithm itself were necessary. It seems
that the modification described above, motivated by limitations
of Elliott-Algol, also improve efficiency, at least for large dimensions of the arrays concerned.

ALGORITHM 174
A POSTERIORI BOUNDS ON A
   ZERO OF A POLYNOMIAL*

ALLAN GIBB

University of Alberta, Calgary, Alberta, Canada

comment The procedures below make use of Algorithm 61, Procedures for Range Arithmetic [*Comm. ACM 4* (1961)]. It is assumed that the procedures below and the range arithmetic procedures are contained in an outer block and, therefore, that the procedures are available as required. Together the procedures make possible an attempt to determine absolute bounds on a zero of a polynomial given an initial estimate of the zero. The procedures below are given for the complex case but may readily be adapted for the real case;

procedure $RngPlyC$ (N, A, Z, P);

comment $RngPlyC$ finds bounds $[P1, P2] + i[P3, P4]$ on the value of an nth degree polynomial $\sum_{k=0}^{n} \{[a_{4k+1}, a_{4k+2}] + i[a_{4k+3}, a_{4k+4}]\}z^k$ with complex range coefficients for a complex range argument $z = [Z1, Z2] + i[Z3, Z4]$;

integer N;   **array** A, Z, P;
begin integer K, J;   **array** $X, Y[1:4]$;
$P[1] := P[2] := P[3] := P[4] := 0$;
for $K := 4 \times N$ **step** -4 **until** 0 **do**
  begin for $J := 1$ **step** 1 **until** 4 **do** $X[J] := A[K+J]$;
  $RNGMPYC$ $(P[1], P[2], P[3], P[4], Z[1], Z[2], Z[3], Z[4], Y[1],$
      $Y[2], Y[3], Y[4])$;
  $RNGSUMC$ $(Y[1], Y[2], Y[3], Y[4], X[1], X[2], X[3], X[4],$
      $P[1], P[2], P[3], P[4])$
  end
end;

procedure $RngAbsC$ (A, C);

comment $RngAbsC$ produces the range absolute value $[C1, C2]$ of the complex range number $[A1, A2] + i[A3, A4]$;

array A, C;
begin array $B[1 : 4]$;
$RANGESQR$ $(A[1], A[2], B[1], B[2])$;
$RANGESQR$ $(A[3], A[4], B[3], B[4])$;
$RANGESUM$ $(B[1], B[2], B[3], B[4], C[1], C[2])$;
$C[1] := sqrt(C[1])$;
$C[2] := sqrt(C[2])$;

comment It is assumed that the accuracy of the $sqrt$ routine used is known and that the maximum error in $sqrt(C)$ is $\pm K \times CORRECTION (C)$. K is to be replaced below by its appropriate numerical value;

$C[1] := C[1] - K \times CORRECTION (C[1])$;
$C[2] := C[2] + K \times CORRECTION (C[2])$
end;

procedure $BndZrPlyC$ $(N, ZOR, ZOJ, A, W,)$;
integer N;   **real** ZOR, ZOJ;   **array** A, W;

comment $BndZrPlyC$ attempts to determine bounds $[W1, W2] + i[W3, W4]$ on a zero of an N-th degree polynomial in z with complex range coefficients. It is assumed that an estimate $ZO = ZOR + iZOJ$ of the zero is available. The following theorem is used. Assume f is regular at z_0 with $f'(z_0) \neq 0$. Let $h_0 = -f(z_0)/f'(z_0)$, let Δ be the region $|z - z_0| \leq r |h_0|$, and

* These procedures were developed under Office of Naval Research Contract Nonr-225(37) at Stanford University. The author wishes to thank Professor George E. Forsythe for assistance with this work.

assume that f is regular in Δ. If, for some $r > 0$, $|f'(z)| \geq (1/r)$. $|f'(z_0)|$ for all $z \in \Delta$ then Δ contains a zero of f (see [1], pp. 29–31);

begin integer I, J;   **array** $B[1:4 \times N]$, $E, F, FP, D[1:4]$, AF, $AFP, G[1:2]$;
real $RH, RHS, NL, NR, R, RNL, RNR$;
for $I := 1$ **step** 1 **until** N **do**
  begin $J := 4 \times I$;
  $RANGEMPY$ $(I, I, A[J+1], A[J+2], B[J-3], B[J-2])$;
  $RANGEMPY$ $(I, I, A[J+3], A[J+4], B[J-1], B[J])$
  end;
$E[1] := E[2] := ZOR$; $E[3] := E[4] := ZOJ$;
$RngPlyC(N, A, E, F)$;
$RngAbsC(F, AF)$;
$RngPlyC(N-1, B, E, FP)$;
$RngAbsC(FP, AFP)$;
$RANGEDVD(AF[1], AF[2], AFP[1], AFP[2], NL, NR)$;
$R := 2$;
1: $RANGEMPY(R, R, NR, NR, RNL, RNR)$;
$RANGESUM(ZOR, ZOR, -RNR, RNR, W[1], W[2])$;
$RANGESUM(ZOJ, ZOJ, -RNR, RNR, W[3], W[4])$;
comment We have replaced the disk of the theorem by a square;
$RngRlyC(N-1, B, W, D)$;
$RngAbsC(D, G)$;
if $G[1] = 0$ **then go to** *failure1*;
comment *failure1* and *failure2* are non-local labels;
$RANGEDVD(AFP[2], AFP[2], R, R, RH, RHS)$;
if $G[1] < RHS$ **then**
  begin $R := 2 \times R$;
  if $R > 1024$ **then go to** *failure2*;
  go to 1
  end
end

comment The following procedure may replace $BndZrPlyC$ above;

procedure $BndZrPlyC2$ (N, ZOR, ZOJ, A, W);
integer N;   **array** A, W;   **real** ZOR, ZOJ;

comment $BndZrPlyC2$ is similar to $BndZrPlyC$ above. The theorem used here follows. If, in the disk $|z - z_0| \leq 2 |h_0|$ we have $|f''(z)| \leq |f'(z_0)|/(2 |h_0|)$, then there is a unique zero in the disk (see [2], pp. 43–50];

begin integer I, J;   **array** $B[1:4 \times N]$, $C[1:4 \times N-4]$, F, D, P, $S[1:4]$, $X, T, Q, Y[1:2]$;   **real** V, VP, R, RL;
for $I := 1$ **step** 1 **until** N **do**
  begin $J := 4 \times I$;
  $RANGEMPY(I, I, A[J+1], A[J+2], B[J-3], B[J-2])$;
  $RANGEMPY(I, I, A[J+3], A[J+4], B[J-1], B[J])$
  end;
for $I := 1$ **step** 1 **until** $N - 1$ **do**
  begin $J := 4 \times I$;
  $RANGEMPY(I, I, B[J+1], B[J+2], C[J-3], C[J-2])$,
  $RANGEMPY(I, I, B[J+3], B[J+4], C[J-1], C[J])$
  end;
$D[1] := D[2] := ZOR$;
$D[3] := D[4] := ZOJ$;
$RngPlyC(N, A, D, F)$;
$RngPlyC(N-1, B, D, P)$;
$RngAbsC(F, T)$;
$RngAbsC(P, X)$;
if $X[1] = 0$ **then go to** *failure1*;

comment *failure*1 and *failure*2 are non-local labels;
$RANGEDVD(T[1], T[2], X[1], X[2], Q[1], Q[2])$;
$RANGEMPY(2, 2, Q[2], Q[2], RL, R)$;
$RNGSUMC(-R, R, -R, R, ZOR, ZOR, ZOJ, ZOJ, W[1], W[2],$
    $W[3], W[4])$;
$RngPlyC(N - 2, C, W, S)$;
$RngAbsC(S, Y)$;
$RANGEDVD(X[1], X[1], R, R, V, VP)$;
if $Y[2] > V$ **then go to** *failure*2
end

References:

1. GIBB, ALLAN. ALGOL procedures for range arithmetic. Tech.
    Report No. 15, Appl. Math. and Statistics Laboratories,
    Stanford University (1961).
2. OSTROWSKI, A. M. *Solution of equations and systems of equations.*
    Academic Press, New York, 1960.

ALGORITHM 175
SHUTTLE SORT
C. J. Shaw and T. N. Trimble
System Development Corporation, Santa Monica, Calif.

```
procedure  shuttle sort (m, Temporary, N);
value m;  integer m;  array N[1:m];
comment  This procedure sorts the list of numbers N[1] through
    N[m] into numeric order, by exchanging out-of-order number
    pairs. The procedure is simple, requires only Temporary as
    extra storage, and is quite fast for short lists (say 25 numbers)
    and fairly fast for slightly longer lists (say 100 numbers). For
    still longer lists, though, other methods are much swifter. The
    actual parameters for Temporary and N should, of course, be
    similar in type;
begin integer i, j;
for i := 1 step 1 until m − 1 do
  begin
  for j := i step −1 until 1 do
    begin
    if N[j] ≤ N[j+1] then go to Test;
Exchange:  Temporary := N[j];  N[j] := N[j+1];
          N[j+1] :=  Temporary;  end of j loop;
Test:  end of i loop
  end shuttle sort
```

CERTIFICATION OF ALGORITHM 175
SHUTTLE SORT [C. J. Shaw and T. N. Trimble, *Comm.
  ACM*, June 1963]
George R. Schubert*
University of Dayton, Dayton, Ohio

* Undergraduate research project, Computer Science Program, Univ. of Dayton.

Algorithm 175 was translated into Balgol and ran successfully on the Burroughs 220. The following actual sorting times were observed:

Number of Items	*Average Time* (sec)
25	1.6
50	6.2
100	25.8
250	181
500	684

The algorithm can be extended so that the sort is made on one array, while retaining a one-to-one correspondence to a second array. This is done by inserting immediately before **end** of the j loop the following:

Temporary := $Y[j]$; $Y[j]$:= $Y[j + 1]$; $Y[j + 1]$:= *Temporary*; where $Y[k]$ is the element to be associated with $N[k]$. Other variations are obviously possible.

REMARK ON ALGORITHM 175
SHUTTLE SORT [C. J. Shaw and T. N. Trimble, *Comm.
  ACM 6*, June 1963]
O. C. Juelich
North American Aviation, Inc., Columbus, Ohio

The authors of this algorithm do well to remind the reader that "Shuttle Sort" is not an efficient procedure, except for lists of items so short that they do not justify the housekeeping apparatus needed by the usual sorting routines.

The algorithm as published is not free from errors. The statement

```
for j := i step − 1 until 1 do
```

should be replaced by either:

```
for j := m − 1 step − 1 until i do
```

or

```
for j := 1 step 1 until m − i do
```

In the former case the process can be visualized as placing the ith smallest element in place on the ith pass; in the latter the ith largest element is put in place on the ith pass.

The label "*Test*" should precede the delimiter "**end** of j loop" rather than the "**end** of i loop". The algorithm can be slightly accelerated by rewriting the body of the procedure·

```
begin integer i, j, j max;
  i := m − 1;
loop:  j max := 1;
  for j := 1 step 1 until i do
    begin
    compare:  if N[j] > N[j + 1] then
      begin Exchange:  Temporary := N[j];
      N[j] := N[j + 1];
      N[j + 1] := Temporary;
      j max := j
    end Exchange;
  end of j loop;
  i := j max;
  if i > 1 then go to loop;
end shuttle sort
```

The revised procedure body will eliminate redundant iterations when some of the data is already ordered.

It was studied in this form by R. L. Boyell and the writer on the Ordvac at Ballistics Research Laboratories, Aberdeen Proving Ground, in 1955. For randomly ordered data the i-loop may be expected to be executed about $m - \sqrt{m}$ times.

REMARK ON ALGORITHM 175
SHUTTLE SORT [C. J. Shaw and T. N. Trimble, *Comm.
  ACM 6* (June 1963), 312; G. R. Schubert, *Comm.
  ACM 6* (Oct. 1963), 619; O. C. Juelich, *Comm. ACM
  6* (Dec. 1963), 739]
Otto C. Juelich (Recd. 18 Dec. 1963)
North American Aviation, 4300 E. Fifth Ave., Columbus, Ohio

The appearance of Schubert's certification has caused me to restudy the algorithm. What I supposed were errors amount to a rearrangement of the order in which the comparisons are carried out. The efficiency of the algorithm is not much affected by the rearrangement, since the number of executions of the statements labeled *Exchange* remains the same.

ALGORITHM 176
LEAST SQUARES SURFACE FIT
T. D. Arthurs
The Boeing Company, Transport Division, Renton, Wash.

```
procedure  SURFIT (F, z, W, m, n) answers: (a, e, rms);
integer m, n;  real rms;  array F, z, W, e;
procedure  Invert, sqrt;
comment Given a set of m ordinates and the corresponding
   values of n prescribed general functions, (fᵢ), of one or more
   linearly independent variables, this procedure fits the points,
   in the least squares sense, with a function of the form a₁f₁ + a₂f₂
   + . . . + aₙfₙ where aᵢ are the unknown coefficients. Also com-
   puted are the vectors of residuals (eⱼ) and their lengths (rms).
   Provision is made for weighting the data points. Essentially, the
   matrix equation FᵀWFa = FᵀWz is solved, where a is the vector
   of unknowns, W is an m × m diagonal matrix of data point
   weights, z is the vector of ordinate values and F is the
   m × n matrix of corresponding function values. The availa-
   bility of a procedure Invert, which replaces a real matrix with
   its inverse, is assumed;
begin integer i, j, k;  real sqsum, g;  array G[1:n, 1:n];
   comment  G is working space for the inversion procedure;
   sqsum := 0;
   for i := 1 step 1 until n do
   for j := 1 step 1 until n do
   begin G[i, j] := 0;
     for k := 1 step 1 until m do
     G[i, j] := G[i, j] + F[k, i] × F[k, j] × W[k]
   end j;
   Invert (G, n);
   for i := 1 step 1 until n do
   begin a[i] := 0;
     for j := 1 step 1 until m do
     begin g := 0;
       for k := 1 step 1 until n do
       g := g + G[i, k] × F[j, k];
       a[i] := a[i] + g × z[j] × W[j]
     end j
   end i;
   for i := 1 step 1 until m do
   begin e[i] = y[i];
     for j := 1 step 1 until n do
     e[i] := e[i] − a[j] × F[i, j];
     sqsum := sqsum + e[i] ↑ 2
   end i;
   rms := sqrt (sqsum/m)
end SURFIT
```

Remark on Algorithm 176 [E2]
Least Squares Surface Fit [T.D. Arthurs, *Comm. ACM 6* (June 1963), 313]

Ernst Schuegraf [Recd. 1 Mar. 1971]
Department of Mathematics. St. Francis Xavier University, Antigonish, Nova Scotia, Canada

Algorithm 176 contains one misprint. The line which reads:
begin $e[i] = y[i]$;

should read:
begin $e[i] = z[i]$;

ALGORITHM 177
LEAST SQUARES SOLUTION WITH CONSTRAINTS
M. J. SYNGE
The Boeing Company, Transport Division, Renton, Wash.

```
procedure  CONLSQ (A, y, w, n, m, r) results: (x) residuals:
    (e, rms);
real rms;  integer n, m, r;  array A, y, w, x, e;  procedure
    abs, SURFIT;
comment  This procedure solves an overdetermined set of n
    simultaneous linear equations in m unknowns, Ax = y. The
    first r equations (r≤m) are satisfied exactly and the remaining
    n − r are satisfied as well as possible by the method of least
    squares. Each equation is assigned a weight from the vector w,
    although the first r weights have no relevance. This procedure
    may be used for curve or surface fitting when the approximating
    function or its derivatives are required to have fixed values at a
    number of points;
begin integer i, j, k, ii, ick;  integer array ic[1:m];
    array B[1:n−r, 1:m−r];  real Amax;
    for i := 1 step 1 until r do
    begin k := 1;  for j := 2 step 1 until m do
        begin if abs (A[i, j]) > abs (A[i, k]) then k := j;  end;
        ic[i] := k;  Amax := A[i, k];  for j := 1 step 1 until m do
        A[i, j] := A[i, j]/Amax;  y[i] := y[i]/Amax;
        for ii := 1 step 1 until r do
        begin if ii = i then go to skip;  Amax := A[ii, k];
            for j := 1 step 1 until m do
            A[ii, j] := A[ii, j] − A[i, j] × Amax;
            y[ii] := y[ii] − y[i] × Amax;
    skip:  end ii
    end i;
    ick := r + 1;  for j := 1 step 1 until m do
    begin k := 1;
repeat:  if j = ic[k] then go to next;
    k := k + 1;  if r ≥ k then go to repeat;
    ic[ick] := j;  ick := ick + 1;
next:  end k;
    for i := r + 1 step 1 until n do
    begin for k := 1 step 1 until r do
        y[i] := y[i] − y[k] × A[i, ic[k]];
        for j := r + 1 step 1 until m do
        begin B[i, j] := A[i, ic[j]];
            for k := 1 step 1 until r do
            B[i, j] := B[i, j] − A[i, ic[k]] × A[k, ic[j]]
        end j
    end i;
    SURFIT  (B, y[r+1:n], w[r+1:n], n − r, m − r, x[r+1:m],
        e[r+1:n], rms);
comment  The procedure SURFIT is called to solve the reduced
    set of n − r simultaneous linear equations in m − r unknowns,
    Bx₂ = y₂', which have no constraints;
    for  j := r + 1 step 1 until m do x[ic[j]] := x[j];
    for  j := 1 step 1 until r do
    begin  x[ic[j]] := y[j];
        for i := r + 1 step 1 until m do
        x[ic[j]] := x[ic[j]] − A[j, ic[i]] × x[ic[i]]
    end j
end CONLSQ
```

REMARK ON ALGORITHM 177
LEAST SQUARES SOLUTION WITH CONSTRAINTS
[Michael J. Synge, Comm. ACM, June 63]
MICHAEL J. SYNGE
The Boeing Co., Transport Division, Renton, Wash.

In row-reducing the constraint equations, *CONLSQ* does not use full pivoting nor does it detect redundancy or inconsistency of the constraints; it was felt that the constraints were likely to be few in number and well-conditioned. However, these omissions may be made good by replacing the statement

$$ick := ick + 1;$$

by

$$done: ick := ick + 1;$$

and substituting the lines below for the first seven lines of the first compound statement of *CONLSQ*. If inconsistency is found, the procedure exits to the nonlocal label *inconsistent*. A roundoff tolerance, *eps*, is used in checking consistency, and some numerical value (e.g. 10^{-6}) should be substituted for it.

```
begin integer i, j, k, ii, ick, mr;  integer array ic[1 : m];
    array B[1 : n−r, 1 : m−r];
    real Amax, Atemp;
    for i := 1 step 1 until r do
    begin k := 1;  mr := i;  Amax := A[i, 1];
        for ii := i step 1 until m do
        begin for j := 1 step 1 until m do
            begin if abs(Amax ≥ abs(A[ii, j]) then go to nogo;
                mr := ii;  k := j;  Amax := A[ii, j];
nogo:  end j
        end ii;
        if abs(Amax) ≥ eps then go to allswell;  mr := i;
test:  if abs(y[mr]) ≥ eps then go to inconsistent else mr := mr + 1;
        if r ≥ mr then go to test else r := i − 1;
        go to done;
allswell:  for j := 1 step 1 until r do
        begin Atemp := A[mr, j];  A[mr, j] := A[i, j];
            A[i, j] := Atemp/Amax
        end j;
        Atemp := y[mr];  y[mr] := y[i];  y[i] := Atemp/Amax:
```

The Algorithm then continues with the line:
```
    for ii := 1 step 1 until r do
```

ALGORITHM 178
DIRECT SEARCH
ARTHUR F. KAUPE, JR.
Westinghouse Electric Corp., Pittsburgh, Penn.

procedure *direct search* (*psi, X DELTA, rho, delta, S*);

value K, *DELTA*, *rho*, *delta*; **integer** K; **array** *psi*;
   real *DELTA*, *rho*, *delta*; **real procedure** S;

comment This procedure may be used to locate the minimum
   of the function S of K variables. A discussion of the use of this
   procedure may be found in: Robert Hooke and T. A. Jeeves,
   'Direct Search' Solution of Numerical and Statistical Problems
   [*J. ACM 8, 2* (1961), 212–229]. The notation is essentially that
   used in Appendix B of the cited paper. The exceptions being the
   spelling of the Greek letters and the introduction of notation to
   distinguish between the process of calculating a value of S and
   the value itself—thus $S(phi)$ and $Sphi$. A modified version of this
   procedure acceptable to the BAC compiler for the Burroughs
   205 and 220 computers has been prepared and run successfully;
begin real SS, $Spsi$, $Sphi$, *theta*; **array** *phi* [1:K]; **integer** K, k;
procedure E; **for** $k := 1$ **step** 1 **until** K **do**
   begin *phi* [k] := *phi* [k] + *DELTA*; $Sphi$:= $S(phi)$;
   if $Sphi < SS$ **then** SS := $Sphi$ **else**
     begin *phi* [k] := *phi* [k] − 2 × *DELTA*; $Sphi$:= $S(phi)$:
     if $Sphi < SS$ **then** SS := $Sphi$ **else** *phi* [k] := *phi* [k] + *DELTA*
   end E;
Start: $Spsi$:= $S(psi)$;
1: SS := $Spsi$;
for $k := 1$ **step** 1 **until** K **do** *phi* [k] := *psi* [k]; E;
if $SS < Spsi$ **then begin**
   2: **for** $k := 1$ **step** 1 **until** K **do begin**
     theta := *psi* [k];
     psi [k] := *phi* [k];
     phi [k] := 2 × *phi* [k] − *theta* **end**;
   $Spsi$:= SS; SS := $Sphi$:= $S(phi)$; E;
   if $SS < Spsi$ **then go to** 2 **else go to** 1 **end**;
3: **if** *DELTA* $\geq$ *delta* **then begin** *DELTA* := *rho* × *DELTA*;
   go to 1 **end end**

REMARK ON ALGORITHM 178 [E4]
DIRECT SEARCH [Arthur F. Kaupe, Jr., *Comm. ACM*
6 (June 1963), 313]
M. BELL AND M. C. PIKE (Recd. 15 Nov. 1965 and 22
Apr. 1966)
Institute of Computer Science, University of London,
London, England, and Medical Research Council's
Statistical Research Unit, London, England

Algorithm 178 has the following syntactical errors:
(1) The parameter list should read
     (*psi,K,DELTA,rho,delta,S*).
(2) The declaration
     integer K,k;
should read
     integer k;
(3) An extra **end** bracket is required immediately before **end** E;.

The algorithm compiled and ran after these modifications had
been made but for a number of problems took a prodigious amount
of computing owing to a flaw in the algorithm caused by rounding
error. This flaw is in **procedure** E and may be illustrated by the
one-dimensional case. Let $S(x) = 1.5 − x$ ($x \leq 1.5$), $3x − 4.5$ ($x >$
1.5), and start at 0 with a step of 1. The first move puts *psi* [1] =
1, *phi* [1] = 2. The second move should then put *phi* [1] = 1 =
psi[1] resulting in a jump to label 1. On many machines, however,
E will put *phi* [1] = 1 + e (e>0 and very small) so that direct
search begins to move away from 1 in very small steps. This is
clearly not desirable and may be avoided by altering the line
     if $SS < Spsi$ **then go to** 2 **else go to** 1 **end**;
to
     if $SS \geq Spsi$ **then go to** 1;
     for $k := 1$ **step** 1 **until** K **do**
     if *abs* (*phi*[k]-*psi*[k]) > 0.5 × *DELTA* **then go to** 2
     end
To accelerate the procedure, direct search should take advantage of its knowledge of the sign of its previous move in each of the
K directions. Take, for example, the one-dimensional case with
starting point zero and the minimum far out and negative; the
pattern moves will arrive there quite efficiently but each first move
of E on the way will be positive whereas the previous experience
of the search should lead it to suspect the minimum to be in the
opposite direction.

Finally, two changes which we have found very useful are (i)
some escape clause in the procedure to enable an exit to be made if
the procedure has not terminated after some given number of
function evaluations *maxeval*, with a Boolean *converge* taking the
value **true** in general but **false** if the procedure has terminated
through exceeding this number of function evaluations; and (ii)
taking $Spsi$ into the parameter list where it is called by name so
that on exit $Spsi$ contains the minimum value of the function.
   With these modifications the procedure now reads:

procedure *direct search* (*psi,K,Spsi,DELTA,rho,delta,S,converge,*
     maxeval);
   value K,*DELTA*,*rho*,*delta*,*maxeval*; **integer** K,*maxeval*;
     array *psi*;
   real *DELTA,rho,delta,Spsi*; **real procedure** S; **Boolean**
     converge;
comment This procedure locates the minimum of the function S of
   K variables. The method used is that of R. Hooke and T. A.
   Jeeves ["Direct search" solution of numerical and statistical
   problems, *J. ACM.* 8 (1961), 212–229] and the notation used is
   theirs except for the obvious changes required by ALGOL. On
   entry: *psi*[1:K] = starting point of the search, *DELTA* =
   initial step-length, *rho* = reduction factor for step-length,
   delta = minimum permitted step-length (i.e. procedure is terminated when step-length < *delta*), *maxeval* = maximum permitted number of function evaluations. On exit: *psi*[1:K] =
   minimum point found and $Spsi$ = value of S at this point,
   converge = **true** if exit has been made from the procedure because a minimum has been found (i.e., step-length < *delta*)
   otherwise *converge* = **false** (i.e. maximum number of function
   evaluations has been reached);
begin integer k,*eval*; **array** *phi,s*[1:K]; **real** $Sphi$,SS,*theta*;
   procedure E;
   for $k := 1$ **step** 1 **until** K **do**
   begin *phi*[k] := *phi*[k] + *s*[k]; $Sphi$:= $S(phi)$; *eval* := *eval*
     + 1;

```
   if Sphi < SS then SS := Sphi else
   begin s[k] := − s[k]; phi[k] := phi[k] + 2.0 × s[k];
     Sphi := S(phi); eval := eval + 1;
     if Sphi < SS then SS := Sphi else
     phi[k] := phi[k] − s[k]
   end
 end E;
Start: for k := 1 step 1 until K do s[k] := DELTA;
 Spsi := S(psi); eval := 1; converge := true;
1: SS := Spsi;
   for k := 1 step 1 until K do phi[k] := psi[k]; E;
   if SS < Spsi then
   begin
2: if eval ≥ maxeval then
   begin converge := false;
     go to EXIT
   end;
   for k := 1 step 1 until K do
   begin if phi[k] > psi[k] ≡ s[k] < 0 then s[k] := −s[k];
     theta := psi[k]; psi[k] := phi[k]; phi[k] := 2.0 × phi[k] −
       theta
   end;
   Spsi := SS; SS := Sphi := S(phi); eval := eval + 1; E;
   if SS ≥ Spsi then go to 1;
   for k := 1 step 1 until K do
   if abs(phi[k]−psi[k]) > 0.5 × abs(s[k]) then go to 2
   end;
3: if DELTA ≥ delta then
   begin DELTA := rho × DELTA;
     for k := 1 step 1 until K do s[k] := rho × s[k]; go to 1
   end;
EXIT:
end direct search
```

REMARK ON ALGORITHM 178 [E4]
DIRECT SEARCH [Arthur F. Kaupe, Jr., *Comm. ACM*
6 (June 1963), 313]
[as revised by M. Bell and M. C. Pike, *Comm. ACM 9*
(Sept. 1966), 684]

R. De Vogelaere (Recd. 4 Dec. 1967)
Department of Mathematics and Computer Center, University of California, Berkeley, Calif. 94720

KEY WORDS AND PHRASES: function minimization, search, direct search
CR CATEGORIES: 5.19

The procedure does not exit, as specified, after *maxeval* (the maximum number of) function evaluations.

The 3 statements *eval := eval +1* should be interchanged with the immediately preceding statement and replaced by a call to the procedure *test eval* defined below. The statement labeled 2 should be deleted.

```
   procedure test eval;
     if eval < maxeval then eval := eval +1
     else begin converge := false;
     go to EXIT
   end test eval
```

REMARK ON ALGORITHM 178 [E4]
DIRECT SEARCH [Arthur F. Kaupe, Jr., *Comm. ACM*
6 (June 1963), 313]; [as revised by M. Bell and M. C.
Pike, *Comm ACM 9* (Sept. 1966), 684]

F. K. Tomlin and L. B. Smith (Recd. 17 May 1968, 9
Sept. 1968 and 30 June 1969)
Stanford Research Institute, Menlo Park, CA 94025, and
CERN, DD Division, Geneva, Switzerland

KEY WORDS AND PHRASES: function minimization, search,
direct search
CR CATEGORIES: 5.19

The procedure *DIRECT SEARCH*, as modified by M. Bell and M. C. Pike [1], does not always provide the determined minimum. In addition, the maximum number of function evaluations permitted is almost always exceeded whenever the step-length is greater than *delta* at the time the number of function evaluations is greater than or equal to *maxeval*. Finally, the label 3 is not used.

To insure that the determined minimum is always provided, the test on the number of evaluations should be moved to a point where the minimum has been properly provided.

In [2] DeVogelaere remarks correctly that the procedure does not exit as specified and gives changes which will indeed cause the procedure to terminate when the number of function evaluations exceeds the specified limit (and not some number of evaluations later). However it is felt that DeVogelaere's solution to this problem causes excessive testing. Therefore the test should be performed after an exploratory move as in [1] but it should also be performed when the step-length is reduced. This method of testing violates the letter of the specified use of *maxeval* but not the intent, which is to provide an escape from excessive calculation.

To obtain the determined minimum, to provide a means for reducing the number of function evaluations when step-length is greater than *delta*, and to eliminate the unused label:

(1) The lines

```
2: if eval ≥ maxeval then
     begin converge := false
       go to EXIT
     end;
```

should be removed.

(2) The line (16th line from the end of the procedure given in [1])

```
for k := 1 step 1 until K do
```

should be changed to

```
2: for k := 1 step 1 until K do
```

(3) The line

```
Spsi := SS; SS := Sphi := S(phi); eval := eval + 1; E;
```

should have the following code inserted after the statement
Spsi := SS;

```
if eval ≥ maxeval then
   begin
3: converge := false;
   go to EXIT
   end;
```

(4) The line

```
3: if DELTA ≥ delta then
```

should be changed to

if $DELTA \geq delta$ then

(5) The line
begin $DELTA := rho \times delta;$

should be changed to

begin if $eval > maxeval$ **then go to** 3 **else**
$DELTA := rho \times delta;$
REFERENCES:
1. BELL, M., AND PIKE, M. C.   Remark on Algorithm 178. *Comm. ACM 9* (Sept. 1966), 684.
2. DEVOGELAERE, R.   Remark on Algorithm 178. *Comm. ACM 11* (July 1968), 498.

REMARK ON ALGORITHM 178 [E4]
DIRECT SEARCH [Arthur F. Kaupe, Jr., *Comm. ACM 6* (June 1963), 313; as revised by M. Bell and M. C. Pike, *Comm. ACM 9* (Sept. 1966), 684]

LYLE B. SMITH* (Recd. 9 Sept. 1968)
Stanford Linear Accelerator Center, Stanford, CA 94305
* Present address. CERN, Data Handling Division, 1211 Geneva 23, Switzerland

KEY WORDS AND PHRASES: function minimization, search, direct search
CR CATEGORIES: 5.19

Algorithm 178, as modified by Bell and Pike [1], has been used successfully by the author on a number of different problems and in a variety of languages (e.g. Burroughs Extended ALGOL on a B5500, SUBALGOL on an IBM 7090, and FORTRAN on the IBM/360 series machines). A modification which has been found to be useful involves tailoring the step size to be meaningful for a wide variation in the magnitudes of the variables.

As currently specified [1], each variable is incremented (or decremented) by $DELTA$ as a minimum is sought. For a function such that the values of the variables differ by several orders of magnitude at the minimum, a universal step size causes some parameters to be essentially ignored during much of the searching process. For example, if a function of two variables has a minimum near $(100.0, 0.1)$, a step size of 10.0 will be useful in minimizing with respect to the first parameter, but it will be meaningless with respect to the second parameter until it has been reduced to near 0.01. On the other hand, a step size of 0.01 would be useful on the second variable but on the first variable it would take an undesirably large number of steps to approach the minimum.

A modification to direct search which circumvents this scaling problem involves the use of a different step size for each variable. This is easily implemented since an array is already used to hold the signed step size for each variable. The change is accomplished by removing the statement labeled *Start* and replacing it by the following statement:

```
Start:   for k := 1 step 1 until K do
            begin s(k) := DELTA × abs (psi(k));
               if s(k) = 0.0 then s(k) := DELTA;
            end;
```

This change sets the step size for each variable to $DELTA$ times the magnitude of the starting value, or if the starting value is 0.0 the step size is set equal to $DELTA$. Thus $DELTA$ is the fraction of the original value of each variable to be used as an initial step size. Subsequent reductions in step size are handled correctly without further modifications to the procedure.

As an example of the usefulness of the above modification, consider the function

$$f(X_1, X_2, X_3) = (X_1 - 0.01)^2 + (X_2 - 1.0)^2 + (X_3 - 100.0)^2$$

with a minimum at $(0.01, 1.0, 100.0)$. The following table shows the results of using direct search on this function with and without the modified step size. The results were computed on an IBM 360/75 computer using single precision with $rho = 0.1$, $delta = 0.001$, $DELTA = 0.2$ for the modified step size (giving 20 percent of initial value for initial step size) and $DELTA =$ [average magnitude of initial guesses for the variables] for the algorithm as published.

TABLE I. $f = (X_1 - 0.01)^2 + (X_2 - 1.0)^2 + (X_3 - 100.0)^2$

	DELTA	Number of function evaluations	Minimum value of f	Final values of the variables		
				X_1	X_2	X_3
For initial values of $(0.0, 0.0, 200.0)$:						
Direct search	66.6667	153	0.841×10^{-7}	0.00999995	0.999995	100.000
Modified direct search	.2	112	0.597×10^{-7}	0.00999998	0.999990	100.000
For initial values of $(0.05, 5.0, 500.0)$:						
Direct search	168.35	174	0.934×10^{-7}	0.0100263	0.998958	99.9999
Modified direct search	.2	75	0.559×10^{-6}	0.00999988	0.999998	99.9992

Note that the modified method will tend to yield the same relative accuracy for each parameter, whereas with a fixed step size direct search will tend to give the same absolute accuracy for all parameters. In most cases a relative accuracy is probably more desirable than an absolute accuracy.

REFERENCES
1. BELL, M., AND PIKE, M. C.   Remark on algorithm 178. *Comm ACM 9* (Sept. 1966), 684.

ALGORITHM 179
INCOMPLETE BETA RATIO*

OLIVER G. LUDWIG

Mathematical Laboratory and Department of Theoretical Chemistry, University of Cambridge, England

* Based in part on work done at Carnegie Institute of Technology, Pittsburgh, Pennsylvania and supported by the Petroleum Research Fund of the American Chemical Society and by the National Science Foundation.

real procedure *incompletebeta* $(x, p, q, epsilon)$;
value x, p, q; **real** $x, p, q, epsilon$;
begin real *finsum, infsum, temp, temp* 1, *term, term* 1, *qrecur, index*;
    Boolean *alter*;
comment This procedure evaluates the ratio $B_x(p, q)/B_1(p, q)$,
    where $B_x(p, q) = \int_0^x t^{p-1}(1-t)^{q-1}\, dt$, with $0 \leq x \leq 1$ and $p, q > 0$,
    but not necessarily integers. It assumes the existence of a non-
    local label, *alarm*, to which control is transferred upon entry to
    the procedure with invalid arguments. Also assumed is a proce-
    dure to evaluate $\int_0^\infty t^p e^{-t}\, dt$ which is called *factorial*(p), (cf. e.g.
    Algorithm 80, March, 1962);
if $x > 1 \lor x < 0 \lor p \leq 0 \lor q \leq 0$ **then go to** *alarm*;
if $x = 0 \lor x = 1$ **then begin** *incompletebeta* := x; **go to** *End* **end**;
comment This part interchanges arguments if necessary to obtain better convergence in the power series below;
if $x \leq 0.5$ **then** *alter* := *false* **else**
    begin *alter* := **true**; *temp* := p; p := q; q := *temp*; x :=
    $1 - x$ **end**;
comment This part recurs on the (effective) q until the power
    series below does not alternate;
finsum := 0; *term* := 1; *temp* := $1 - x$; *qrecur* := *index* := q;
for *index* := *index* $- 1$ **while** *index* > 0 **do**
    begin *qrecur* := *index*;
      term := *term* $\times$ $(qrecur+1)/(temp\times(p+qrecur))$;
      finsum := *finsum* $+$ *term*
    end;
comment This part sums a power series for non-integral effective q and yields unity for integer q;
infsum := *term* := 1; *index* := 0;
comment In the following statement the convergence criterion
    might well be altered to *term* $>$ *epsilon*, since *infsum* > 1 al-
    ways, thus saving one divide per cycle at the cost, perhaps, of a
    few more cycles;
for *index* := *index* $+ 1$ **while** $(term/infsum) >$ *epsilon* **do**
    begin *term* := *term* $\times$ $x \times$ $(index\text{-}qrecur) \times$ $(p+index-1)/$
    $(index\times(p+index))$; *infsum* := *infsum* $+$ *term*
    end;
comment This part evaluates most of the necessary factorial
    functions, minimizing the number of entries into the factorial
    procedure;
temp := *temp* 1 := *factorial* $(qrecur-1)$;
term := *term* 1 := *factorial* $(qrecur+p-1)$;
for *index* := *qrecur* **step** 1 **until** $(q-0.5)$ **do**
    begin *temp* 1 := *temp* 1 $\times$ *index*;
      term 1 := *term* 1 $\times$ $(index+p)$
    end;
comment This part combines the partial results into the final
    one;
temp := $x \uparrow p \times$ $(infsum\times term/(p\times temp)+finsum\times term$ $1\times$
    $(1-x) \uparrow q/(q\times temp$ $1))/factorial$ $(p-1)$;

incompletebeta := **if** *alter* **then** $1-temp$ **else** *temp*;
end: **end** *incompletebeta*

REMARK ON ALGORITHM 179 [S 14]
INCOMPLETE BETA RATIO [Oliver G. Ludwig, *Comm. ACM 6* (June 1963), 314]

M. C. PIKE AND I. D. HILL (Recd. 8 Oct. 1965 and 12 Jan. 1966)

Medical Research Council's Statistical Research Unit, University College Hospital Medical School, London, England

Algorithm 179 has the following two typographical errors:

(1) the line

    if $x \leqslant 0.5$ **then** *alter* := *false* **else**

should read

    if $x \leqslant 0.5$ **then** *alter* := **false else**

(2) the line

    end:**end** *incompletebeta*

should read

    End:**end** *incompletebeta*

With these changes Algorithm 179 ran successfully on the ICT Atlas computer using Algorithm 221 [Walter Gautschi, *Comm. ACM 7* (Mar. 1964), 143], to evaluate the factorials required. A minor improvement might be to call *epsilon* by value.

As the algorithm stands, the permitted range of p and q is dictated by overflow problems associated with finding the values of factorials. For most machines this will mean that $p+q$ will have to be less than about 70. In the statistical applications of this algorithm which we describe below this restriction is very serious. However, these factorials appear essentially only in the form of ratios, and by making use of this fact the permitted range of p and q can be enormously extended. This is most simply accomplished by using the real procedure *loggamma* [Algorithm 291, M. C. Pike and I. D. Hill, *Comm. ACM 9* (Sept. 1966), 684] modifying Algorithm 179 as follows: replace the instructions

    temp := *temp*1 := *factorial*$(qrecur-1)$;

to

    temp := $x \uparrow p \times$ $(infsum\times term/(p\times temp)+finsum\times term1\times$
    $(1-x) \uparrow q/(q\times temp1))/factorial(p-1)$;

inclusive, by

    temp := $x \uparrow p \times$ $(infsum\times exp(loggamma(qrecur+p)-loggamma$
    $(qrecur)-loggamma(p+1.0))+finsum\times(1.0-x) \uparrow q$
    $\times exp(loggamma(p+q)-loggamma(p)-loggamma(q+1.0)))$;.

This also means that the declarations of *temp*1 and *term*1 are not required. For even moderately large values of p or q this will also have the effect of speeding up the algorithm [see Remark on Algorithm 291, M. C. Pike and I. D. Hill, *Comm. ACM 9* (Sept. 1966), 685].

The following real procedures use this algorithm to evaluate three of the most frequently required statistical distribution functions.

real procedure $Ftail$ (k, f1, f2, epsilon);
value k, f1, f2, epsilon;   **real** k, f1, f2, epsilon;
comment $Ftail$ evaluates the probability that a random variable following an F distribution, on f1 and f2 degrees of freedom, exceeds a positive constant k;
  $Ftail := incompletebeta(f2/(f2+f1\times k), 0.5\times f2, 0.5\times f1, epsilon);$

real procedure $Student$(k, f, epsilon);
value k, f, epsilon;   **real** k, f, epsilon;
comment $Student$ evaluates the probability that the absolute value of a random variable following a t distribution, on f degrees of freedom, exceeds a positive constant k;
  $Student := incompletebeta(f/(f+k \uparrow 2), 0.5\times f, 0.5, epsilon);$
real procedure $Binomial$(k, n, p, epsilon);
value k, n, p, epsilon;   **real** k, n, p, epsilon;
comment $Binomial$ evaluates the probability that a random variable following a binomial distribution, with parameters n and p, takes a value greater than or equal to k;
  $Binomial := incompletebeta(p, k, n-k+1.0, epsilon);$

Remark on Algorithm 179 [S14]
Incomplete Beta Ratio
[Oliver G. Ludwig, *Comm. ACM* 6 (June 1963), 314]

Nancy E. Bosten and E.L. Battiste [Recd. 1 Sept. 1972 and 15 Mar. 1973]
IMSL, Suite 510/6200 Hillcroft, Houston, TX 77036

Description

Algorithm 179 (modified to include the remark by M.C. Pike and I.D. Hill [1]) computes the Incomplete Beta Ratio using this equation

$$I_x(p,q) = \frac{INFSUM \cdot x^p \cdot \Gamma(PS+p)}{\Gamma(PS)\cdot\Gamma(p+1)} + \frac{x^p\cdot(1-x)^q\cdot\Gamma(p+q)}{\Gamma(p)\cdot\Gamma(q+1)} FINSUM$$

$INFSUM$ and $FINSUM$ represent two series summations defined as follows:

$$INFSUM = \sum_{i=0}^{\infty} \frac{(1-PS)_i\cdot p}{p+i}\ \frac{x^i}{i!}, \text{ where}$$

$$(1-PS)_i = 1 \qquad\qquad [i=0]$$
$$= (1-PS)\cdot(2-PS)\cdots(i-PS) = \frac{\Gamma(1+i-PS)}{\Gamma(1-PS)} \quad [i>0]$$

and $$FINSUM = \sum_{i=1}^{[q]} \frac{q\cdot(q-1)\cdots(q-i+1)}{(p+q-1)(p+q-2)\cdots(p+q-i)}\ \frac{1}{(1-x)^i},$$

where $[q]$ is equal to the largest integer less than q. If $[q] = 0$, then $FINSUM = 0$. PS is defined as

$PS = 1$, if q is an integer; otherwise
  $= q - [q]$.

By rearranging Algorithm 179 so that scaling can be introduced, the argument range of p and q can be extended and accuracy can be improved.

Since $I_x(p, q)$ is a probability and, therefore, bounded [0, 1], and $INFSUM$ and $FINSUM$ are series having only positive terms, we see that $I_x(p, q)$ is a collection of terms all of which are positive and bounded in the range [0, 1] if: (1) each term of $INFSUM$ is multiplied by $(x^p \cdot \Gamma(PS + p))/(\Gamma(PS) \cdot \Gamma(p + 1))$; and (2) each term of $FINSUM$ is multiplied by $(x^p \cdot (1-x)^q \cdot \Gamma(p+q))/(\Gamma(p) \cdot \Gamma(q+1))$.

Knowing this fact, we can apply a scaling procedure to the algorithm. $INFSUM$ is a decreasing series. If the product of the first term of $INFSUM$ and its multiplicative factor would underflow, then the sum of this series could be set to zero and all calculations involving underflow could be avoided. This is handled in the modification of the algorithm given below. However, since $INFSUM$ is a decreasing series, underflows may occur later in the calculations. No attempt has been made to handle them here.

The second summation is more complicated. The series is decreasing if $q/((q + p - 1)(1 - x))$ is less than 1. If an individual term becomes less than $1.E-6$ times the previous sum, calculation can be legitimately terminated since no additivity is apparent. If a term of the decreasing series is less than an arbitrarily small constant ($EPS2$), calculation is also terminated. This is done to prevent underflows in the later terms.

If the series is increasing, the first terms may underflow. In this case a power of ϵ_1 (machine precision $- 1.E-78$ on the IBM 360/370) may be factored from each term in $FINSUM$ (times its multiplier). These terms cannot be added to the sum since they are less than machine precision; however, they are useful in retaining the accuracy of the initial terms, which are then used recursively. By the nature of the problem, we know that any term in $FINSUM$, times its multiplier, must be less than or equal to 1, but we have factored out powers of ϵ_1. Therefore, if a term of $FINSUM$ becomes greater than 1, we know that rescaling, by multiplying the term by ϵ_1, is in order.

Testing on the IBM 360/195 has shown that, by rearranging the calculations of the original Algorithm 179, and thus including scaling, the input range of the algorithm can be greatly extended with a high degree of accuracy.

$MDBETA$ requires a double precision function $DLGAMA$ which computes the log of the gamma function. ACM Algorithm 291 may be used. $MDBETA$ was tested against the SSP routine $BDTR$ given in the manual *System/360 Scientific Subroutine Package (360A-CM-03X) Version III Programmer's Manual*, H20-0205. $MDBETA$ ran 3.5 times faster than $BDTR$ with greater accuracy. For example, in the case $x = .5, p = 2000$ and $q = 2000$, $MDBETA$ gave the correct result, .5, while $BDTR$ gave an answer of .497026. The IMSL subroutine, $MDBIN$, was used for an additional comparison when p and q are integers. $MDBIN$ maintains IBM 370/360 single precision accuracy (approximately six significant digits). Over the tests performed the maximum difference occurred in the fifth significant digit when p and q were less than 200. Three to four significant digits of accuracy can be expected with p and q as large as 2000.

Acknowledgments. The above ideas are the application of ideas learned from the late Hirondo Kuki. Routine $MDBETA$ originated from a code which resides in IMSL Library 1. We thank Wayne Fullerton, from the University of California, Los Alamos Scientific Laboratory, for refereeing the paper.

Algorithm

```
      SUBROUTINE MDBETA(X, P, Q, PROB, IER)
C FUNCTION                  - INCOMPLETE BETA PROBABILITY
C                             DISTRIBUTION FUNCTION
C USAGE       - CALL MDBETA (X,P,Q,PROB,IER)
C PARAMETERS
C   X      - VALUE TO WHICH FUNCTION IS TO BE INTEGRATED. X
C             MUST BE IN THE RANGE (0,1) INCLUSIVE.
C   P      - INPUT (1ST) PARAMETER (MUST BE GREATER THAN 0)
C   Q      - INPUT (2ND) PARAMETER (MUST BE GREATER THAN 0)
C   PROB - OUTPUT PROBABILITY THAT A RANDOM VARIABLE FROM A
C             BETA DISTRIBUTION HAVING PARAMETERS P AND Q
C             WILL BE LESS THAN OR EQUAL TO X.
C   IER  - ERROR PARAMETER.
C             IER = 0 INDICATES A NORMAL EXIT
C             IER = 1 INDICATES THAT X IS NOT IN THE RANGE
C                (0,1) INCLUSIVE.
C             IER = 2 INDICATES THAT P AND/OR Q IS LESS THAN
C                OR EQUAL TO 0.
      DOUBLE PRECISION PS, PX, Y, P1, DP, INFSUM, CNT, WH, XB,
     * DQ, C, EPS, EPS1, ALEPS, FINSUM, PQ, D4, EPS2, DLGAMA
C DOUBLE PRECISION FUNCTION DLGAMA
C MACHINE PRECISION
      DATA EPS/1.D-6/
C SMALLEST POSITIVE NUMBER REPRESENTABLE
      DATA EPS1/1.D-78/
```

```
C NATURAL LOG OF EPS1
      DATA ALEPS/-179.6016D0/
C ARBITRARILY SMALL NUMBER
      DATA EPS2/1.D-50/
C CHECK RANGES OF THE ARGUMENTS
      Y = X
      IF ((X.LE.1.0) .AND. (X.GE.0.0)) GO TO 10
      IER = 1
      GO TO 140
   10 IF ((P.GT.0.0) .AND. (Q.GT.0.0)) GO TO 20
      IER = 2
      GO TO 140
   20 IER = 0
      IF (X.GT.0.5) GO TO 30
      INT = 0
      GO TO 40
C SWITCH ARGUMENTS FOR MORE EFFICIENT USE OF THE POWER
C SERIES
   30 INT = 1
      TEMP = P
      P = Q
      Q = TEMP
      Y = 1.D0 - Y
   40 IF (X.NE.0. .AND. X.NE.1.) GO TO 60
C SPECIAL CASE - X IS 0. OR 1.
   50 PROB = 0.
      GO TO 130
   60 IB = Q
      TEMP = IB
      PS = Q - FLOAT(IB)
      IF (Q.EQ.TEMP) PS = 1.D0
      DP = P
      DQ = Q
      PX = DP*DLOG(Y)
      PQ = DLGAMA(DP+DQ)
      P1 = DLGAMA(DP)
      C = DLGAMA(DQ)
      D4 = DLOG(DP)
      IF (Y.GT.EPS) GO TO 70
C SPECIAL CASE - X IS CLOSE TO 0. OR 1.
      XB = PX + PQ - D4 - P1 - C
      IF (XB.LE.ALEPS) GO TO 50
      PROB = DEXP(XB)
      GO TO 130
C DLGAMA IS A FUNCTION WHICH CALCULATES THE DOUBLE
C PRECISION LOG GAMMA FUNCTION
   70 XB = PX + DLGAMA(PS+DP) - DLGAMA(PS) - D4 - P1
C SCALING
      IB. = XB/ALEPS
      INFSUM = 0.D0
C FIRST TERM OF A DECREASING SERIES WILL UNDERFLOW
      IF (IB.NE.0) GO TO 90
      INFSUM = DEXP(XB)
      CNT = INFSUM*DP
C CNT WILL EQUAL DEXP(TEMP)*(1.D0-PS)I*P*Y**I/FACTORIAL(I)
      WH = 0.0D0
   80 WH = WH + 1.D0
      CNT = CNT*(WH-PS)*Y/WH
      XB = CNT/(DP+WH)
      INFSUM = INFSUM + XB
      IF (XB/EPS.GT.INFSUM) GO TO 80
C DLGAMA IS A FUNCTION WHICH CALCULATES THE DOUBLE
C PRECISION LOG GAMMA FUNCTION
   90 FINSUM = 0.D0
      IF (DQ.LE.1.D0) GO TO 120
      XB = PX + DQ*DLOG(1.D0-Y) + PQ - P1 - DLOG(DQ) - C
C SCALING
      IB = XB/ALEPS
      IF (IB.LT.0) IB = 0
      C = 1.D0/(1.D0-Y)
      CNT = DEXP(XB-FLOAT(IB)*ALEPS)
      PS = DQ
      WH = DQ
      P1 = (PS*C)/(DP+WH-1.D0)
      XB = P1*CNT
      IF (XB.LE.EPS2 .AND. P1.LE.1.D0) GO TO 120
  100 WH = WH - 1.D0
      IF (WH.LE.0.0D0) GO TO 120
      IF (P1.LE.1.D0 .AND. CNT/EPS.LE.FINSUM) GO TO 120
      CNT = (PS*C*CNT)/(DP+WH)
      IF (CNT.LE.1.D0) GO TO 110
C RESCALE
      IB = IB - 1
      CNT = CNT*EPS1
  110 PS = WH
      IF (IB.EQ.0) FINSUM = FINSUM + CNT
      GO TO 100
  120 PROB = FINSUM + INFSUM
  130 IF (INT.EQ.0) GO TO 140
      PROB = 1.0 - PROB
      TEMP = P
      P = Q
      Q = TEMP
  140 RETURN
      END
```

ACM Transactions on Mathematical Software, Vol. 2, No. 2, June 1976, Pages 207–208.

REMARK ON ALGORITHM 179

Incomplete Beta Ratio [S14]
[O. G. Ludwig, *Comm. ACM 6*, 6(June 1963), 314]

Malcolm C. Pike and Jennie SooHoo [Recd 5 March 1975 and 11 September 1975]
School of Medicine, University of Southern California, 1840 North Soto Street,
Los Angeles, CA 90032

N. E. Bosten and T. J. Aird, International Mathematical and Statistical Libraries,
Inc., 7500 Bellaire Blvd., Houston, TX 77036.

This work was supported by the Virus Cancer Program, National Cancer Institute, Bethesda, Md., under Grant PO ICA 17054-01 and Contract N01-CP-53500.

Algorithm 179 (*MDBETA*) can be improved as shown.

1. Remove *EPS*2 from the double precision statement.

2. Remove the data statement and comment:

```
C       ARBITRARILY SMALL NUMBER
        DATA EPS2/1.D-50/
```

3. Remove the three statements preceding statement number 100:

```
P1 = (PS*C)/(DP+WH-1.D0)
XB = P1*CNT
IF(XB.LE.EPS2.AND.P1.LE.1.D0) GO TO 120
```

4. After statement number 100, replace the following statements:

```
IF(P1.LE.1.D0.AND. CNT/EPS.LE.FINSUM) GO TO 120
CNT = (PS*C*CNT)/(DP+WH)
```

with the following:

```
         PX = (PS*C)/(DP+WH)
         IF(PX.GT.1.0D0) GO TO 105
         IF(CNT/EPS.LE.FINSUM .OR. CNT.LE.EPS1/PX) GO TO 120
   105   CNT = CNT*PX
```

The above changes eliminate the occurrence of underflow in the computation of *FINSUM* and decrease the execution time of the algorithm with no apparent change in accuracy.

5. Remove the statements and comment:

```
         IF(Y.GT.EPS) GO TO 70
   C     SPECIAL CASE — X IS CLOSE TO 0. OR 1.
         XB = PX+PQ-D4-P1-C
         IF(XB.LE.ALEPS) GO TO 50
         PROB = DEXP(XB)
         GO TO 130
```

6. Remove the statement number from:

```
   70   XB = PX+DLGAMMA(PS+DP)-DLGAMMA(PS) — D4-P1
```

ALGORITHM 180
ERROR FUNCTION—LARGE X
HENRY C. THACHER, JR.*
Argonne National Laboratory, Argonne, Ill.
  * Work supported by the U. S. Atomic Energy Commission.

real procedure $erfL(x)$; **value** x; **real** x;
comment This procedure evaluates the error function of real

argument, $erf(x) = (2/\sqrt{\pi})\int_0^x e^{-u^2}\,du$ by the Laplace continued

fraction for the complementary error function: $erf(x) = 1 - (1/(1+v/(1+2v/(1+3v/(1+\cdots)))))/(\sqrt{\pi}x\ e^{x^2})$ where $v = 1/(2x^2)$. Successive even convergents of the continued fraction are evaluated, using an algorithm suggested by Maehly, until the full accuracy of the arithmetic being used is attained.

  The continued fraction converges for all $x > 0$. For small x, however, convergence may be excessively slow, and overflow may occur. In this region, the Taylor series converges satisfactorily, and algorithms such as No. 123 are suitable.

  For $x \leq 0$, the procedure calls the global procedure alarm.

  The body of this procedure has been checked on the LGP-30 computer, using the Dartmouth Self Contained Algol Processor. The program was used to tabulate $erf(x)$ from 0.9(.1)5.0. The maximum error was 2×10^{-6}, which is explainable by roundoff errors. The number of convergents calculated ranged from 36 for $x = 0.9$ to 2 for $x \geq 3.3$. Overflow occurred for $x = 0.87$;

```
begin integer m;  real B min 2, B min 3, P, R, T, v, v2;
if x ≤ 0 then alarm;
v := x × x;
T := −0.56418958/x/exp(v);
```
comment The constant $0.56418958\cdots = \pi^{-1/2}$, and should be given to the full accuracy required of the procedure;
```
v := 0.5/v;
P = v × T;
v2 := v × v;
T := T + 1;
m := 0;
R := B min 3 := B min 2 := 1;
for m := m + 1 while T ≠ R do
  begin R := T;
  B min 3 := v × (m−1) × B min 3 + B min 2
  T := B min 2;
  B min 2 := v × m × B min 2 + B min 3;
  T := R − P/B min 2/T;
  P := m × (m+1) × v2 × P
  end while;
  erfL := T
end
```

REMARKS ON:
ALGORITHM 123 [S15]
REAL ERROR FUNCTION, ERF(x)
  [Martin Crawford and Robert Techo *Comm. ACM 5* (Sept. 1962), 483]

ALGORITHM 180 [S15]
ERROR FUNCTION—LARGE X
  [Henry C. Thacher Jr. *Comm. ACM 6* (June 1963), 314]

ALGORITHM 181 [S15]
COMPLEMENTARY ERROR FUNCTION—
LARGE X
  [Henry C. Thacher Jr. *Comm. ACM 6* (June 1963), 315]

ALGORITHM 209 [S15]
GAUSS
  [D. Ibbetson. *Comm. ACM 6* (Oct. 1963), 616]

ALGORITHM 226 [S15]
NORMAL DISTRIBUTION FUNCTION
  [S. J. Cyvin. *Comm. ACM 7* (May 1964), 295]

ALGORITHM 272 [S15]
PROCEDURE FOR THE NORMAL DISTRIBUTION
FUNCTIONS
  [M. D. MacLaren. *Comm. ACM 8* (Dec. 1965), 789]

ALGORITHM 304 [S15]
NORMAL CURVE INTEGRAL
  [I. D. Hill and S. A. Joyce. *Comm. ACM 10* (June 1967), 374]

I. D. HILL AND S. A. JOYCE (Recd. 21 Nov. 1966)
Medical Research Council,
Statistical Research Unit, 115 Gower Street, London
  W.C.1., England

These algorithms were tested on the ICT Atlas computer using the Atlas ALGOL compiler. The following amendments were made and results found:

ALGORITHM 123
  (i) **value** x; was inserted.
  (ii) $abs(T) \leqslant {}_{10}-10$ was changed to $Y - T = Y$
      both these amendments being as suggested in [1].
  (iii) The labels 1 and 2 were changed to $L1$ and $L2$, the **go to** statements being similarly amended.
  (iv) The constant was lengthened to 1.12837916710.
  (v) The extra statement $x := 0.707106781187 \times x$ was made the first statement of the algorithm, so as to derive the normal integral instead of the error function.

  The results were accurate to 10 decimal places at all points tested except $x = 1.0$ where only 2 decimal accuracy was found, as noted in [2]. There seems to be no simple way of overcoming the difficulty [3], and any search for a method of doing so would hardly be worthwhile, as the algorithm is slower than Algorithm 304 without being any more accurate.

ALGORITHM 180
  (i) $T := -0.56418958/x/exp(v)$ was changed to
      $T := -0.564189583548 \times exp(-v)/x$. This is faster and also has the advantage, when v is very large, of merely giving 0 as the answer instead of causing overflow.

(ii) The extra statement $x := 0.707106781187 \times x$ was made as in (v) of Algorithm 123.

(iii) **for** $m := m + 1$ was changed to **for** $m := m + 2$. $m+1$ is a misprint, and gives incorrect answers.

The greatest error observed was 2 in the 11th decimal place.

ALGORITHM 181

(i) Similar to (i) of Algorithm 180 (except for the minus sign).

(ii) Similar to (ii) of Algorithm 180.

(iii) m was declared as **real** instead of **integer**, as an alternative to the amendment suggested in |4].

The results were accurate to 9 significant figures for $x \leqslant 8$, but to only 8 significant figures for $x = 10$ and $x = 20$.

ALGORITHM 209

No modification was made. The results were accurate to 7 decimal places.

ALGORITHM 226

(i) $10 \uparrow m/(480 \times sqrt(2 \times 3.14159265))$ was changed to $10 \uparrow m \times 0.000831129750836$.

(ii) **for** $i := 1$ **step** 1 **until** $2 \times n$ **do** was changed to $m := 2 \times n$; **for** $i := 1$ **step** 1 **until** m **do**.

(iii) $-(i \times b/n) \uparrow 2/8$ was changed to $-(i \times b/n) \uparrow 2 \times 0.125$.

(iv) **if** $i = 2 \times n - 1$ was changed to **if** $i = m - 1$

(v) $b/(6 \times n \times sqrt(2 \times 3.14159265))$ was changed to $b/(15.0397696478 \times n)$.

Tests were made with $m = 7$ and $m = 11$ with the following results:

x	Number of significant figures correct		Number of decimal places correct	
	$m = 7$	$m = 11$	$m = 7$	$m = 11$
-0.5	7	11	7	11
-1.0	7	10	7	10
-1.5	7	10	8	10
-2.0	7	9	8	10
-2.5	6	9	8	11
-3.0	6	7	8	9
-4.0	5	7	10	11
-6.0	2	1	12	10
-8.0	0	0	11	9

Perhaps the comment with this algorithm should have referred to decimal places and not significant figures. To ask for 11 significant figures is stretching the machine's ability to the limit, and where 10 significant figures are correct, this may be regarded as acceptable.

ALGORITHM 272

The constant .99999999 was lengthened to .9999999999.

The accuracy was 8 decimal places at most of the points tested, but was only 5 decimal places at $x = 0.8$.

ALGORITHM 304

No modification was made. The errors in the 11th significant figure were:

$abs(x)$	$x > 0 \equiv upper$	$x > 0 \not\equiv upper$
0.5	1	1
1.0	1	2
1.5	21[a] (5)	2
2.0	25[a] (0)	4
3.0	0	0
4.0	2	3
6.0	6	0
8.0	14	0
10.0	23	0
20.0	35	0

[a] Due to the subtraction error mentioned in the comment section of the algorithm. Changing the constant 2.32 to 1.28 resulted in the figures shown in brackets.

To test the claim that the algorithm works virtually to the accuracy of the machine, it was translated into double-length instructions of Mercury Autocode and run on the Atlas using the EXCHLF compiler (the constant being lengthened to 0.398942280401432677939946). The results were compared with hand calculations using Table II of [5]. The errors in the 22nd significant figure were:

$abs(x)$	$x > 0 \equiv upper$	$x > 0 \not\equiv upper$
1.0	2	3
2.0	7	1
4.0	2	0
8.0	8	0

Timings. Timings of these algorithms were made in terms of the Atlas "Instruction Count," while evaluating the function 100 times. The figures are not directly applicable to any other computer, but the relative times are likely to be much the same on other machines.

INSTRUCTION COUNT FOR 100 EVALUATIONS

$abs(x)$	Algorithm number							
	123	180	181	209	226 $m = 7$	272	304[a]	304[b]
0.5	58			8	97	24	25	24
1.0	65[c]			8	176	24	29	29
1.5	164	128	127	9	273	25	35	35
2.0	194	78	90	8	387	24	39	39
2.5	252	54	68	10	515	24	131	44
3.0		42	51	9	628	25	97	50
4.0		27	39	9	900[d]	25	67	44
6.0		15	30	6	1400[d]	16	49	23
8.0		9	28	7	2100[d]	18	44	11
10.0		10	25	5	2700[d]	16	38	11
20.0		9	22	5	6500[d]	16	32	11
30.0		9	9	5	10900[d]	16	11	11

[a] Readings refer to $x > 0 \equiv upper$.
[b] Readings refer to $x > 0 \not\equiv upper$.
[c] Time to produce incorrect answer. A count of 120 would fit a smooth curve with surrounding values.
[d] 100 times Instruction Count for 1 evaluation.

Opinion. There are advantages in having two algorithms available for normal curve tail areas. One should be very fast and reasonably accurate, the other very accurate and reasonably fast. We conclude that Algorithm 209 is the best for the first requirement, and Algorithm 304 for the second.

Algorithms 180 and 181 are faster than Algorithm 304 and may be preferred for this reason, but the method used shows itself in Algorithm 181 to be not quite as accurate, and the introduction of this method solely for the circumstances in which Algorithm 180 is applicable hardly seems worth while.

Acknowledgment. Thanks are due to Miss I. Allen for her help with the double-length hand calculations.

REFERENCES:

1. THACHER, HENRY C. JR. Certification of Algorithm 123. *Comm. ACM 6* (June 1963), 316.

2. IBBETSON, D. Remark on Algorithm 123. *Comm. ACM 6* (Oct. 1963), 618.

3. BARTON, STEPHEN P., AND WAGNER, JOHN F. Remark on Algorithm 123. *Comm. ACM 7* (Mar. 1964), 145.

4. CLAUSEN, I., AND HANSSON, L. Certification of Algorithm 181. *Comm. ACM 7* (Dec. 1964), 702.

5. SHEPPARD, W. F. *The Probability Integral.* British Association Mathematical Tables VII, Cambridge U. Press, Cambridge, England, 1939.

ALGORITHM 181
COMPLEMENTARY ERROR FUNCTION—
LARGE X
HENRY C. THACHER, JR.*
Argonne National Laboratory, Argonne, Ill.
* Work supported by the U. S. Atomic Energy Commission.

real procedure $erfcL(x)$; **value** x; **real** x;
comment This procedure evaluates the complementary error
function, $erfc(x) = 1 - erf(x) = (2/\sqrt{\pi}) \int_x^\infty exp(-u^2)du$ by
the Laplace continued fraction:

$$erfc(x) = (1/(1+v/(1+2v/(1+3v/(1+\cdots)))))/(\sqrt{\pi}x\, e^{x^2})$$

where $v = 1/(2x^2)$. Successive even convergents of the continued
fraction are evaluated, using an algorithm suggested by Maehly,
until the full accuracy of the arithmetic being used is attained.

The continued fraction converges for all $x > 0$. For small x,
however, convergence may be excessively slow, and overflow
and round-off accumulation may occur. In this region, the
Taylor series converges satisfactorily.

For $x \leq 0$, the procedure calls the global procedure alarm.

The body of this procedure has been checked on the LGP-30
Computer, using the Dartmouth Self Contained Algol Processor,
for $x = 1.2(0.1)5.0$. Results were generally correct to 1 in the
6th significant digit, although a few errors were as large as 6
in that digit. The errors are believed to be due to round-off
only. The number of convergents calculated ranged from 46
for $x = 1.2$ to 10 for $x = 5.0$.

Overflow occurred for $x = 1.183$;

```
begin integer m;  real B min 2, B min 3, P, R, T, v, v2;
if x ≤ 0 then alarm;
v := x × x;
T := 0.56418958/x/exp(v);
comment  The constant 0.56418958 ··· = π^(-1/2), and should be
   given to the full accuracy required of the procedure;
v := 0.5/v;
v2 := v × v;
P := v × T;
m := R := 0;
B min 3 := B min 2 := 1;
for  m := m + 2 while R ≠ T do
   begin  R := T;
   B min 3 := v × (m−1) × B min 3 + B min 2;
   T := B min 2;
   B min 2 := v × m × B min 2 + B min 3;
   T := R − P/B min 2/T;
   P := m × (m+1) × v2 × P
   end while;
   erfc L := T
end
```


CERTIFICATION OF ALGORITHM 181 [S15]
COMPLEMENTARY ERROR FUNCTION—LARGE
X   [Henry C. Thacher, Jr., *Comm. ACM 6* (June 1963),
315]
I. CLAUSEN AND L. HANSSON (Recd. 20 Aug. 1964)
DAEC, Risø, Denmark.

The procedure $erfcL$ was tested in GIER-ALGOL with 29 signifi-
cant bits and the number-range $abs(x) < 2 \uparrow 512$ (approx. $1.3_{10}154$).
The statement $m := R := 0$; was corrected to $m := 0$; $R :=$
0; [Because m and R are of different type; cf. Sec. 4.2.4 of the
ALGOL Report, *Comm. ACM 6* (Jan. 1963), 1–17.—Ed.] After this
the tests were successful. The procedure was checked a.o. for
$x = 1.19\ (-0.01)\ 0.72$. The differences from table values increased
from $_{10}-8$ at $x = 1.1$ to $7_{10}-8$ at $x = 0.75$. Overflow occurred at
$x = 0.71$.

REMARKS ON:
ALGORITHM 123 [S15]
REAL ERROR FUNCTION, ERF(x)
   [Martin Crawford and Robert Techo *Comm. ACM 5*
   (Sept. 1962), 483]

ALGORITHM 180 [S15]
ERROR FUNCTION—LARGE X
   [Henry C. Thacher Jr. *Comm. ACM 6* (June 1963), 314]

ALGORITHM 181 [S15]
COMPLEMENTARY ERROR FUNCTION—
LARGE X
   [Henry C. Thacher Jr. *Comm. ACM 6* (June 1963), 315]

ALGORITHM 209 [S15]
GAUSS
   [D. Ibbetson. *Comm. ACM 6* (Oct. 1963), 616]

ALGORITHM 226 [S15]
NORMAL DISTRIBUTION FUNCTION
   [S. J. Cyvin. *Comm. ACM 7* (May 1964), 295]

ALGORITHM 272 [S15]
PROCEDURE FOR THE NORMAL DISTRIBUTION
FUNCTIONS
   [M. D. MacLaren. *Comm. ACM 8* (Dec. 1965), 789]

ALGORITHM 304 [S15]
NORMAL CURVE INTEGRAL
   [I. D. Hill and S. A. Joyce. *Comm. ACM 10* (June
   1967), 374]

I. D. HILL AND S. A. JOYCE (Recd. 21 Nov. 1966)
Medical Research Council,
Statistical Research Unit, 115 Gower Street, London
   W.C.1., England

These algorithms were tested on the ICT Atlas computer using
the Atlas ALGOL compiler. The following amendments were made
and results found:

ALGORITHM 123
   (i) **value** x; was inserted.
   (ii) $abs(T) \leqslant _{10}-10$ was changed to $Y - T = Y$
      both these amendments being as suggested in [1].

(iii) The labels 1 and 2 were changed to $L1$ and $L2$, the **go to** statements being similarly amended.

(iv) The constant was lengthened to 1.12837916710.

(v) The extra statement $x := 0.707106781187 \times x$ was made the first statement of the algorithm, so as to derive the normal integral instead of the error function.

The results were accurate to 10 decimal places at all points tested except $x = 1.0$ where only 2 decimal accuracy was found, as noted in [2]. There seems to be no simple way of overcoming the difficulty [3], and any search for a method of doing so would hardly be worthwhile, as the algorithm is slower than Algorithm 304 without being any more accurate.

ALGORITHM 180

(i) $T := -0.56418958/x/exp(v)$ was changed to $T := -0.564189583548 \times exp(-v)/x$. This is faster and also has the advantage, when v is very large, of merely giving 0 as the answer instead of causing overflow.

(ii) The extra statement $x := 0.707106781187 \times x$ was made as in (v) of Algorithm 123.

(iii) **for** $m := m + 1$ was changed to **for** $m := m + 2$. $m+1$ is a misprint, and gives incorrect answers.

The greatest error observed was 2 in the 11th decimal place.

ALGORITHM 181

(i) Similar to (i) of Algorithm 180 (except for the minus sign).

(ii) Similar to (ii) of Algorithm 180.

(iii) m was declared as **real** instead of **integer**, as an alternative to the amendment suggested in |4|.

The results were accurate to 9 significant figures for $x \leqslant 8$, but to only 8 significant figures for $x = 10$ and $x = 20$.

ALGORITHM 209

No modification was made. The results were accurate to 7 decimal places.

ALGORITHM 226

(i) $10 \uparrow m/(480 \times sqrt(2 \times 3.14159265))$ was changed to $10 \uparrow m \times 0.000831129750836$.

(ii) **for** $i := 1$ **step** 1 **until** $2 \times n$ **do** was changed to $m := 2 \times n$; **for** $i := 1$ **step** 1 **until** m **do**.

(iii) $-(i \times b/n) \uparrow 2/8$ was changed to $-(i \times b/n) \uparrow 2 \times 0.125$.

(iv) **if** $i = 2 \times n - 1$ was changed to **if** $i = m - 1$

(v) $b/(6 \times n \times sqrt(2 \times 3.14159265))$ was changed to $b/(15.0397696478 \times n)$.

Tests were made with $m = 7$ and $m = 11$ with the following results:

x	Number of significant figures correct		Number of decimal places correct	
	$m = 7$	$m = 11$	$m = 7$	$m = 11$
-0.5	7	11	7	11
-1.0	7	10	7	10
-1.5	7	10	8	10
-2.0	7	9	8	10
-2.5	6	9	8	11
-3.0	6	7	8	9
-4.0	5	7	10	11
-6.0	2	1	12	10
-8.0	0	0	11	9

Perhaps the comment with this algorithm should have referred to decimal places and not significant figures. To ask for 11 significant figures is stretching the machine's ability to the limit, and where 10 significant figures are correct, this may be regarded as acceptable.

ALGORITHM 272

The constant .99999999 was lengthened to .9999999999.

The accuracy was 8 decimal places at most of the points tested, but was only 5 decimal places at $x = 0.8$.

ALGORITHM 304

No modification was made. The errors in the 11th significant figure were:

$abs(x)$	$x > 0 \equiv upper$	$x > 0 \not\equiv upper$
0.5	1	1
1.0	1	2
1.5	21[a](5)	2
2.0	25[a](0)	4
3.0	0	0
4.0	2	3
6.0	6	0
8.0	14	0
10.0	23	0
20.0	35	0

[a] Due to the subtraction error mentioned in the comment section of the algorithm. Changing the constant 2.32 to 1.28 resulted in the figures shown in brackets.

To test the claim that the algorithm works virtually to the accuracy of the machine, it was translated into double-length instructions of Mercury Autocode and run on the Atlas using the EXCHLF compiler (the constant being lengthened to 0.398942280401432677939946). The results were compared with hand calculations using Table II of [5]. The errors in the 22nd significant figure were:

$abs(x)$	$x > 0 \equiv upper$	$x > 0 \not\equiv upper$
1.0	2	3
2.0	7	1
4.0	2	0
8.0	8	0

Timings. Timings of these algorithms were made in terms of the Atlas "Instruction Count," while evaluating the function 100 times. The figures are not directly applicable to any other computer, but the relative times are likely to be much the same on other machines.

INSTRUCTION COUNT FOR 100 EVALUATIONS

$abs(x)$	Algorithm number							
	123	180	181	209	226 $m = 7$	272	304[a]	304[b]
0.5	58			8	97	24	25	24
1.0	65[c]			8	176	24	29	29
1.5	164	128	127	9	273	25	35	35
2.0	194	78	90	8	387	24	39	39
2.5	252	54	68	10	515	24	131	44
3.0		42	51	9	628	25	97	50
4.0		27	39	9	900[d]	25	67	44
6.0		15	30	6	1400[d]	16	49	23
8.0		9	28	7	2100[d]	18	44	11
10.0		10	25	5	2700[d]	16	38	11
20.0		9	22	5	6500[d]	16	32	11
30.0		9	9	5	10900[d]	16	11	11

[a] Readings refer to $x > 0 \equiv upper$.
[b] Readings refer to $x > 0 \not\equiv upper$.
[c] Time to produce incorrect answer. A count of 120 would fit a smooth curve with surrounding values.
[d] 100 times Instruction Count for 1 evaluation.

Opinion. There are advantages in having two algorithms available for normal curve tail areas. One should be very fast and reasonably accurate, the other very accurate and reasonably fast. We conclude that Algorithm 209 is the best for the first requirement, and Algorithm 304 for the second.

Algorithms 180 and 181 are faster than Algorithm 304 and may be preferred for this reason, but the method used shows itself in Algorithm 181 to be not quite as accurate, and the introduction of this method solely for the circumstances in which Algorithm 180 is applicable hardly seems worth while.

Acknowledgment. Thanks are due to Miss I. Allen for her help with the double-length hand calculations.

REFERENCES:

1. THACHER, HENRY C. JR. Certification of Algorithm 123. *Comm. ACM 6* (June 1963), 316.

2. IBBETSON, D. Remark on Algorithm 123. *Comm. ACM 6* (Oct. 1963), 618.

3. BARTON, STEPHEN P., AND WAGNER, JOHN F. Remark on Algorithm 123. *Comm. ACM 7* (Mar. 1964), 145.

4. CLAUSEN, I., AND HANSSON, L. Certification of Algorithm 181. *Comm. ACM 7* (Dec. 1964), 702.

5. SHEPPARD, W. F. *The Probability Integral.* British Association Mathematical Tables VII, Cambridge U. Press, Cambridge, England, 1939.

ALGORITHM 182
NONRECURSIVE ADAPTIVE INTEGRATION
W. M. McKeeman and Larry Tesler
Stanford University, Stanford, Calif.

```
real procedure  Simpson(F) limits : (a, b) tolerance : (eps);
real procedure  F;  real a, b, eps;  value a, b, eps;
begin comment  A nonrecursive translation of Algorithm 145.
  Note that the device used here can be used to simulate recursion
  for a wide class of algorithms;
  integer lvl;
  switch  return := r1, r2, r3;
  real array  dx, epsp, x2, x3, F2, F3, F4, Fmp, Fbp,
  est2, est3 [1:30], pval[1:30, 1:3];
  integer array  rtrn [1:30];
  real  absarea, est, Fa, Fm, Fb, da, sx, est1, sum, F1;
  comment  the parameter setup for the initial call;
  lvl := absarea := est := 0;  da := b − a;
  Fa := F(a);  Fm := 4.0 × F((a+b)/2.0);  Fb := F(b);
  recur:
    lvl := lvl + 1;  dx[lvl] := da/3.0;
    sx := dx[lvl]/6.0;  F1 := 4.0 × F(a+dx[lvl]/2.0);
    x2[lvl] := a + dx[lvl];  F2[lvl] := F(x2[lvl]);
    x3[lvl] := x2[lvl] + dx[lvl];  F3[lvl] := F(x3[lvl]);
    epsp[lvl] := eps;  F4[lvl] := 4.0 × F(x3[lvl]+dx[lvl]);
    Fmp[lvl] := Fm;  est1 := (Fa+F1+F2[lvl]) × sx;
    Fbp[lvl] := Fb;  est2[lvl] := (F2[lvl]+F3[lvl]+Fm) × sx;
                     est3[lvl] := (F3[lvl]+F4[lvl]+Fb) × sx;
    sum := est1 + est2[lvl] + est3[lvl];
    absarea := absarea − abs(est) + abs(est1) + abs(est2[lvl]) +
      abs (est3[lvl]);
    if  (abs(est−sum) ≤ epsp[lvl] × absarea)  ∨ (lvl≥30) then
    begin comment  done on this level;
      up : lvl := lvl − 1;
      pval[lvl, rtrn[lvl]] := sum;
      go to  return [rtrn[lvl]]
    end;
    rtrn[lvl] := 1;  da := dx[lvl];  Fm := F1;
    Fb := F2[lvl];  eps := epsp[lvl]/1.7;  est := est1;
    go to  recur;  r1:
    rtrn[lvl] := 2;  da := dx[lvl];  Fa := F2[lvl];
    Fm := Fmp[lvl];  Fb := F3[lvl];  eps := epsp[lvl]/1.7;
    est := est2[lvl];  a := x2[lvl];  go to  recur;  r2:
    rtrn[lvl] := 3;  da := dx[lvl];  Fa := F3[lvl];
    Fm := F4[lvl];  Fb := Fbp[lvl];  eps := epsp[lvl]/1.7;
    est := est3[lvl];  a := x3[lvl];  go to  recur;  r3:
    sum := pval[lvl, 1] + pval[lvl, 2] + pval[lvl, 3];
    if  lvl > 1 then go to  up;
  Simpson  :=  sum
end  Simpson
```

CERTIFICATION OF ALGORITHM 182
NONRECURSIVE ADAPTIVE INTEGRATION [W.
 M. McKeeman and Larry Tesler, *Comm. ACM 6* (June
 1963), 315]
Harold S. Butler (Recd 8 Nov. 1963; rev. 6 Dec. 1963)
Stanford Linear Accelerator Center, Stanford, Calif.

A Balgol transliteration of *Simpson* has been prepared at
Stanford by its authors and it has been used in a number of problems involving numerical integration. Its value was most strik-
ingly displayed when it was utilized in a triple integral in which
the final integration was over a strongly peaked function that
spanned seven orders of magnitude. *Simpson* effectively minimized
the number of evaluations and completed the integration five
times faster than alternate schemes to subdivide the region of
interest. The values of the integral agreed with independent
calculations well within the required tolerance.

The following changes should be made to the published
algorithm:

Line 13 should be changed to:
  $lvl := 0$; $absarea := est := 1.0$; $da := b−a$;
Line 17 should read:
  $sx := dx[lvl]/6.0$; $F1 := 4.0 × F(a + dx[lvl]/2.0)$;
Line 20 should read:
  $epsp[lvl] := eps$; $F4[lvl] := 4.0 × F(x3[lvl] + dx[lvl]/2.0)$;
The condition of line 27 should be changed to:
  if $((abs(est−sum) ≤ epsp[lvl] × absarea) ∧ (est ≠ 1.0)) ∨$
  $(lvl ≥ 30)$ **then**

ALGORITHM 183

REDUCTION OF A SYMMETRIC BANDMATRIX
TO TRIPLE DIAGONAL FORM

H. R. Schwarz

Swiss Federal Institute of Technology, Zürich, Switzerland

```
procedure  bandred(a, n, m);
  value n, m;  integer n, m;  array a;
  comment  bandred reduces a real and symmetric matrix of band
    type (order n, a[i, k]=0 for |i−k|>m) by a sequence of orthog-
    onal similarity transformations to triple diagonal form. The
    procedure represents a generalization of the algorithm m21 by
    H. Rutishauser. Due to symmetry only the upper part of the
    band matrix must be given and these elements are denoted for
    convenience in the following way: a[i, 0] (i=1, 2, ···,  n) repre-
    sents the diagonal element in the ith row, and a[i, k] (i=1, 2,
    ···, n−k and k=1, 2, ···, m) represents the generally nonzero
    element in the ith row and the kth position to the right of the
    diagonal. After completion of the reduction, the elements of the
    symmetric triple diagonal matrix are given by a[i, 0] (i=1, 2,
    ···, n) and a[i, 1] (i=1, 2, ···, n−1);
begin  integer r, k, i, j, p, rr;  real  b, g, c, s, c2, s2, cs, u, v;
  for  r := m step −1 until 2 do
  begin
    for k := 1 step 1 until n−r do
    begin
      for j := k step r until n−r do
      begin
        comment  This compound statement describes the rota-
          tion involving the ith and (i+1)st rows and columns
          in order to reduce either a[j, r] or the off-band element
          g to zero, respectively. This rotation produces a new
          off-band element g (in general different from zero) pro-
          vided i + r < n;
        if  j = k then
        begin if  a[j, r] = 0 then go to endk;
          b := −a[j, r−1]/a[j, r]
        end
        else
        begin if  g = 0 then go to endk;
          b := −a[j−1, r]/g
        end;
        s := 1/sqrt(1 + b×b);  c := b × s;
        c2 := c × c;  s2 := s × s; cs := c × s;
        i := j + r − 1;
cross elements:
        u := c2 × a[i, 0] − 2 × cs × a[i, 1] + s2 × a[i + 1, 0];
        v := s2 × a[i, 0] + 2 × cs × a[i, 1] + c2 × a[i+1, 0];
        a[i, 1] := cs × (a[i, 0] − a[i+1, 0]) + (c2−s2) ×a[i, 1];
        a[i, 0] := u;  a[i+1, 0] := v;
column rotation:
        for p := j step 1 until i − 1 do
        begin
          u := c × a[p, i−p] − s × a[p, i−p+1];
          a[p, i−p+1] := s × a[p, i−p] + c × a[p, i−v+1];
          a[p, i−p] := u
        end  p;
        if  j ≠ k then
          a[j−1, r] := c × a[j−1, r] − s × g;
row rotation:
        rr := if r ≤ n − i then r else n − i;
        for  p := 2 step 1 until rr do
        begin
          u := c × a[i, p] − s × a[i+1, p−1];
          a[i+1, p−1] := s × a[i, p] + c × a[i+1, p−1];
          a[i, p] := u
        end  p;
        if  i + r < n then
new g:  begin  g := −s × a[i+1, r];
          a[i+1, r] := c × a[i+1, r]
        end
      end  j;
endk: end k
    end r
end  bandred
```

ALGORITHM 184
ERLANG PROBABILITY FOR CURVE FITTING
A. COLKER
U. S. Steel Applied Research Laboratory
Monroeville, Penn.

```
procedure ERLANG (X, XO, M, VARS, C, FACTORIAL, P);
value XO, M, VARS, C;  integer C;  real array X, P;
integer procedure FACTORIAL;
```

comment Computes the Erlang probability for the ith interval by $\int_0^{x_i} f(x)dx - \int_0^{x_{i-1}} f(x)dx$ where $f(x) = + [(K\mu)^K/(K-1)!] \cdot (x-x_0)^{K-1} e^{-K\mu(x-x_0)}$ where $\mu = 1/M$, $K = (M-X_0)^2 VARS$ is the upper boundary for the class intervals, X_0 is the lower boundary of the first class interval, M is the mean of the Erlang, $VARS$ is the variance corrected by Sheppard's correction, C is the number of class intervals and P_i is the calculated probability;

```
begin
  integer I, J, K, F;  real array XE[0 : C];
    for I := 1 step 1 until C do
      XE[I] := X[I] − XO;
    XE[0] := 0;
    ME := M − XO;
    K := 0.5 + (ME↑2)/VARS;
    U := K/ME;
    SP := 0;
    for I := 1 step 1 until C do
    begin
        SUM1 := 0;
        SUM2 := 0;
        for J := 0 step 1 until K − 1 do
        begin
          F := FACTORIAL (J);
          Z1 := U × XE[I−1];
      SUM1 := SUM1 + (Z1↑J)/F;
          Z2 := U × XE[I];
          SUM2 := SUM2 + (Z2↑J)/F;
        end J;
        P[I] := SUM1 × (EXP(−U×XE[I−1])) − SUM2
          × (EXP(−U×XE[I]));
        SP := SP + P[I];
    end I;
      P[C+1] := 1.0 − SP;
end Erlang
```

ALGORITHM 185

NORMAL PROBABILITY FOR CURVE FITTING

A. Colker

U. S. Steel Applied Research Laboratory

Monroeville, Penn.

procedure *NORMAL* (*X*, *M*, *VARS*, *C*, *HASTINGS*, *P*);
value *M*, *VARS*, *C*; **integer** *C*; **real array** *X*, *P*;
real procedure *HASTINGS*;
comment Computes the normal probabilities for the ith interval
by $\int_0^{x_i} f(x)dx - \int_0^{x_{i-1}} f(x)dx$ where $f(x)$ is Hastings' approximation to the normal interval. Hastings' formula is

$$\phi(X_{ni}) = \tfrac{1}{2}[1 - (1 + a_1 X_{ni} + a_2 X_{ni}^2 + a_3 X_{ni}^3 + a_4 X_{ni}^4 + a_5 X_{ni}^5)^{-8}]$$

where $a_1 = 0.09979268$, $a_2 = 0.04432014$, $a_3 = 0.00969920$,
$a_4 = -0.00009862$, and $a_5 = 0.00058155$. The X_{ni} are normalized
boundary values of X_i where $X_{ni} = (X_i - M)/\sqrt{VARS}$, where
M is the mean and $VARS$ is the variance corrected by Sheppard's
correction, C is the number of class intervals and P_i the calculated probability;
begin
  integer *I*; **real array** *XN*[1 : *C*];
  for *I* := 1 **step** 1 **until** *C* **do** *XN*[*I*] := (*X*[*I*]−*M*)/*SQRT*(*VARS*);
  P[1] := 0.5 − *HASTINGS* (*ABS*(*XN*[1]));
  for *I* := 2 **step** 1 **until** *C* **do**
  begin
   if
XN[*I*] < 0 **then**
P[*I*] := *HASTINGS* (*ABS*(*XN*[*I*−1])) − *HASTINGS*
  (*ABS*(*XN*[*I*])); **else**
   begin
    if (*XN*[*I*]>0) ∧ (*XN*[*I*−1]<0)
    then *P*[*I*] := *HASTINGS* (*XN*[*I*]) + *HASTINGS*
     (*ABS*(*XN*[*I*−1])); **else**
    P[*I*] := *HASTINGS* (*XN*[*I*]) − *HASTINGS* (*XN*[*I*−1]);
   end;
  end *I*;
  P[*C*+1] := 0.5 − *HASTINGS* (*XN*[*C*]);
end *NORMAL*

ALGORITHM 186
COMPLEX ARITHMETIC
R. P. van de Riet
Mathematical Centre, Amsterdam, Holland

procedure *Complex arithmetic* (a, b, R, r); **value** a, b; **array** a, b, R, r;

comment This procedure assigns the value $a^2 + b^2$ to R and the value $(a+ib)/(a-ib)$ to r, where a, b, R and r are complex numbers. These two arithmetic expressions are of course fully arbitrary. They serve only to demonstrate the use of the procedures P, Q, S, T, J and U. With them one can build up any arithmetic expression with complex variables, as easily as one can form them with real variables in Algol 60 (As one sees immediately these procedures can easily be extended for use in quaternion arithmetic or general vector and tensor calculus). We focus attention to the **value** call of the procedure-parameters, which is essential. Furthermore, we notice that the depth or height of the accumulator H is the number of right-handed brackets placed one after another not counting the brackets which occur in parameter-delimeters. It is perhaps superfluous to mention that this procedure was tested on the X1 computer of the Mathematical Centre.;

begin integer i, k; **array** $H[1{:}4,1{:}2]$;

integer procedure $P(i, j)$; **value** i, j; **integer** i, j;
comment P forms the product of the ith and jth element of H;
begin real a; $k := k - 1$; $a := H[i, 1] \times H[j, 1] - H[i, 2] \times H[j, 2]$; $H[k, 2] := H[i, 1] \times H[j, 2] + H[i, 2] \times H[j, 1]$; $H[k, 1] := a$; $P := k$
end;

integer procedure $Q(i, j)$; **value** i, j; **integer** i, j;
comment Q forms the quotient of the ith and jth element of H;
begin real a, b; $k := k - 1$; $b := H[j, 1] \uparrow 2 + H[j, 2] \uparrow 2$;
$a := (H[i, 1] \times H[j, 1] + H[i, 2] \times H[j, 2])/b$;
$H[k, 2] := (H[i, 2] \times H[j, 1] - H[i, 1] \times H[j, 2])/b$;
$H[k, 1] := a$; $Q := k$
end;

integer procedure $S(i, j)$; **value** i, j; **integer** i, j;
comment S forms the sum of the ith and jth element of H;
begin $k := k - 1$; $H[k, 1] := H[i, 1] + H[j, 1]$;
$H[k, 2] := H[i, 2] + H[j, 2]$; $S := k$
end;

integer procedure $T(a)$; **array** a;
comment T assigns to the $k+1$th element of H the complex variable a;
begin $k := k + 1$; $H[k, 1] := a[1]$; $H[k, 2] := a[2]$; $T := k$
end;

integer procedure $J(i, expi)$; **integer** i; **real** $expi$;
comment J assigns to the $(k+1)$th element of H a complex variable which is decomposed in real and imaginary part;
begin $k := k + 1$; $i := 1$; $H[k, 1] := expi$; $i := 2$;
$H[k, 2] := expi$;
$J := k$
end;

procedure $U(i, R)$; **value** i; **integer** i; **array** R;
comment U assigns to R the ith element of H;
begin $R[1] := H[i, 1]$; $R[2] := H[i, 2]$; $k := 0$ **end**;
$k := 0$; $U(S(P(T(a))\text{times}{:}(T(a)))\text{plus}{:}(P(T(b))\text{times}{:}$
$(T(b))), R)$;

comment $(a \times a) + (b \times b) =: R$; $U(Q(S(T(a))$ plus:
$(P(J(i, i-1))$ times: $(T(b))))$ divided by: $(S(T(a))$
plus: $(P(J(i, 1-i))$ times: $(T(b)))), r)$;
comment $(a+(i \times b))/(a+(-i \times b)) =: r$;
end *Complex Arithmetic*;

The contents of this Algorithm are published in the Technical Note TN 27, Mathematical Centre, Nov. 1962.

ALGORITHM 187
DIFFERENCES AND DERIVATIVES
R. P. VAN DE RIET
Mathematical Centre, Amsterdam, Holland

```
begin  real h;  integer i, k;  array A[1 : 50];
comment  This program calculates, only to demonstrate the
    procedures DELTA and DER, the third derivative of the expo-
    nential function with a sixth order difference scheme. We do
    not propose to use these procedures in actual calculations, for
    as we observed with the X1 computer of the Math. Centre, they
    work, but very slowly as a consequence of the strong recursive-
    ness of the procedures. In actual programming one has to take
    the trouble to write out the well-known formula of Gregory, or
    for higher derivatives to multiply this formula a number of
    times by itself, then one has to collect the same function-values.
    All this trouble is taken over by the computer if one uses the
    procedures described below. My purpose, however, in publishing
    these procedures lies not in the numerical use but in a demon-
    stration of the flexibility of ALGOL 60, if one uses the recursive-
    ness property of procedures.;
real procedure SUM(i, h, k, ti);  value k;  integer i, k, h;
    real ti;
begin  real s;  s := 0;  for i := h step 1 until k do s := s + ti;
        SUM := s
end;
real procedure DELTA (N, k, k0, fk);  value N, k0;  real fk;
integer N, k, k0;
comment  N is the order of the forward difference which is
    calculated from a set of function-values with equidistant
    parameter-values;
begin  integer i;
        DELTA := if N = 1
        then SUM (k, k0, k0+1, (−1)↑(k+1−k0)×fk)
        else DELTA (1, i, k0, DELTA (N−1, k, i, fk))
end;
real procedure DER (OR, N, h, k, k0, fk);  value OR, N, h, k0;
    real fk, h;
integer OR, N, k, k0;
comment  OR is the order of the derivative, calculated from a
    given set of function-values f(k), with equidistant parameter-
    values, the error is of the order h ↑ (N+1−OR), where h is the
    steplength. k0 is the point where the derivative is calculated;
begin  integer i;
        DER := if OR = 1
        then SUM(i, 1, N, DELTA(i, k, k0, fk)
        ×(−1)↑(i+1)/i)/h
        else DER(1, N+1−OR, h, i, k0, DER(OR−1, N−1, h,
        k, i, fk))
end;
for i := 1 step 1 until 50 do A[i] := exp(i/50);
for i := 1 step 1 until 25 do A[i] := DER(3, 6, .02, k, i, A[k])
end
```

The contents of this Algorithm are published in the Technical
Note TN 27, Mathematical Centre, Nov. 1962.

ALGORITHM 188
SMOOTHING 1.
F. Rodriguez-Gil
Central University, Caracas, Venezuela

```
procedure Smooth 13(n, x);
integer n;
real array x;
comment This procedure uses Gram's first-degree three-point
    formulas, as described in Hildebrand's "Introduction to Nu-
    merical Analysis," Ch. 7, to smooth a series of n equally spaced
    values. If the procedure is entered with less than three points,
    control is transferred to a nonlocal label error;
begin real array xp[1 : n];  integer i;
    if n < 3 then go to error;
    for i := 1 step 1 until n do xp[i] := x[i];
    x[1] := 0.83333333 × xp[1] + 0.33333333 × xp[2] − 0.16666667
        × xp[3];
    for i := 2 step 1 until n − 1 do x[i] :=  (xp[i−1]+xp[i]
        + xp[i+1]) × 0.33333333;
    x[n] := − 0.16666667 × xp[n−2] + 0.33333333 × xp[n−1]
        + 0.83333333 × xp[n]
end  Smooth 13
```

ALGORITHM 189
SMOOTHING 2
F. Rodriguez Gil
Central University, Caracas, Venezuela

procedure *Smooth* 35(n, x);
integer n;
real array x;
comment This procedure is similar to *Smooth* 13, except that
  Gram's third-degree five-point formulas are used, and that a
  minimum of five points is needed for a successful application;
begin real array $xp[1 : n]$; **integer** i;
  if $n < 5$ **then go to** *error*;
  for $i := 1$ **step** 1 **until** n **do** $xp[i] := x[i]$;
  $x[1] := 0.98571429 \times xp[1] + 0.05714286 \times (xp[2]+xp[4])$
    $- 0.08571429 \times xp[3] - 0.01428571 \times xp[5]$;
  $x[2] := 0.05714286 \times (xp[1]+xp[5]) + 0.77142857 \times xp[2]$
    $+ 0.34285714 \times xp[3] - 0.22857143 \times xp[4]$;
  for $i := 3$ **step** 1 **until** $n - 2$ **do** $x[i] := - 0.08571429 \times (xp[i-2]$
    $+xp[i+2]) + 0.34285714 \times (xp[i-1]+xp[i+1]) + 0.48571429$
    $\times xp[i]$;
  $x[n-1] := 0.05714286 \times (xp[n-4]+xp[n]) - 0.22857143$
    $\times xp[n-3] + 0.34285714 \times xp[n-2] + 0.77142857 \times xp[n-1]$;
  $x[n] := - 0.01428571 \times xp[n-4] + 0.05714286 \times (xp[n-3]$
    $+xp[n-1]) - 0.08571429 \times xp[n-2] + 0.98571429 \times xp[n]$
end *Smooth* 35

ALGORITHM 190
COMPLEX POWER
A. P. Relph
The English Electric Co. Ltd., Whetstone, England

procedure *Complex power* (a, b, c, d, n, x, y); **value** a, b, c, d, n;
real a, b, c, d, x, y; **integer** n;
comment This procedure calculates $(x+iy) = (a+ib) \uparrow (c+id)$
where i is the root of -1. In the complex plane, with a cut along
the real axis from 0 to $-\infty$, p is the sum of the principal value
of the argument of $(a+ib)$ and $2n\pi$ (n is positive, negative or
zero depending on the solution required). *arctan* is assumed to
be in the range $-\pi/2$ to $\pi/2$. The case $n = 0$, $d = 0$ is given by
Algorithm 106;
begin   **real** p, r, v, w;
        if $a = 0$ **then begin if** $b = 0$ **then begin** $x := y := 0$;
                                    go to L **end**
                          else $p := 1.57079633 \times$
                                  $(sign(b)+4\times n)$
                  end
            else begin $p := 6.28318532 \times n + arctan(b/a)$;
                  if $a < 0$ **then begin if** $b \geqq 0$ **then**
                              $p := p + 3.14159265$
                        else
                              $p := p - 3.14159265$
                        end
                end;
        $r := .5 \times ln(a\uparrow 2+b\uparrow 2)$; $v := c \times p + d \times r$;
        $w := exp(c\times r-d\times p)$;
        $x := w \times cos(v)$; $y := w \times sin(v)$;
L: **end**

ALGORITHM 191
HYPERGEOMETRIC
A. P. RELPH
The English Electric Co. Ltd., Whetstone, England

procedure *Hypergeometric* (a1, a2, b1, b2, c1, c2, z1, z2) Results:
(s1, s2); **value** a1, a2, b1, b2. c1, c2, z1, z2; **real** a1, a2, b1, b2,
c1, c2, z1, z2, s1, s2;
begin comment calculates the hypergeometric function
$1F2(a, b, c, z)$ with complex parameters $(a=a1+ia2,$
etc);
    real $d, y1, y2$; **integer** n;
    procedure *comp mult* (a1, a2, b1, b2, c1, c2); **value** a1,
    a2, b1, b2; **real** a1, a2, b1, b2, c1, c2;
        begin comment calculates the product of the two
            complex numbers $(a1+ia2)$ and $(b1+ib2)$
            where i is the root of -1;
                $c1 := a1 \times b1 - a2 \times b2$; $c2 := a2 \times b1 +$
                $a1 \times b2$
        end;
    $s1 := y1 := 1$; $s2 := y2 := 0$;
    for $n := 1$ **step** 1 **until** 100 **do**
    begin $d := n \times ((c1+n-1)\uparrow2+c2\uparrow2)$;
        comp mult (a1+n-1, a2, y1/d, y2/d, y1, y2);
        comp mult (y1, y2, b1+n-1, b2, y1, y2);
        comp mult (y1, y2, c1+n-1, -c2, y1, y2);
        comp mult (y1, y2, z1, z2, y1, y2);
        if $s1 = s1 + y1 \wedge s2 = s2 + y2$ **then go to** L;
        $s1 := s1 + y1$; $s2 := s2 + y2$
    end;
L: **end**

CERTIFICATION OF ALGORITHMS 191 AND 192
HYPERGEOMETRIC AND CONFLUENT HYPER-
  GEOMETRIC [A. P. Relph, *Comm. ACM* 6 (July
  1963), 388]
HENRY C. THACHER, JR.* (Recd 2 Dec. 1963)
Argonne National Laboratory, Argonne, Ill
  * Work supported by the U.S. Atomic Energy Commission.

The bodies of these two procedures were transcribed for the
Dartmouth SCALP processor for the LGP-30 computer. No syntactical errors were found, and the programs gave results agreeing
within roundoff (7D) with tabulated values for the following special cases: $_2F_1(0.5, 0.5; 1; k^2) = (2/\pi) K(k)$; $_2F_1(0.5, -0.5; 1; k^2) =$
$(2/\pi)E(k)$ where K and E are complete elliptic integrals of the first
and second kinds; $_1F_1(.5; 1; iy) = J_0(x)$, and with $_1F_1(-1; 0.1; x)$;
$_1F_1(-0.5; 0.1; x)$, and $_1F_1(-0.5; 0.5; x)$.

It should be observed that the function calculated by 191 is
$_2F_1(a, b; c; z)$, not $_1F_2(a, b; c; z)$ as stated in the comment. These
programs evaluate the functions by direct summation of the hypergeometric series. They are, therefore, relatively general, but
inefficient. Precautions must also be taken against attempting to
compute outside the range of effective convergence of the series.

Certification and Remark on Algorithm 191 [S22]
Hypergeometric [A.P. Relph, *Comm. ACM* 6 (July
1963), 388]

Henk Koppelaar (Recd. 14 Sept. 1973) Physical Laboratory, Division: Atom Physics, Utrecht State University,
Utrecht, The Netherlands

The following changes were made in the algorithm:
(a) The subroutine for complex multiplication was erased.
(b) In accordance with (a) and with the standard notation $_2F_1$, the
heading and end of the procedure were changed to read
real procedure *hyp2geom1* (a, b, c, z);
end *hyp2geom1*.
(c) Erasing the subroutine for complex multiplication caused us to
modify the algorithm further as follows.
  real procedure *hyp2geom1* (a, b, c, z);
    value $a, b\ c, z$; **real** a, b, c, z;
  begin
    real s, y; **integer** i;
    $s := y := 1$;
    for $i := 0$ **step** 1 **until** 100 **do**
    begin
        $y := y \times ((a+i)/(c+i)) \times (b+i) \times z/(i+1)$;
        if $s = s + y$ **then go to** *exit*; $s := s + y$
    end;
exit:
    $hyp2geom1 := s$
  end *hyp2geom1*
The inefficiency of the original algorithm for real arguments, as
mentioned by Henry C. Thacher Jr. in his certification of it, is
largely reduced by these modifications, because they make a considerable reduction in the computational costs.

With these modifications the algorithm was translated for the
CDC-6500 using the Control Data Algol 3 compiler and ran only
partially satisfactorily, as reported below.

The following two tests (a) and (b) were performed, using identities 1 and 2 and Algorithm 160 [1] for the combinatorial $\binom{m}{n}$.
The identities are:

1. $(1/B(a\ b))\ (z^a/a)\,_2F_1(a, 1-b; a+1; z) = I_z(a, b)$,

2. $a \times B(a, b) = \binom{a+b-1}{a}^{-1}$, where

$$B(a, b) = \int_0^1 t^{a-1}(1-t)^{b-1}dt$$

is the complete beta function and

$$I_z(a, b) = (1/B(a, b)) \int_0^z t^{a-1}(1 - t)^{b-1}dt$$

is the incomplete beta function, $a \geq 1, b \geq 1$, and a, b integer.

Test (a). Computing $\binom{a+b-1}{a} \times hyp2geom1$ $(a, 1-b, a+1,$
$z) \times z \uparrow a$, gave correct results to 7D., according [2], for the
following values of a, b and z.

$b = 7, a = 7(1)9$ and z 0.61(0.01)0.97;
$b = 17, a = 17$ and $z = 0.13$
$b = 17, a = 17, 18$ and $z = 0.14$
$b = 17, a = 17(1)19$ and $z = 0.15$
$b = 17, a = 17(1)34$ and $z = 0.50$

The total computation time for this test using combinatorial and
hyp2geom1 was less than 10 sec on the Control Data 6500 computer.

Test (b). The same test as test (a) was performed for the following values of a, b, and z.

$b = 17$, $a = 17(1)34$ and $z = 0.51(0.01)0.60$.

The algorithm gave correct results to $5D$, according [2].

For the values

$b = 17$, $a = 17(1)34$ and $z = 0.61(0.01)0.89$, the results became worse with increasing z. This error is due to slower convergence of the series

$$_2F_1(a, b; c; z) = \sum_{n=0}^{\infty} (a)_n (b)_n / (c)_n \, z^n / n!$$

with increasing z and increasing b.

Pochhammer's symbol means

$(a)_n = \Gamma(a+n) / \Gamma(a)$.

More precisely we see

$(1/B(a, b))(z^a/a) \, _2F_1(a, 1-b; a+1; z) =$

$$\sum_{n=0}^{\infty} \binom{a+b-1}{a} ((a)_n (1-b)_n / (a+1)_n) \, z^{n+a} / n!$$

From this expression it is clear that the rate of convergence of the expansion by and large is dominated by the values of b and z. This explains why in test (a) for $b = 7$ the results remained accurate to $7D$ for increasing z, and also it explains why in test (b) for $b = 17$ the results became worse for increasing z.

References
1. Wolfson, M.L., and Wright, H.V. Combinatorial of M things taken N at a time. *Comm. ACM 6*, 4 (Apr. 1963), 161.
2. Pearson, K. (Ed.). *Tables of the Incomplete Beta Function.* Cambridge U. Press, Cambridge, England, 1948.

ALGORITHM 192
CONFLUENT HYPERGEOMETRIC
A. P. RELPH
The English Electric Co. Inc., Whetstone, England

```
procedure Confluent hypergeometric (a1, a2, c1, c2, z1, z2)
  Result : (s1, s2);  value a1, a2, c1, c2, z1, z2;
  real a1, a2, c1, c2, z1, z2, s1, s2;
begin  comment calculates the confluent hypergeometric func-
        tion 1F1(a, c, z) with complex parameters
        (a=a1+ia2, etc);
        real d, y1, y2;  integer n;
        procedure comp mult (a1, a2, b1, b2, c1, c2);
          value a1, a2, b1, b2;  real a1, a2, b1, b2, c1, c2;
          begin  comment calculates the product of the two
                  complex numbers (a1+ia2) and (b1+ib2)
                  where i is the root of −1;
                c1 := a1 × b1 − a2 × b2;
                c2 := a2 × b1 + a1 × b2
          end;
        s1 := y1 := 1;  s2 := y2 := 0;
        for n := step 1 until 100 do
        begin  d := n × ((c1+n−1)↑2+c2↑2);
                comp mult (a1+n−1, a2, y1/d, y2/d, y1, y2);
                comp mult (y1, y2, c1+n−1, −c2, y1, y2);
                comp mult (y1, y2, z1, z2, y1, y2);
                if s1 = s1 + y1 ∧ s2 = s2 + y2 then go to L;
                s1 := s1 + y1;  s2 := s2 + y2
        end;
L:  end
```


CERTIFICATION OF ALGORITHMS 191 AND 192
HYPERGEOMETRIC AND CONFLUENT HYPER-
    GEOMETRIC [A. P. Relph, *Comm. ACM* 6 (July
1963), 388]
HENRY C. THACHER, JR.* (Recd 2 Dec. 1963)
Argonne National Laboratory, Argonne, Ill.
    * Work supported by the U.S. Atomic Energy Commission.

The bodies of these two procedures were transcribed for the
Dartmouth SCALP processor for the LGP-30 computer. No syntactical errors were found, and the programs gave results agreeing
within roundoff (7D) with tabulated values for the following spe-
cial cases: $_2F_1(0.5, 0.5; 1; k^2) = (2/\pi) K(k)$; $_2F_1(0.5, -0.5; 1; k^2) = (2/\pi)E(k)$ where K and E are complete elliptic integrals of the first
and second kinds; $_1F_1(.5; 1; iy) = J_0(x)$, and with $_1F_1(-1; 0.1; x)$;
$_1F_1(-0.5; 0.1; x)$, and $_1F_1(-0.5; 0.5; x)$.

It should be observed that the function calculated by 191 is
$_2F_1(a, b; c; z)$, not $_1F_2(a, b; c; z)$ as stated in the comment. These
programs evaluate the functions by direct summation of the hypergeometric series. They are, therefore, relatively general, but
inefficient. Precautions must also be taken against attempting to
compute outside the range of effective convergence of the series.

ALGORITHM 193
REVERSION OF SERIES
HENRY E. FETTIS
Aeronautical Research Laboratories, Wright-Patterson Air
  Force Base, Ohio

```
procedure SERIESRVRT (A, B, N);
value A, N;  array A, B;  integer N;
comment This procedure gives the coefficients B[i] for the series
  x = y + ΣB[i] × y ↑ i (i=2, 3, ⋯ , n) when the coefficients
  A[i] of the series y = x + ΣA[i] × x ↑ i are given. The procedure
  uses successive approximations after writing y_{L+1} = x − ΣB[i] ×
  y_L ↑ i (i=2, 3, ⋯ , L+2 and L=0, 1, ⋯ , N−2) starting with
  y_0 = x;
begin integer i, j, k, m;
  array Q, R [0 : N];
  real s;
  A[1] := B[0] := 0;
  B[1] := 1;
  for k := 1 step 1 until N − 1 do
  begin B[k+1] := 0;
    for i := 0 step 1 until k + 1 do
    R[i] := 0;
    for j := k + 1 step − 1 until 1 do
    begin Q[0] := R[0] − A[j];
      for i := 1 step 1 until k + 1 do
      Q[i] := R[i];
      for i := 0 step 1 until k + 1 do
      begin s := 0;
        for m := 0 step 1 until i do
        s := s + B[m] × Q[i−m];
        R[i] := s
      end for i;
    end for j;
    for i := 2 step 1 until k + 1 do B[i] := R[i]
  end for k;
end SERIESRVRT
```


CERTIFICATION OF ALGORITHM 193
REVERSION OF SERIES [Henry E. Fettis, *Comm.
  ACM 5*, 1962]
HENRY C. THACHER, JR.*
Argonne National Laboratory, Argonne, Ill.
  * Work supported by the U. S. Atomic Energy Commission

The body of Algorithm 193 was tested on the LGP-30 using the
ALGOL 60 translator developed by the Dartmouth College Computer Center. No syntactical errors were found. The program successfully found the first four coefficients for the series for $ln(1+y)$
from the first four coefficients of the series for $y = e^x − 1$.

ALGORITHM 194
ZERSOL
CARLOS DOMINGO
Universidad Central, Caracas, Venezuela

procedure $ZERSOL$ $(h, YI, m, epsi, F, f, Z)$; **real** h, $epsi$, f;
  array YI, Z; **integer** m; **procedure** F;
comment $ZERSOL$ finds the simple zeros of the solution $Y1(Y0)$
  of the set of m first order differential equations $Yj = Fj(Y0,$
  $Y1, \cdots, Ym)$. h is the step of integration, $epsi$ the error with
  which the zeros are to be determined (assuming no error in the
  process of integration). $F(YS, j, v)$ is a procedure which calculates the functions Fj, taking the arguments from the array
  YS and leaving the results in v. The search for zeros stops
  when $Y0 > f$. The zeros are stored as elements of the array Z.
  MR is a 4×4 matrix with the coefficients of a Runge-Kutta
  method. For example MR may be row-wise $0.5, 1, 0.5, 0, 1 - a,$
  $1 - a, 1 - a, 0.5, 1 + a, 1 + a, 1 + a, 0, \frac{1}{6}, \frac{1}{3}, 0.5, 0.5$, where
  $a = sqrt(2)$;
begin real v, r, d; **integer** j, s, n, k; **array** $Q[1:m]$, $YS[0:m]$,
  $YAL[0:m]$, $YT[1:m]$, $MR[1:4,1:4]$; **switch** $S := NOZ, ZER$;
  $n := 1$;
for $d := h$ **while** $YI[0] \leq f$ **do**
  begin $s := 1$;
    $R1$: **for** $j := 1$ **step** 1 **until** m **do**
      begin $Q[j] := 0.0$; $YS[j] := YI[j]$; $YT[j] := YI[j]$ **end**;
      $YS[0] := YI[0]$;
    $R2$: **for** $k := 1$ **step** 1 **until** 4 **do**
      begin $YS[0] := YS[0] + MR[k, 4] \times d$;
        for $j := 1$ **step** 1 **until** m **do**
        begin $F(YS,j,v)$; $v := v \times d$;
          $r := MR[k,1] \times v - MR[k,2] \times Q[j]$;
          $YT[j] := YT[j] + r$;
          $Q[j] := Q[j] + 3.0 \times r - MR[k,3] \times v$
        end;
        for $j := 1$ **step** 1 **until** m **do** $YS[j] := YT[j]$
      end;
      go to $S(s)$;
  NOZ: **if** $sign(YI[1]) \neq sign(YS[1])$ **then go to** IT;
   TR: **for** $j := 0$ **step** 1 **until** m **do** $YI[j] := YS[j]$; **go to** $R2$;
   IT: $s := 2$;
      for $j := 0$ **step** 1 **until** m **do** $YAL[j] := YS[j]$;
  ZER: $d := d/2$;
    if $d \leq epsi$ **then go to** STZ;
    if $sign(YI[1]) = sign(YS[1])$ **then go to** TR **else go to** $R1$;
  STZ: $Z[n] := YI[0] := YI[0] + d$; $n := n + 1$;
      for $j := 0$ **step** 1 **until** m **do** $YI[j] := YAL[j]$
  end;
end

ALGORITHM 195
BANDSOLVE
DONALD H. THURNAU
Marathon Oil Co., Littleton, Colo.

procedure $BANDSOLVE$ (C,N,M,V); **value** N,M; **integer** N,M; **real array** C,V;

comment $BANDSOLVE$ is effective in solving the matrix equation $AX = B$ when the matrix A is of large order and sparse such that a narrow band centered on the main diagonal includes all the non-zero elements. Parameter N is the order of A, and M is the width of the band, necessarily an odd number of elements. $BANDSOLVE$ is very efficient because it operates only on the band portion of the matrix A, given in the N by M array C. The band elements of a given row of A appear in the same row of C but shifted such that element $A[i,j]$ becomes $C[i,j-i+(M+1)/2]$. All band elements whether zero or non-zero must be given. The values of undefined elements of C, such as $C[1,1]$ or $C[N,M]$, are irrelevant. The array V initially contains the vector B. After solution, the array V contains the answer vector X. The contents of array C are destroyed during solution which is done by Gauss elimination with row interchanges, followed by back substitution;

```
begin integer JM,LR,I,PIV,R,J;  real T;
  LR := (M+1) ÷ 2;
  for R := 1 step 1 until LR − 1 do
    for I := 1 step 1 until LR − R do
      begin for J := 2 step 1 until M do
        C[R,J−1] := C[R,J];
        C[R,M] := C[N+1−R,M+1−I] := 0
      end of row shifting and zero placement;
  for I := 1 step 1 until N − 1 do
    begin PIV := I;
      for R := I + 1 step 1 until LR do
        if abs(C[R,1])>abs(C[PIV,1]) then
          PIV := R;
      if PIV ≠ I then
        begin T := V[I];
          V[I] := V[PIV];
          V[PIV] := T;
          for J := 1 step 1 until M do
            begin T := C[I,J];
              C[I,J] := C[PIV,J];
              C[PIV,J] := T
            end J
        end of row interchange;
      V[I] := V[I]/C[I,1];
      for J := 2 step 1 until M do
        C[I,J] := C[I,J]/C[I,1];
      for R := I + 1 step 1 until LR do
        begin T := C[R,1]
          V[R] := V[R] − T × V[I];
          for J := 2 step 1 until M do
            C[R,J−1] := C[R,J] − T × C[I,J];
          C[R,M] := 0
        end R;
      if LR ≠ N then LR := LR + 1
    end of triangularization;
```

```
  V[N] := V[N]/C[N,1];
  JM := 2;
    for R := N − 1 step −1 until 1 do
      begin for J := 2 step 1 until JM do
        V[R] := V[R] − C[R,J] × V[R−1+J];
      if JM ≠ M then JM := JM + 1
      end of back solution
end BANDSOLVE
```

Remark on Algorithm 195 [F4]
BANDSOLVE [Donald H. Thurnau, *Comm. ACM 6* (Aug. 1963), 441]

Ernst Schuegraf [Recd. 1 Mar. 1971]
Department of Mathematics, St. Francis Xavier University, Antigonish, Nova Scotia, Canada

Algorithm 195 was transliterated into Fortran IV for the IBM 360/50. Various matrices with different values of N and M were used. The execution time was recorded and the accuracy of the results was checked.

Execution time [sec]

	$M = 11$	$M = 15$	$M = 21$	$M = 25$
$N = 50$	.2	.7	1.1	1.9
$N = 100$	.6	1.6	2.5	4.2

The execution time shows the expected proportionality to $((M − 1) ÷ 2)^2 \cdot N$. (Note the definition of M!) When checking the results, it was found that the algorithm failed for singular and near singular matrices. To protect against this, it is recommended to introduce a tolerance *eps* for a test on singularity and a label *fail*. This requires the following changes in the procedure declaration:

procedure $BANDSOLVE$ $(C,N,M,V,eps,fail)$;

It is necessary to insert the following statements in the blockhead of the procedure:
real *eps*; **label** *fail*;

After the statement *piv* := *r*; insert:
if *abs* $(C[piv, 1]) < eps$ **go to** *fail*;

ALGORITHM 196
MULLER'S METHOD FOR FINDING ROOTS OF
AN ARBITRARY FUNCTION
ROBERT D. RODMAN
Burroughs Corp., Pasadena, Calif.

```
procedure MULLER (p1, p2, p3, mxm, nrts, ep1, ep2, sw1, sw2,
    sw3, swr, rrt, irt);
value p1, p2, p3, mxm, nrts, ep1, ep2, sw1, sw2, sw3, swr;
integer mxm, nrts;  boolean sw1, sw2, sw3, swr;
real p1, p2, p3, ep1, ep2;  array rrt, irt;
begin comment  procedure MULLER finds real and complex
    roots of an arbitrary function. p1, p2, and p3 are starting values.
    Roots nearest these points are found first. mxm is the maximum
    number of iterations to be made in finding any one root. ep1 and
    ep2 are specified as tolerance parameters. If ABS((X[i+1]−X[i])/
    X[i+1]) < ep1 or if the function value and modified function value
    are both less than ep2, a root has been found. If sw1 is true,
    then each iterant of each root is printed. If sw2 is true, the
    value of each root found is printed. If sw3 is true, then, when
    applicable, the complex conjugate of each root found is admitted
    as a root. If swr is true, only real roots are found. rrt and irt
    contain the real, and imaginary parts of each root found. Proce-
    dure function is the function generator and procedure com-
    plex performs necessary complex operations;
boolean bool;  integer c1, rtc, i, itc;  real rx1, rx2, rx3, ix1,
    ix2, ix3, rroot, iroot, rdnr, idnr, t1, it1, frroot, firoot, rfx1, rfx2,
    rfx3, ifx1, ifx2, ifx3, rh, ih, rlam, ilam, rdel, idel, t2, it2, t3, it3, t4,
    it4, rg, ig, rden, iden, rfunc, ifunc;
switch j := m2, m3, m4, m7, m11;
procedure function (reale, imag, reval, ieval);
value reale, imag;  real reale, imag, reval, ieval;
begin comment  Coding for this procedure must be inserted at
    compile time. reale and imag are the real and imaginary parts
    of the dependent variable. reval and ieval, the real and imaginary
    parts of the function;
end function;
procedure complex (a, ia, b, ib, k, c, ic);
value a, ia, b, ib, k;  integer k;
real a, ia, b, ib, c, ic;
begin real temp;  switch j := mpy, dvd, sqt;
        go to j[k];
mpy:  c := a × b − ia × ib;  ic := a × ib + ia × b; go to exit;
dvd:  if (ib=0) ∧ (b=0) then begin ic := 0; c := 1;
        go to exit end; temp := b ↑ 2 + ib ↑ 2;
    c := (a×b+ia×ib)/temp;  ic := (ia×b−a×ib)/temp;
        go to exit;
sqt:  if (ia=0) ∧ (a<0) then
        begin c := 0;  ic := sqrt (−a) end
        else if ia = 0 then
        begin c := sqrt(a);  ic := 0 end
        else begin temp := sqrt (a↑2+ia↑2);
            c := sqrt ((temp + a)/2);
            ic := if (temp − a) < 0 then 0
                else sqrt ((temp − a)/2) end;
        if ((b+c) ↑ 2 + (ib+ic) ↑ 2) < ((b−c) ↑ 2 + (ib−ic) ↑ 2)
        then begin c := b − c;  ic := ib − ic end
        else begin c := b + c;  ic := ib + ic end;
exit:  end of complex;
start: for i := 1 step 1 until nrts do rrt [i] := irt [i] := 0; rtc := 0;

m0:  ix1 := ix2 := ix3 := cl, := iroot := itc := 0;
    rroot := p1;  bool := false;
m1:  cl; := cl + 1;  rdnr := 1;  idnr := 0;
    for i := 1 step 1 until rtc do
        begin
            complex (rdnr, idnr, rroot-rrt [i], iroot-irt [i], 1, t1, it1);
                rdnr := t1;  idnr := it1
        end;
    function (rroot, iroot, t1, it1);
    complex (t1, it1, rdnr, idnr, 2, frroot, firoot);
    go to j[cl];
m2:  rfx1 := frroot;  ifx1 := firoot;  rroot := p2;
    go to m1;
m3:  rfx2 := frroot;  ifx2 := firoot;  rroot := p3;
    go to m1;
m4:  rfx3 := frroot;  ifx3 := firoot;  rx1 := p1;
    rx2 := p2;  rx3 := p3;  rh := rx3 − rx2;
    ih := ix3 − ix2;
    complex (rh, ih, rx2− rx1, ix2− ix1, 2, rlam, ilam);
    rdel := rlam + 1;  idel := ilam;
m9:  if (rfx1=rfx2) ∧ (rfx2=rfx3) ∧ (ifx1=ifx2) ∧ (ifx2=ifx3)
    then begin rlam := 1;  ilam := 0;  go to m8 end;
    complex (rfx1, ifx1, rlam, ilam, 1, t1, it1);
    complex (rfx2, ifx2, rdel, idel, 1, t2, it2);
    t1 := t1 − t2 + rfx3;  it1 := it1 − it2 + ifx3;
    complex (rdel, idel, rlam, ilam, 1, t2, it2);
    complex (t1, it1 t2, it2, 1, t3, it3);
    complex (rfx3, ifx3, t3, it3, 1, t1, it1);
    t1 := −4 × t1;  it1 := −4 × it1;
    complex (rfx3, ifx3, rlam+rdel, ilam+idel, 1, t2, it2);
    complex (rdel ↑ 2−idel ↑ 2, 2×rdel×idel, rfx2, ifx2, 1, t3, it3);
    complex (rlam ↑ 2−ilam ↑ 2, 2×rlam×ilam, rfx1, ifx1, 1,
        t4, it4);
    rg := t4 − t3 + t2;  ig := it4 − it3 + it2;
    if swr ∧ ((rg ↑ 2+t1)<0) then
        begin rden := rg;  iden := ig := 0 end
    else complex (rg ↑ 2−ig ↑ 2+t1, 2×rg×ig+it1, rg, ig, 3,
        rden, iden);
    complex (−2×rfx3, −2×ifx3, rdel, idel, 1, t1, it1);
    complex (t1, it1, rden, iden, 2, rlam, ilam);
m8:  itc := itc + 1;
    rx1 := rx2;  rx2 := rx3;  rfx1 := rfx2;  rfx2 := rfx3;
    ix1 := ix2;  ix2 := ix3;  ifx1 := ifx2;  ifx2 := ifx3;
    complex (rlam, ilam, rh, ih, 1, t1, it1);
    rh := t1;  ih := it1;
m6:  rdel := rlam + 1;  idel := ilam;  rx3 := rx2 + rh;
    ix3 := ix2 + ih;  cl := 3;  rroot := rx3;
    iroot := ix3;  go to m1;
m7:  rfx3 := frroot;  ifx3 := firoot;
    function (rx3, ix3, rfunc, ifunc);
    complex (rfx3, ifx3, rfx2, ifx2, 2, t1, it1);
    if (t1 ↑ 2+it1 ↑ 2) > 100 then
        begin rlam := rlam/2;  rh := rh/2;  ilam := ilam/2;
        ih := ih/2;  go to m6 end;
    if sw1 then . . .
comment  option to output iterant and associated function
    values;
    t1 := rx3 − rx2;  it1 := ix3 − ix2;
    complex (t1, it1, rx2, ix2, 2, t2, it2);
```

```
    if sqrt (t2↑2+it2↑2) ≦ ep1 then go to fin1;
    if (sqrt (rfx3↑2+ifx3↑2)≦ep2) ∧
    (sqrt (rfunc↑2+ifunc↑2)≦ep2) then go to fin 2;
        go to if itc ≧ mxm then fin3 else m9;
fin1: if sw2 then . . .
comment   option to output root;   go to m12;
fin2: if sw2 then . . .
comment   option to output root;   go to m12;
fin3: if sw2 then . . .
comment   no convergence, option to output last iterant;
    bool := true;
m12: rtc := rtc + 1;   rrt[rtc] := rx3;   irt[rtc] := ix3;
        if rtc ≧ nrts then go to exit;
        if (ABS(ix3)>ep1) ∧ sw3 ∧ ¬ bool then
            begin ix3 := −ix3;  function (rx3, ix3, rfunc, ifunc);
            rroot := rx3;   iroot := ix3;   c1 := 4;
            go to m1;
m11: if sw2 then . . .
comment   the complex conjugate of the last root found is accept-
    able. Option to output this root;
        rtc := rtc + 1;   rrt[rtc] := rx3;   irt[rtc] := ix3
            end else go to m0;
        if rtc < nrts then go to m0;
exit:  end of procedure MULLER
```

CERTIFICATION OF ALGORITHM 196 [C5]
MULLER'S METHOD FOR FINDING ROOTS OF
AN ARBITRARY FUNCTION [Robert D. Rodman,
   Comm. ACM 6 (Aug. 1963), 442]
Virginia W. Whitley (Recd. 11 Oct. 1966, 24 Feb. 1967
   and 8 Sept. 1967)
Environmental Research Corp., Alexandria, Va.

KEY WORDS AND PHRASES: equation roots, function zeros
CR   CATEGORY: 5.15

The Algorithm. Algorithm 196 has been compiled in Fortran IV on the CDC-3600 and the IBM-7090 both in single and in
double precision. The single precision versions used the system-
supplied complex arithmetic subroutines; the double precision
versions used subroutines agreeing as closely as possible with those
described in the IBJOB Manual [4]. Thus, the algorithm tested
differs from the published algorithm only in the complex square
root subroutine.

There are five remarks to be made about Algorithm 196.

(1) As the Algorithm stands, if one of the values $P1$, $P2$, or $P3$
is a root of the equation and if more than one root is to be found,
then the procedure will fail with a 0/0 form in the computation of

$$F_r(z) =) f(z)/\prod_{i=1}^{rtc}(z-z_i).$$

Our decision was to terminate the procedure with a message to the
user. The referee has suggested an alternative:

```
m1:  rdnr := 1;   idnr := 0;
        for i := 1 step 1 until rtc do
        begin
        if rroot = rrt [i] ∧ iroot < irt [i] then
        begin
            if c1 = 0 then P1 := rroot := 2 × rroot + ep1 else
            if c1 = 1 then P2 := rroot := 2 × rroot + ep1 else
            if c1 = 2 then P3 := rroot := 2 × rroot + ep1 else
            begin comment   we have converged to a zero found pre-
                viously, we accept it without any test;
```

```
            go to fin1;
        end;
            go to m1
        end;
        complex (rdnr, idnr, rroot − rrt [i], iroot − irt [i], 1, t1,
        it1); rdnr := t1;   idnr := it1
    end;
    c1 := c1 + 1;
    function (rroot, iroot, t1, it1);
```

(2) The logical variable *bool* should be called by name in the
procedure statement and should be an array of the same dimension
as *rrt* and *irt*. Otherwise, unless *sw2* is *true*, the user will not know
which of the "roots" satisfies the convergence criteria and which
was returned by default.

(3) The statement *fin2* is unnecessary, since it is identical to
fin1.

(4) Frank [1] states, "This procedure (Muller's method) works
readily for functions having simple roots. On the other hand, if
a function possesses multiple roots, ξ, then $F_r(z)$ is indeterminate
when z approaches a ξ which may already have been found. However, even in this case the process has never failed. In fact, roots
of multiplicity six or more have been found successfully. This is
primarily due to the fact that multiple roots are found to much
less accuracy than simple roots and behave, in effect, like clustered
roots." In a private communication, Frank explained that the subroutine described in [1] included steps which perturbed roots
already found, forcing them to behave like clustered roots. Frank's
remark is not true for Algorithm 196, as a simple test with $(z-2)^2$,
requesting two roots, will demonstrate.

(5) The complex square root in procedure *complex* contains at
least two errors, not the least of which is that it fails to take ac-
count of the location of the complex number whose root is being
computed. The referee has pointed out a second error: "The 5th
line after the line labelled *sqt* should be

$$c := sqrt((temp+a)/2) \times sign(ia).$$

The construct **if** $(temp-a) < 0$ **then** 0 is unnecessary as the
Boolean expression cannot be true. Moreover, even with corrections the complex square root included is unsatisfactory because,
when *ia* is small, either $temp - a$ or $temp + a$ will be a difference
of two nearly equal numbers and loss of significance will occur."

It is suggested that the four lines beginning four lines below the
label *sqt* be replaced by

```
else begin temp := sqrt((abs(a)+sqrt(a↑2+ia↑2))/2);
    ic := .5 × ia/temp;
    if a ≥ 0 then c := temp
    else begin c := abs(ic);
        ic := sign(ia) × temp
    end;
end;
if ((b+c)↑2+(ib+ic)↑2) < · · ·
```

Under some systems, the case $ia = -0$ might cause problems.
With the possible exception of the case $ia = -0$, this coding will
choose the square root whose real part is positive.

Modifications to the Algorithm. Both the single and
double precision versions have been altered as suggested by Traub
[3, p. 212]:

$$z_{i+1} = z_i - \frac{2f_i}{\rho_i}, \quad f_i = f(z_i)$$

$$\rho_i = \omega_i \pm \{\omega_i^2 - 4f_i f[z_i, z_{i-1}, z_{i-2}]\}^{\frac{1}{2}}$$

$$\omega_i = f[z_i, z_{i-1}] + (z_i - z_{i-1})f[z_i, z_{i-1}, z_{i-2}]$$

$$f[z_i, z_{i-1}, z_{i-2}] = \frac{f[z_i, z_{i-1}] - f[z_{i-1}, z_{i-2}]}{z_i - z_{i-2}}$$

$$f[z_i, z_{i-1}] = \frac{f_i - f_{i-1}}{z_i - z_{i-1}}$$

Both Algorithm 196 and Traub's iteration function choose the sign of the square root to maximize the modulus of the denominator.

Although the two iteration functions are equivalent, Traub's requires fewer operations (8 additions, 5 multiplications and 3 divisions compared to Muller's 10 additions, 15 multiplications and 2 divisions), less storage and less computing time.

The behavior of the coded version of Traub's method differed little from that of Algorithm 196. Given the same starting values, both methods converged in the same number of iterations, even though the first iterates were sometimes different. Example 1 compares Muller and Traub in single precision with double precision. The difference between the first iterates in single precision is the result of roundoff due to the fact that the initial function values, $f(P1)$, $f(P2)$, and $f(P3)$, are very close together. In double precision the two methods agreed to 17 significant figures (the CDC-3600 carries approximately 24 decimal digits in double precision).

Comparison of double versus single precision results on the same machine and double versus double (single versus single) on two

---

Example 1. $f(z) = z^{20} - 1$

$P1 = .1875$ $f(P1) = -.99999999999999997115799543$ $ep1 = .5_{10} -6$
$P2 = .375$ $f(P2) = -.99999999969756966621957785$ $ep2 = .5_{10} -6$
$P3 = .5$ $f(P3) = -.99999990463246834693750000$ $nrts = 2$

	CDC-3600 Iteration		z		$f(z)$	
1	s.p.	Alg.196	8.959044683	−001	−8.890227259	−001
		Traub	8.959064917	−001	−8.890177130	−001
	d.p.		8.9590562323582572	−001	−8.8901986473910256	−001
2	s.p.	Alg.196	1.009184101	+000	2.006266970	−001
		Traub	1.009182792	+000	2.005955638	−001
	d.p.		1.0091833539604292	+000	2.0060892685123016	−001
3	s.p.	Alg.196	9.915438059	−001	−1.562027260	−001
		Traub	9.915451528	−001	−1.561798028	−001
	d.p.		9.9154457471920591	−001	−1.5618964171373523	−001
4	s.p.	Alg.196	9.996899988	−001	−6.181798671	−003
		Traub	9.996900961	−001	−6.179863674	−003
	d.p.		9.9969005435603457	−001	−6.1806941810976143	−003
5	s.p.	Alg.196	9.999986299	−001	−2.740146010	−005
		Traub	9.999986307	−001	−2.738516196	−005
	d.p.		9.9999863036901786	−001	−2.7392263226776617	−005
6	s.p.	Alg.196	1.000000000	+000	3.492459655	−009
		Traub	1.000000000	+000	3.492459655	−009
	d.p.		1.0000000001972048	+000	3.9440963812509629	−009

Example 2. Acoustic Waveguide Function

$$f(z) = P \frac{\sin (2\pi S)}{S} + \rho \cos (2\pi S), \quad \rho = 2.50$$

$P = \sqrt{(k^2 - z^2)}, \quad \mathrm{Re}(P) < 0; \quad S = \sqrt{[(z/\epsilon)^2 - k^2]}, \quad \mathrm{Re}(s) < 0;$
$k = 0.0, \quad \epsilon = 0.288; \quad P1 = 0.0, \quad P2 = 0.02, \quad P3 = 0.04$

Iteration	$z = x + iy$		$f(z) = u + iv$	
	x	y	u	v
1	0.67672247−01	0.44008261−02	0.24001462−00	0.49014466−01
2	0.71677143−01	0.51274250−02	0.23259824−01	0.98558515−02
3	0.72102452−01	0.53056952−02	0.20655826−03	0.19331276−07
4	0.72106262−01	0.53092607−02	0.39539112−07	0.40978193−07

Example 3. The 20 Roots of Unity

Starting values at approximately 19, 27, and 35 degrees. Conjugates accepted as roots. $ep1 = ep2 = .5_{10} - 7$. Roots are at $e^{ir\pi/10}$, $r = 0, \cdots, 19$.

Root	r	Number of iterations
0	3	5
1	17	*
2	4	15
3	16	*
4	7	22
5	13	*
6	9	14
7	11	*
8	5	14
9	15	*
10	1	10
11	19	*
12	2	7
13	18	*
14	6	10
15	14	*
16	0	5
17	8	8
18	12	*
19	10	2

* Asterisk in column 3 indicates conjugate taken; i.e., $z_1 = \bar{z}_0$, $z_3 = \bar{z}_2$, etc.

different machines have been made. In every case differences can be satisfactorily explained by (1) different BCD-binary conversion on different machines, (2) different word lengths, or (3) differences in library subroutines on the various machines.

Miscellaneous Comments. Both versions of Muller's method (Algorithm 196 and Traub's iteration function) have the advantage over other one-point-with-memory methods in that it is possible to locate complex zeros using real starting values. There is, of course, the possibility of spending unnecessary time doing complex arithmetic. As a general purpose library routine, unless one is looking for complex zeros, there are other iterative functions that require less space and have the same order; there are also others requiring less space and enjoying higher order, although they usually involve computing derivatives. For our purposes, the Traub version of Muller's method has proved quite satisfactory.

Example 2 is included to indicate the behavior of the Algorithm using a non-algebraic function. With this function both Algorithm

196 and the Traub modification agreed exactly except for a difference in the last digit in the first iterate.

Example 3 is included to show the order in which the 20 roots of unity are found and the number of iterations required to attain the specified accuracy. Values were checked against the NBS *Tables of Sines and Cosines to 15 Decimal Places*. All but two roots were correct to 9 significant figures; the remaining two were correct to 8.

Acknowledgment. The support of this work by the Atomic Energy Commission, Contract AT(29-2)-1163, is gratefully acknowledged.

REFERENCES:
1. FRANK, WERNER L. Finding zeros of arbitrary functions. *J. ACM 5* (1958), 154–160.
2. MULLER, DAVID E. A method for solving algebraic equations using an automatic computer. *MTAC 10* (1956), 208–215.
3. TRAUB, J. F. *Iterative Methods for the Solution of Equations.* Prentice-Hall, Englewood Cliffs, N. J., 1964.
4. IBM Systems Reference Library. File No. 7090-27, Form C28-6389-2. IBM 7090/7094 IBSYS Operating System. Version 13. IBJOB Processor.

ALGORITHM 197
MATRIX DIVISION
M. WELLS
University of Leeds, Leeds, England

procedure *Pos Div* (*b, c, m, n, solve*);
value *m, n, solve*; **array** *b, c*; **integer** *m, n*; **Boolean** *solve*;
comment The matrix *c*, with *m* rows and *n* columns, is divided
by the positive definite matrix *b*, of order *m*, by the square root
method (see Fadeeva, V. N., Computational Methods of Linear
Algebra, Chap 2, §10). The upper triangle of *b* is replaced by
an upper triangular matrix *N* such that $N^t N = b$. The other
elements of *b* are undisturbed. The matrix *c* is replaced by $b^{-1}c$.
The **Boolean** *solve* is used as a switch. If its value is **true**, then
it is assumed that an earlier entry to *Pos Div* has left the matrix
N in place, and a further division of *c* by *b* takes place;
begin integer *i, j, k*;
　real procedure *dot* (*a, b, p, q*);
　value *q*; **real** *a, b*; **integer** *p, q*;
　comment This is innerproduct, modified to define a function
　　designator;
　begin real *s*; *s* := 0;
　for *p* := 1 **step** 1 **until** *q* **do** *s* := *s* + *a* × *b*;
　dot := *s* **end** *dot*;
　Start of program: **if** *solve* **then go to** *back substitution*;
　for *i* := **step** 1 **until** *m* **do**
　begin *b* [*i, i*] := *sqrt* (*b*[*i, i*] − *dot* (*b*[*j, i*]↑2, 1, *j*, *i* − 1));
　　for *j* := *i* + 1 **step** 1 **until** *m* **do**
　　b[*i,j*] := (*b*[*i,j*] − *dot* (*b*[*k,i*], *b*[*k,j*], *k*, *i*−1))/*b*[*i,i*]
　end formation of upper triangular matrix;
back substituton: **for** *i* := 1 **step** 1 **until** *n* **do**
　begin for *j* := 1 **step** 1 **until** *m* **do**
　c[*i,j*] := (*c*[*i,j*] − *dot* (*b*[*k,j*], *c*[*i,k*], *k*, *j*−1))/*b*[*j,j*];
　　for *j* := *m* **step** −1 **until** 1 **do**
　c[*i,j*] := (*c*[*i,j*] − *dot* (*b*[*j,m*+1−*k*], *c*[*i,m*+1−*k*], *k*, *m*−*j*))/
　　b[*j,j*]
　end of double back substitution
end of *Pos Div*

CERTIFICATION OF ALGORITHM 197
MATRIX DIVISION [M. Wells, *Comm. ACM* 6 (Aug.
　1963), 443]
M. WELLS (Recd 18 Nov. 63)
University of Leeds, Leeds, England

The procedure was tested on a Ferranti Pegasus, using the
ALGOL compiler developed by the de Havilland Aircraft Company
at Hatfield. The line after the one labelled 'start of program'
should read

　　　for *i* := 1 **step** 1 **until** *m* **do**

(the first 1 was omitted).
　The statement labelled *back substituton* is incorrect, and should
read

back substitution: **for** *j* := 1 **step** 1 **until** *n* **do**
begin **for** *i* := 1 **step** 1 **until** *m* **do**
　c[*i,j*] := (*c*[*i,j*] − *dot* (*b*[*k,i*], *c*[*k,j*], *k*, *i*−1))/*b*[*i, i*];

for *i* := *m* **step** −1 **until** 1 **do**
c[*i,j*] := (*c*[*i,j*] − *dot* (*b*[*i,m*+1−*k*], *c*[*m*+1−*k,j*], *k,m*−*i*))/*b*[*i,i*]
end *of double back substitution*

With these changes the program was operated successfully on
a number of small test problems. The procedure is only applicable
to symmetric positive definite matrices, and no systematic attempt has yet been made to assess the accuracy of the results.
　The word 'symmetric' should be inserted before 'positive
definite' in the comment.
　It is interesting to note that the original, incorrect version of
the procedure will divide one symmetric matrix by another, and
so can be used for matrix inversion.

ALGORITHM 198
ADAPTIVE INTEGRATION AND MULTIPLE
INTEGRATION
WILLIAM MARSHALL MCKEEMAN
Stanford University, Stanford, Calif.

```
begin comment  This program illustrates the declaration and
  call of a procedure used to numerically approximate definite
  integrals and multiple integrals. The integrand is an expression
  substituted for the first formal parameter and must be a func-
  tion of the simple variable replacing the second formal pa-
  rameter. Multiple integration is accomplished by substituting
  a complete call of Integral for the first formal parameter. Note
  that in this case that the limits of integration on the inside calls
  may be functions of the variable of integration on the outer
  call. The parameter rule selects a Newton-Cotes formula which
  matches a polynomial of degree = rule to the function in the
  interval of integration. (See Hamming, Numerical Methods for
  Scientists and Engineers, Sec. 12.2). In any case, the procedure
  integral adapts its step size to the function in seeking to mini-
  mize the number of function evaluations. The program has been
  tested and run on a variety of functions using the ALGOL com-
  piler on the Burroughs B-5000.;
real procedure  Integral (F) a function of the real variables: (x)
  between limits: (a,b) polynomial degree: (rule) tolerance: (eps);
value a, b, rule, eps;  integer rule;
real F, x, a, b, eps;
begin comment  set up the parameters for the recursion before
  calling the procedure NC;
switch nct := R1, R2, R3, R4, R5, R6, R7;
real array cf, fn [1:rule+1];
integer k;  real da, ab;
real procedure NC(F,x,a,da,fn,k,cf,rule,eps,es,ab,1v1);
value a, da, rule, eps, es, 1v1;  real array cf;
integer k, rule, 1v1;  real F, x, a, da, fn, eps, es, ab;
begin comment  NC is the adaptive heart of Integral;
  real array fc[1:rule+1,1:rule+1], est, xx[1:rule+1];
  integer i, j;  real dx, int, ep;
  real procedure SUM(term, index, upperlimit);
  real term;  integer index, upperlimit;
  begin real t;  t := 0;
    for index := 1 step 1 until upperlimit do
      t := t + term;
    SUM := t
  end of SUM;
  comment begin the integration by evaluating F on the mesh
    points;
  for k := 1 step 1 until rule + 1 do fc[k,k] := fn;
  dx := da/(rule×(rule+1));
  x := a;
  for i := 1 step 1 until rule + 1 do
    for j := 1 step 1 until rule do
      begin
        if j = 1 then xx[i] := x;
        if i ≠ j then fc[i,j] := cf[j] × F;
        x := x + dx;
      end having done all necessary function evaluations;
  for i := 1 step 1 until rule do
    fc[i, rule+1] := fc[i+1,1];
  ep := eps/sqrt(rule+1);
```

```
  comment  eps/(rule + 1) is the value to give an absolute
    error bound of eps in the final answer. It proves too strict in
    practice;
  dx := dx × rule;
  comment  compute the integrals of the subintervals;
  for i := 1 step 1 until rule + 1 do
    est[i] := SUM(fc[i,j],j,rule+1)×dx;
  ab := ab − abs(es) + SUM(abs(est[i]),i,rule+1);
  comment  ab is the area under abs(F). It is used in computing
    the relative error upon which to terminate;
  int := SUM(est[i],i,rule+1);
  if 1v1 ≥ 100/(rule+1) then go to error;
  NC := if abs(es−int) ≤ eps × ab ∧ es ≠ 1.0 then int
    else SUM(NC(F,x,xx[i],dx,fc[i,j],j,cf,rule,ep,est[i], ab,1v1+1),
    i,rule+1);
  go to return;
  error;  NC := int;
    comment  abs(es − int) is the approximate error caused by
      terminating the recursion. In most cases, termination at
      this level will not adversely affect the accuracy of the result;
  return:
end of NC;
comment  now initialize the Newton-Cotes coefficients;
go to nct [rule];
R1:  cf[1] := cf[2] := 1.0/2.0;  go to compute;
R2:  cf[1] := cf[3] := 1.0/6.0;  cf[2] := 4.0/6.0;
comment  R1 is trapezoidal rule, R2 is Simpson's rule;
go to compute;
R3:  cf[1] := cf[4] := 1.0/8.0;
     cf[2] := cf[3] := 3.0/8.0;  go to compute;
R4:  cf[1] := cf[5] := 7.0/90.0;
     cf[2] := cf[4] := 32.0/90.0;
     cf[3] := 12.0/90.0;  go to compute;
R5:  cf[1] := cf[6] := 19.0/288.0;
     cf[2] := cf[5] := 75.0/288.0;
     cf[3] := cf[4] := 50.0/288.0;  go to compute;
R6:  cf[1] := cf[7] := 41.0/840.0;
     cf[2] := cf[6] := 216.0/840.0;
     cf[3] := cf[5] := 27.0/840.0;
     cf[4] := 272.0/840.0;  go to compute;
R7:  cf[1] := cf[8] := 75.1/1728.0;
     cf[2] := cf[7] := 357.7/1728.0;
     cf[3] = cf[6] := 134.3/1728.0;
     cf[4] := cf[5] := 298.9/1728.0;
compute: da := b − a;
for k := 0 step 1 until rule do
begin
  x := a + k × da/rule;
  fn[k+1] := F × cf[k+1];
end;
ab := 1.0;
  Integral := NC(F,x,a,da,fn[k],k,cf,rule,eps,1.0,ab,0);
  end of Integral;
  comment  Now evaluate the integral of 1.0/sqrt(abs(x+y))
    on the unit disk in the x,y-plane;
  real x, y, answer;
  answer := Integral(Integral(1.0/sqrt(abs(x+y)), x,
    −sqrt(1.0−y↑2), sqrt(1.0−y↑2), 7, 0.001),y,−1.0,1.0,3,0.001);
end of program;
```

ALGORITHM 199
CONVERSIONS BETWEEN CALENDAR DATE
AND JULIAN DAY NUMBER
ROBERT G. TANTZEN
Air Force Missile Development Center, Holloman AFB,
New Mex.

procedure $JDAY$ (d,m,y,j);
integer d,m,y,j;
comment $JDAY$ converts a calendar date, Gregorian calendar,
to the corresponding Julian day number j. From the given day
d, month m, and year y, the Julian day number j is computed
without using tables. The procedure is valid for any valid
Gregorian calendar date. When transcribing $JDAY$ for other
compilers, be sure that integers of size 3×10^6 can be handled;
  begin integer c, ya;
    if $m > 2$ **then** $m := m - 3$
      else begin $m := m + 9$; $y := y - 1$ **end**;
    $c := y \div 100$; $ya := y - 100 \times c$;
    $j := (146097 \times c) \div 4 + (1461 \times ya) \div 4 + (153 \times m + 2) \div 5 + d + 1721119$
  end $JDAY$

procedure $JDATE$ (j,d,m,y);
integer j,d,m,y;
comment $JDATE$ converts a Julian day number j to the corresponding calendar date, Gregorian calendar. Since j is an integer
for this procedure, it is correct astronomically for noon of the
day. $JDATE$ computes the day d, month m, and year y, without
using tables. The procedure is valid for any valid Gregorian
calendar date. When transcribing $JDATE$ for other compilers,
be sure that integers of size 3×10^6 can be handled:
  begin $j := j - 1721119$;
    $y := (4 \times j - 1) \div 146097$; $j := 4 \times j - 1 - 146097 \times y$;
    $d := j \div 4$;
    $j := (4 \times d + 3) \div 1461$; $d := 4 \times d + 3 - 1461 \times j$;
    $d := (d + 4) \div 4$;
    $m := (5 \times d - 3) \div 153$; $d := 5 \times d - 3 - 153 \times m$;
    $d := (d + 5) \div 5$;
    $y := 100 \times y + j$; **if** $m < 10$ **then** $m := m + 3$
      else begin $m := m - 9$; $y := y + 1$ **end**;
  end $JDATE$

procedure $KDAY$ (d,m,ya,k);
integer d,m,ya,k;
comment $KDAY$ converts a calendar date, Gregorian calendar,
to the corresponding serial day number k. From the given day
d, month m, and the last two decimals of the year, ya, the serial
day number k is computed without using tables. The procedure
is valid from 1 March 1900 $(k=1)$ to 31 December 1999
$(k = 36465)$. To obtain the Julian day number j (valid at noon)
use $j = k + 2415079$;
  begin if $m > 2$ **then** $m := m - 3$
    else begin $m := m + 9$; $ya := ya - 1$ **end**;
    $k := (1461 \times ya) \div 4 + (153 \times m + 2) \div 5 + d$
  end

procedure $KDATE$ (k,d,m,ya);
integer k,d,m,ya;

comment $KDATE$ converts a serial day number k to the corresponding calendar date, Gregorian calendar. It computes day d,
month m, and the last two decimals of the year, ya, without
using tables. The procedure is valid from $k = 1$ (1 March 00) to
$k = 36465$ (31 December 99) for any one century. For the 20th
Century the relation between k and the Julian day number j
(at noon) is $j = k + 2415079$;
begin $ya := (4 \times k - 1) \div 1461$; $d := 4 \times k - 1 - 1461 \times ya$;
  $d := (d + 4) \div 4$; $m := (5 \times d - 3) \div 153$;
    $d := 5 \times d - 3 - 153 \times m$;
  $d := (d + 5) \div 5$;
  if $m < 10$ **then** $m := m + 3$
    else begin $m := m - 9$; $ya := ya + 1$ **end**;
end $KDATE$

CERTIFICATION OF ALGORITHM 199 [Z]
CONVERSIONS BETWEEN CALENDAR DATE AND
  JULIAN DAY NUMBER [Robert G. Tartzen, *Comm.
  ACM 8* (Aug. 1963), 444].
DAVID K. OPPENHEIM (Recd. 10 Jul. 64 and 27 Jul. 64)
System Development Corp., Santa Monica, Calif.

Algorithm 199 was translated into JOVIAL J3 and tested on the
Philco 2000. Input was generated with a random number generator
that produced uniformly distributed dates between the years
1583 and 2583. The results were checked for 50 different dates in
that range.

The procedures as written place unnecessary restrictions on
some of the parameters. Expressions cannot always be used as
inputs to the procedures. Also, the original input to $JDAY$,
$JDATE$ and $KDAY$ will be modified during the operation of the
respective procedures. It should also be noted that in many im-
plementations of ALGOL the use of parameters called by name may
be more expensive than those called by value. The call by name
is a far more powerful tool than is necessary for most of the parameters of these procedures. For these reasons the following
changes are suggested:

1. In **procedure** $JDAY$
    change: **integer** d, m, y, j;
    to: **value** d, m, y; **integer** d, m, y, j;
2. In **procedure** $JDATE$
    change: **integer** j, d, m, y; to: **value** j; **integer** j, d, m, y;
3. In **procedure** $KDAY$
    change: **integer** d, m, ya, k;
    to: **value** d, m, ya; **integer** d, m, ya, k;
4. In **procedure** $KDATE$
    change: **integer** k, d, m, ya;
    to: **value** k; **integer** k, d, m, ya;

ALGORITHM 200
NORMAL RANDOM
RICHARD GEORGE*
Argonne National Laboratory, Argonne, Ill.
* Work supported by United States Atomic Energy Commission.

```
real procedure NORMAL RANDOM (Mean, Sigma n);
  procedure  Random;
  real Mean, Sigma;
  integer n;
  comment  Random is assumed to be a real procedure which
    generates a random number uniform on the interval (−1, +1).
    The value of n should be greater than 10, in order to approxi-
    mate the normal distribution with accuracy. However, very
    large values of n will increase the running time. The use of
    Mean and Sigma should be obvious. Reference: R. W. Ham-
    ming, Numerical Methods for Scientists and Engineers;
  begin
    integer i;  real sum;
    sum := 0;
    for i := step 1 until n do
      sum := sum + Random;
    NORMAL RANDOM := Mean + Sigma × sum × sqrt (3.0/n)
  end NORMAL RANDOM
```

CERTIFICATION OF ALGORITHM 200 [G5]
NORMAL RANDOM
[Richard George, *Comm. ACM 6* (Aug. 1963), 444]
M. C. PIKE (Recd. 3 May 1965)
Statistical Research Unit of the Medical Research Coun-
cil, U. College Hospital Medical School, London.

Algorithm 200 has the following errors:
(1) The line

    real procedure *NORMAL RANDOM* (*Mean, Sigma n*);

should be changed to

    real procedure *NORMAL RANDOM* (*Random, Mean,
Sigma, n*);

(2) The line

    procedure *Random*;

should be changed to

    real procedure *Random*;

With these corrections *NORMAL RANDOM* has been run successfully on the ICT Atlas computer with the Atlas ALGOL compiler.

ALGORITHM 201
SHELLSORT
J. BOOTHROYD
English Electric-Leo Computers, Kidsgrove, Staffs, England

procedure *Shellsort* (a, n); **value** n; **real array** a; **integer** n;
comment $a[1]$ through $a[n]$ of $a[1:n]$ are rearranged in ascending
  order. The method is that of D. A. Shell, (A high-speed sorting
  procedure, *Comm. ACM* 2 (1959), 30–32) with subsequences
  chosen as suggested by T. N. Hibberd (An empirical study of
  minimal storage sorting, SDC Report SP-982). Subsequences
  depend on m_1 the first operative value of m. Here $m_1 = 2^k - 1$
  for $2^k \leq n < 2^{k+1}$. To implement Shell's original choice of $m_1 =$
  $[n/2]$ change the first statement to $m := n$;
begin integer i, j, k, m; **real** w;
  for $i := 1$ **step** i **until** n **do** $m := 2 \times i - 1$;
  for $m := m \div 2$ **while** $m \neq 0$ **do**
    begin $k := n - m$;
      for $j := 1$ **step** 1 **until** k **do**
        begin for $i := j$ **step** $-m$ **until** 1 **do**
          begin if $a[i+m] \geq a[i]$ **then go to** 1;
            $w := a[i]$; $a[i] := a[i+m]$; $a[i+m] := w$;
          end i;
      1: **end** j
    end m
end *Shellsort*;

CERTIFICATION OF ALGORITHM 201
SHELLSORT [J. BOOTHROYD, *Comm. ACM 6* (Aug.
  1963), 445]
M. A. BATTY   (Recd 27 Jan. 1964)
English Electric Co., Whetstone, Nr. Leicester, England

  This algorithm has been tested successfully using the DEUCE
ALGOL Compiler. When the first statement of the algorithm was
replaced by the statement

$$m := n;$$

to implement Shell's original choice of $m_1 := n/2$, a slight increase
in sorting time was observed with most of the cases tested.

REMARK ON ALGORITHM 201 [M1]
SHELLSORT [J. Boothroyd, *Comm. ACM 6* (Aug. 1963),
  445]
J. P. CHANDLER AND W. C. HARRISON* (Recd. 19 Sept.
  1969)
Department of Physics, Florida State University, Tallahassee, FL 32306

  * This work was supported in part by AEC Contract No. AT-
(40-1)-3509. Computational costs were supported in part by
National Science Foundation Grant GJ 367 to the Florida State
University Computing Center.

KEY WORDS AND PHRASES: sorting, minimal storage sorting, digital computer sorting
CR CATEGORIES: 5.31

  Hibbard [1] has coded this method in a way that increases the
speed significantly. In SHELLSORT, each stage of each sift con-
sists of successive pair swaps. The modification replaces each set
of n pair swaps by one "save," n − 1 moves, and one insertion.
  Table I gives timing information for ALGOL, FORTRAN, and
COMPASS (assembly language) versions of SHELLSORT and the

TABLE I. SORTING TIMES IN SECONDS FOR 10,000 RANDOMLY
ORDERED NUMBERS ON THE CDC 6400 COMPUTER

Algorithm	*Source Language*		
	ALGOL	FORTRAN	COMPASS
SHELLSORT	53.40	7.18	2.38
SHELLSORT2	36.56	5.98	1.87

modified version (called SHELLSORT2), for the CDC 6400 computer. The savings in time achieved by the modification are 32%,
17%, and 21%, respectively. The savings are greater than this
when vectors of more than one word each are being sorted.
  The comparative execution times of the ALGOL and FORTRAN
versions, for these compilers, are quite interesting.
REFERENCES:
1. HIBBARD, T. N. An empirical study of minimal storage sort-
    ing. *Comm. ACM 6* (May 1963), 206.

ALGORITHM 202
GENERATION OF PERMUTATIONS IN LEXICO-
GRAPHICAL ORDER
Mok-Kong Shen
Postfach 74, München 34, Germany

procedure *PERLE* (S, N, I, E);
integer array S; **integer** N; **Boolean** I; **label** E;
comment If the array S contains a certain permutation of the
N digits $1, 2, \cdots, N$ before call, the procedure will replace
this with the lexicographically next permutation. If initialization is required set the Boolean variable I equal **true**, which
will be changed automatically to **false** through the first call,
otherwise set I equal **false**. If no further permutation can be
generated, exit will be made to E. For reference see *BIT 2*
(1962), 228–231;
begin integer j, u, w;
if I **then begin for** $j := 1$ **step** 1 **until** N **do** $S[j] := j$;
       $I :=$ **false**; **go to** *Rose*
       end;
$w := N$;
Lilie: **if** $S[w] < S[w-1]$ **then**
       begin if $w = 2$ **then go to** E;
        $w := w - 1$; **go to** *Lilie*
       end;
$u := S[w-1]$;
for $j := N$ **step** -1 **until** w **do**
begin if $S[j] > u$ **then**
       begin $S[w-1] := S[j]$;
        $S[j] := u$; **go to** *Tulpe*
       end
end;
Tulpe: **for** $j := 0$ **step** 1 **until** $(N-w-1)/2 + 0.1$ **do**
       begin $u := S[N-j]$;
        $S[N-j] := S[w+j]$; $S[w+j] := u$
       end;
Rose:
end *PERLE*

CERTIFICATION OF ALGORITHM 202 [G6]
GENERATION OF PERMUTATIONS IN LEXICO-
GRAPHICAL ORDER
[Mok-Kong Shen, *Comm. ACM 6* (Sept. 1963), 517]
Roger W. Elliott (Recd. 5 May 1965)
The University of Texas, Austin

The equal sign in the second line after the comment should be
replaced by a replacement operator. With this minor correction,
PERLE was translated into Algol for the CDC 1604. The following times for generating all of the $n!$ permutations of a given vector
of length n and the following values of $r_n = t_n/nt_{n-1}$ [See *Comm.
ACM 5* (Apr. 1962), 209] were observed.

n	5	6	7	8
t_n(sec)	.168	1.01	7.08	56.75
r_n	1.0	1.00	1.00	1.00

REMARKS ON:

ALGORITHM 87 [G6]
PERMUTATION GENERATOR
[John R. Howell, *Comm. ACM 5* (Apr. 1962), 209]
ALGORITHM 102 [G6]
PERMUTATION IN LEXICOGRAPHICAL ORDER
[G. F. Schrak and M. Shimrat, *Comm. ACM 5* (June
(1962), 346]
ALGORITHM 130 [G6]
PERMUTE
[Lt. B. C. Eaves, *Comm. ACM 5* (Nov. 1962), 551]
ALGORITHM 202 [G6]
GENERATION OF PERMUTATIONS IN
LEXICOGRAPHICAL ORDER
[Mok-Kong Shen, *Comm. ACM 6* (Sept. 1963), 517]

R. J. Ord-Smith (Recd. 11 Nov. 1966, 28 Dec. 1966 and
17 Mar. 1967)
Computing Laboratory, University of Bradford, England

A comparison of the published algorithms which seek to generate
successive permutations in lexicographic order shows that Algorithm 202 is the most efficient. Since, however, it is more than twice
as slow as transposition Algorithm 115 [H. F. Trotter, Perm,
Comm. ACM 5 (Aug. 1962), 434], there appears to be room for improvement. Theoretically a "best" lexicographic algorithm
should be about one and a half times slower than Algorithm 115.
See Algorithm 308 [R. J. Ord-Smith, Generation of Permutations
in Pseudo-Lexicographic Order, *Comm. ACM 10* (July 1967), 452]
which is twice as fast as Algorithm 202.

ALGORITHM 87 is very slow.

'ALGORITHM 102 shows a marked improvement.

ALGORITHM 130 does not appear to have been certified before.
We find that, certainly for some forms of vector to be permuted,
the algorithm can fail. The reason is as follows.

At execution of $A[f] := r$; on line prior to that labeled *schell*, f
has not necessarily been assigned a value. f has a value if, and
only if, the Boolean expression $B[k] > 0 \wedge B[k] < B[m]$ is **true** for
at least one of the relevant values of k. In particular when matrix
A is set up by $A[i] := i$; for each i the Boolean expression above is
false on the first call.

ALGORITHM 202 is the best and fastest algorithm of the
exicographic set so far published.

A collected comparison of these algorithms is given in Table I.
t_n is the time for complete generation of $n!$ permutations. Times
are scaled relative to t_8 for Algorithm 202, which is set at 100.
Tests were made on an ICT 1905 computer. The actual time t_8
for Algorithm 202 on this machine *was* 100 seconds. r_n has the
usual definition $r_n = t_n/(n \cdot t_{n-1})$.

comment 3. In the next part—the compound tail—the calls for $ATIVE$ and $STEP$ are organized. The values 1.5 and .5 of the factors of p are not very important. During the iteration p gets an optimal value, which slowly varies. Only at the end p rapidly tends to 0. The programme was tested on the functions $\frac{y^2+1}{x^2+1}$ and $\frac{(x-y)^2-2}{(x+y)^2+2}$, the latter being the first one except for a rotation of the xy-plane over $\pi/4$ radians. In the first case a "gutter" coincides with the x-axis, while for $x > 0$ and $|y| \gtrless 1 \frac{\partial f}{\partial x} \lessgtr 0$.

In the second case, where the gutter is along the line $x=y$, the relaxation is especially interesting, because with $relax = 0$ (and $pmax=100$) the iteration follows the gutter in an unstable way. With starting values $x=-14$ and $y=21$ from $x=y=26$ about 300 steps were taken along the gutter with p about .01. With $relax = .35$ and $pmax=.5$ we had about 150 steps from $x=y=23$. In the gutter itself $relax = .85$ gave the best results, but in that case the gutter was reached at $x=y=63$.

Other parameter values were: $zeta = psi = 1$, $dxmax=100$, $eta = 10^{-7}$ with $eps = 10^{-8}$ gave $fmin$ in 10 figures correctly and $xmin[i]$ in 4 to 6 figures for various starting values of $xs[i]$;

```
p := 1;
for j := 1 step 1 until n do
begin xmin[j] := xs[j]; dfpr[j] := 0 end; fmin := FUNK
  (xmin);
deriv: ATIVE;
next: STEP;
    alpha := FUNK (xstep);
if alpha < fmin then
  begin fmin := alpha;  p := 1.5 × p;
      if p > pmax then p := pmax;
      for j := 1 step 1 until n do xmin[j] := xstep[j];
      go to deriv end;
p := .5 × p;
if p > eps then go to next;
```

comment As p has become smaller than eps this is the end of $STEEP1$. The program $ATIVE$ takes up rather a lot of computer time by the way it chooses a value for $dx(i)$. A thorough simplification is obtained by taking $dx(i)$ as $10 \uparrow - 3 \times abs(xmin[i])$, where again $xmin[i]$ may be replaced by $zeta$. Further, at the cost of some loss of accuracy, computing time is saved by taking $\frac{f(x+h)-f(x)}{h}$ as an estimate for the derivative. This program, as far as it differs from $STEEP1$, is described in algorithm 204, $STEEP2$. An interesting compromise between the two methods is obtained by interchanging the computation of dx and $dfdx$ in $ATIVE$ of $STEEP1$ and omitting the iteration on dx. This routine $ATIVE$, which has to be used in $STEEP1$, is given by J. G. A. Haubrich in algorithm 205;

end $STEEP1$

CERTIFICATION OF ALGORITHM 203 [E4]
STEEP1 [E. J. Wasscher, *Comm. ACM 9* (Sept. 1963), 517]

Philip Wallack (Recd. 25 May 1964)
Republic Aviation Corp., Farmingdale, L. I., N. Y.

$STEEP1$ was translated into Fortran IV and run on the IBM 7094. The program was tested on the function $x^4 + y^4 - 1$, with starting values $x = y = 1.5$. Other parameter values were those suggested in the body of the algorithm. After 17 steps the values of the variables were $x = .0180$, $y = .0191$, and the function value $fmin = -.9999999$.

I feel that good programming practice requires that a count be kept of the number of steps taken in $STEEP1$ and the number of iterations in $ATIVE$, with running checks on both these quantities to control looping. Counters were set up for this purpose in the version of the program I ran.

CERTIFICATION OF ALGORITHM 203 [E4]
STEEP1 [E. J. Wasscher, *Comm. ACM 6* (Sept. 1963), 517; *Comm. ACM 7* (Oct. 1964), 585]

J. M. Varah (Recd. 30 July 1964)
Computation Center, Stanford University, Stanford, Calif.

Algorithm 203 was run on the B5000 at Stanford with the necessary modifications for Burroughs' Extended Algol. After some testing, the following errors were found.

1. There is an extra **begin** in procedure $ATIVE$. The first statement after the comment in this procedure should be changed from

  begin $ATIVE$: $lambda := 0$;

to

  $lambda := 0$;

[It was the author's original intention that this **begin** be not in bold-face but that it should be part of the label $begin$ $ATIVE$ inserted to clarify the program.—Ed.]

Also, there is a missing semicolon in procedure $ATIVE$ at the end of the line preceding $comp$: and procedure $STEP$ has an unnecessary **begin-end** block.

2. Because the domain of definition of the function $FUNK$ is bounded by the rectangular hyperbox $lb[j] \leq x[j] \leq ub[j]$, $j = 1, 2, \cdots, n$, before giving a new direction in which to proceed, the value of $xmin$ is checked (in $ATIVE$, under $large$:). If, for any j, $xmin[j]$ is within $dx[j]$ of the boundary, $xmin[j]$ is changed so that it is exactly $dx[j]$ from the boundary. However, if the minimum value of $FUNK$ occurs at just such a place (say right at the boundary), then a step will be made from this new position back to the boundary. Then the new $xmin[j]$ will again be within $dx[j]$ of the boundary, so it is moved away, and so on forming a loop. To correct this, the old value of $xmin[j]$ should be saved (in $xstep[j]$, for example) and below, when A is tested, the function value set equal to the minimum of values at $xmin$ and $xstep$. The author, when A was true (i.e. when such a shift had been made), merely set the function equal to the value at $xmin$.

Specifically, this means changing the lines following $large$: to

```
A := B := false;  if xmin[j] + dx[j] > ub[j] then
begin
  xstep[j] := xmin[j];
  xmin[j] := ub[j] - dx[j];  A := true
end
else if xmin[j] - dx[j] < lb[j] then
begin
  xstep[j] := xmin[j];
  xmin[j] := lb[j] + dx[j];  A := true
end;
```

and the conditional statement involving A (3rd line after $small$:) to

```
if A then
begin
  gamma := FUNK(xmin);
  if fmin ≤ gamma then xmin[j] := xstep[j]
  else fmin := gamma
end;
```

3. Also in $ATIVE$, under $comp$:, the derivative approximations are all normalized after the **for** loop by division by $lambda$. However, $lambda$ will be zero if all $dfdx[j]$ are zero to working accuracy. So we should only divide by $lambda$ when it is not zero.

Specifically, this means inserting the line
$$\textbf{if } lambda \neq 0 \textbf{ then}$$
before the third line from the end of procedure $ATIVE$.

With these corrections, the algorithm did run successfully. It should also be mentioned that procedures $ATIVE$ and $STEP$ could just as well be blocks with labels $ATIVE$ and $STEP$ rather than procedures, with the calls on them changed to **go to** $ATIVE$ and **go to** $STEP$.

ALGORITHM 204
STEEP2
E. J. Wasscher
Philips Research Laboratories
N. V. Philips' Gloeilampenfabrieken
Eindhoven-Netherlands

```
procedure  STEEP2 (lb, xs, ub, dx, xmin, fmin, n, eps, relax
    dxmax, pmax, zeta, FUNK);
value dx, n, eps, relax, dxmax, pmax, zeta;
integer n;
real dx, fmin, eps, relax, dxmax, pmax, zeta;
array lb, xs, ub, xmin;  real procedure FUNK;
comment  dx should now be taken about 10↑ − 3, dxmax could
    be taken equal to 1. As the program is equal to STEEP1 after
    the declaration of the procedure ATIVE, the Algol description
    is cut off there;
begin integer j;  real alpha, p;
  array xstep, dfdx, dfpr [1:n];
procedure ATIVE;
begin real beta, lambda;  lambda := 0;
  for j := 1 step 1 until n do
begin alpha := dx × (if abs(xmin[j]) < dxmax
  then dxmax else abs (xmin[j]));
  if xmin[j] + alpha > ub[j] then alpha := −alpha;
  xmin[j] := xmin[j] + alpha;  beta := FUNK (xmin);
  xmin[j] := xmin[j] − alpha;
  dfdx[j] := (beta − fmin)/alpha;
  lambda := lambda + dfdx[j] ↑2
end for;  lambda := sqrt (lambda);
  for j := 1 step 1 until n do dfdx[j] := dfdx[j]/lambda;
end procedure ATIVE
```

ALGORITHM 205
ATIVE
J. G. A. HAUBRICH
Philips Research Laboratories
N. V. Philips' Gloeilampenfabrieken
Eindhoven-Netherlands

```
procedure ATIVE;
begin real beta, lambda;  Boolean A;
comment  This routine may replace ATIVE in STEEP1. The
  significance of eta has slightly changed;
lambda := 0;
for j := 1 step 1 until n do
begin A := false;  alpha := dx[j];

if xmin[j] + alpha > ub[j] then
begin xmin[j] := ub[j] − alpha;  A := true end
else if xmin[j] − alpha < lb[j] then
begin xmin[j] := lb[j] + alpha;  A := true end;
      xmin[j] := xmin[j] + dx[j];  alpha := FUNK(xmin);
      xmin[j] := xmin[j] − 2 × dx[j];  beta := FUNK(xmin);
      xmin[j] := xmin[j] + dx[j];  if A then fmin := FUNK
        (xmin);
      dfdx[j] := (alpha−beta)/(2 × dx[j]);
      lambda := lambda+dfdx[j] ↑ 2;
if alpha − fmin > 0 ∧ beta − fmin > 0 then go to end;
beta := abs((alpha−beta)/(if abs(fmin)<psi then psi else fmin));
if beta > eta then dx[j] := .3 × dx[j] else
begin dx[j] := × d3x[j];  if dx[j] > dxmax then dx[j]: = dxmax end;
end:  end for;
lambda := sqrt (lambda);
for j := 1 step 1 until n do dfdx[j] := dfdx[j]/lambda
end procedure ATIVE
```


REMARK ON ALGORITHM 205 [E4]
ATIVE [J. G. A Haubrich, *Comm. ACM 6* (Sept. 1963),
  519]
E. J. WASSCHER (Recd. 23 Nov. 1964)
Philips Computer Center, N. V. Philips' Gloeilampen-
  fabrieken, Eindhoven, Netherlands

There is a misprint in this Algorithm. The first statement in
the fifth line from the end of the procedure *ATIVE* should read:
$$dx[j] := 3 \times dx[j];$$

ALGORITHM 206
ARCCOSSIN
Misako Konda

Japan Atomic Energy Research Institute, Tokai, Ibaraki, Japan

procedure *ARCCOSSIN* (x) Result:$(arccos, arcsin)$;
value x;
real $x, arccos, arcsin$;
comment This procedure computes $arccos(x)$ and $arcsin(x)$ for $-1 \leq x \leq 1$. The constant 2^{-27} depends on the word length and relative machine precision, and may be replaced by a variable identifier. *Alarm* is the procedure which messages that x is invalid.

The approximation formula used here was coded for MUSA-SINO-1 in its own language at the Electrical Communication Laboratory Tokyo. This algorithm was translated into FAP and successfully ran on an IBM 7090;

```
begin real A, x1, x2, a;  integer r;
  if abs(x) > 1
  then go to Alarm
  else if abs(x) > 2 ↑ (−27)
    then go to L1
    else begin arccos := 1.5707963;  go to L3
      end;
L1:  if x = 1
    then begin arccos := 0;  go to L3
      end
    else if x = − 1
      then begin arccos := 3.1415926;  go to L3
        end
      else begin A := 0;  x1 := x;
          for r := 0 step 1 until 26 do
          begin if x1<0
            then begin a := 1;  x2 := 1−2 × x1 ↑2 end
            else begin a := 0;  x2 := 2 × x1 ↑2 − 1 end;
            A := A + a × 2 ↑(−r−1);
            x1 := x2
          end;
        arccos := 3.1415926 × A;
        end;
L3:  arcsin := 1.570963 − arccos;
    end ARCCOSSIN
```

REMARK ON ALGORITHM 206 [B1]
ARCCOSSIN [Misako Konda, *Comm. ACM 6* (Sept. 1963), 519]

Henry J. Bowlden (Recd. 30 Sept. 1964 and 5 Nov. 1964)
Westinghouse Electric Corp., R&D Ctr., Pittsburgh, Pa.

Algorithm 206 was transcribed into Burroughs Extended Algol after correcting one typographical error, namely the value of $\pi/2$ in the statement labeled $L3$, which should be 1.5707963.

Results were obtained for a selection of values of the argument between 0 and 1. Accuracy is about 7+ decimal digits over the entire range, by comparison with the tables of inverse sines in [*Handbook of Mathematical Functions*, National Bureau of Standards Applied Mathematics Series #55, U.S. Government Printing Off., Washington, D.C., June 1964, 203–212]. Average execution time was 43 milliseconds.

The efficiency of the procedure could be significantly improved by avoiding the computation of $a \times 2 \uparrow (−r−1)$. Powers of 0.5 may be accumulated within the loop, and the modification of A may be skipped entirely when $a = 0$. Actually, if efficiency is important, procedures using the intrinsic *arctan* and the common trigonometric identities are preferable. Such routines, on the B-5000, give full machine accuracy (11+ significant figures) in about 2 milliseconds execution time.

ALGORITHM 207
STRINGSORT
J. Boothroyd
English Electric-Leo Computers, Ltd.
Staffordshire, England

Number of Items	Time in Seconds
10	0.03
20	0.05
50	0.20
100	0.38
200	1.03
500	3.22
1000	6.43
2000	12.85
5000	38.72
10000	90.72

procedure *stringsort* (a, n); **comment** elements $a[1] \cdots a[n]$ of $a[1:2n]$ are sorted into ascending sequence using $a[n+1] \cdots a[2n]$ as auxiliary storage. Von Neumann extended string logic is employed to merge input strings from both ends of a sending area into output strings which are sent alternately to either end of a receiving area. The procedure takes advantage of naturally occurring ascending or descending order in the original data;

value n; **integer** n; **array** a;
begin integer d, i, j, m, u, v, z; **integer array** $c[-1:1]$;

```
    switch p := jz1, str i;   switch q := merge, jz2;
oddpass:  i := 1;   j := n;   c[-1] := n + 1;   c[1] := 2 × n;
allpass:  d := 1;   go to firststring;
merge:  if a[i] ≧ a[z]
            then begin go to p[v];
              jz1:  if a[j] ≧ a[z]
                        then ij:  begin if a[i] ≧ a[j]
                                        then str j: begin a[m] := a[j];
                                        j := j − 1 end
                                        else str i: begin a[m] := a[i]:
                                        i := i + 1 end
                                end
                        else begin v := 2;   go to str i end
              end
            else begin u := 2;
              jz2:  if a[j] ≧ a[z]
                        then go to str j
                        else begin d := −d;   c[d] := m;
                            firststring:   m := c[−d];
                            v := u := 1;
                            go to ij
                        end
              end;
z := m;   m := m + d;   if j ≧ i then go to q[u];
if m > n + 1 then begin comment evenpass;   i := n + 1;
              j := 2 × n;   c[−1] := 1;   c[1] := n;   go to
          allpass end
          else if m < n + 1 then go to oddpass
end stringsort;
```

CERTIFICATION OF ALGORITHM 207 [M1]
STRINGSORT [J. Boothroyd, *Comm. ACM* 6 (Oct. 1963),
   615]
Charles R. Blair (Recd. 31 Jul. 1964)
Department of Defense, Washington 25, D. C.

STRINGSORT compiled and ran successfully without correction on the Aldap translator for the CDC 1604A. The following sorting times were observed.

ALGORITHM 208
DISCRETE CONVOLUTION
William T. Foreman, Jr.
Collins Radio Co.
Newport Beach, Calif.

```
procedure Discrete Convolution (m, n, prs) result: (Conv);
integer m, n;  real procedure prs;  real array conv;
comment This procedure finds the probability distribution of
   the sum of m independent variables, each with a known distribu-
   tion over the nonnegative integers. A real procedure prs with
   results pr[k] is assumed to find each probability distribution in
   succession. The maximum sum for which probabilities are
   computed must be fixed by the user. The number of iterations
   is roughly m²n/2. The procedure prs will in general depend on
   additional parameters and should include the read-in of the
   parameters for that distribution. It may include the selection
   of one function from a set;
begin integer i, j, k, ix1, ix2;
real array prob [1:2, 0:m], pr[0:m];
i := 1;  ix1 := 1;  ix2 := 2;  prs (m) result: (pr);
for j := 0 step 1 until m do prob[ix1, j] := pr[j];
for i := 2 step 1 until n do
   begin
      if ix1 = 1 then begin ix2 := 1;  ix1 := 2 end
         else begin ix2 := 2;  ix1 := 1 end
      prs (m) result:  (pr);
      for j := 0 step 1 until m do
      begin
         prob[ix1, j] := 0;
         for k := 0 step 1 until j do
            prob[ix1, j] := prob[ix1, j] + pr[k] × prob[ix2, j−k]
      end j
   end i;
for j := 0 step 1 until m do conv[j] := prob[ix1, j]
end Discrete Convolution
comment The convolution of discrete probability series is
   isomorphic to the multiplication of polynomials. A useful vari-
   ation is to omit the parameters i, n and have prs recognize
   the end of input. A FORTRAN program using this procedure has
   been run on the IBM 7090 to find the sum of queue lengths in a
   teletype switching center, where messages arrived according
   to the Poisson distribution and message lengths were distributed
   negative-exponentially. The following was used as the prob-
   ability procedure;
procedure prs (m) result:  (pr);
value m;  procedure read;
real array pr;  integer m;
begin real trafficrate, linespeed, rho;  integer j;
   read (trafficrate, linespeed);
   rho := trafficrate/linespeed;
   pr[0] : 1 − rho;
   for j := 1 step until m do pr[j] := rho × pr[j−1]
end prs
```

ALGORITHM 209
GAUSS
D. Ibbetson,
Elliott Brothers (London) Ltd.,
Elstree Way, Borehamwood, Herts., England

real procedure $Gauss(x)$; **value** x; **real** x;
comment Gauss calculates $(1/\sqrt{2\pi})\int_{-\infty}^{x} exp(-\frac{1}{2}u^2)\,du$ by means
of polynomial approximations due to A. M. Murray of Aberdeen
University;
begin real y, z, w;
if $x = 0$ **then** $z := 0$
else
begin $y := abs(x)/2$;
  if $y \geq 3$ **then** $z := 1$
  else if $y < 1$ **then**
  begin $w := y \times y$;
    $z := (((((((((0.000124818987 \times w$
      $-0.001075204047) \times w +0.005198775019) \times w$
      $-0.019198292004) \times w +0.059054035642) \times w$
      $-0.151968751364) \times w +0.319152932694) \times w$
      $-0.531923007300) \times w +0.797884560593) \times y \times 2$
  end
  else
  begin $y := y - 2$;
    $z := (((((((((((((-0.000045255659 \times y$
      $+0.000152529290) \times y -0.000019538132) \times y$
      $-0.000676904986) \times y +0.001390604284) \times y$
      $-0.000794620820) \times y -0.002034254874) \times y$
      $+0.006549791214) \times y -0.010557625006) \times y$
      $+0.011630447319) \times y -0.009279453341) \times y$
      $+0.005353579108) \times y -0.002141268741) \times y$
      $+0.000535310849) \times y +0.999936657524$
  end
end;
  $Gauss := $ **if** $x > 0$ **then** $(z+1)/2$ **else** $(1-z)/2$
end $Gauss$;

CERTIFICATION OF ALGORITHM 209
GAUSS [D. Ibbetson, *Comm. ACM 6* (Oct. 1963), 616]
(Pvt.) G. W. Gladfelter (Recd 4 Nov. 63)
RA17667701, 1st Inf. Battle Group U.S. Military Academy
(9822), West Point, N.Y.

The algorithm was translated into Fortran for the GE 225 and
used to publish a table of the error function. No errors were found
in the algorithm and the table produced agreed with the published
tables at hand (6 significant figures).

CERTIFICATION OF ALGORITHM 209 [S15]
GAUSS [D. Ibbetson, *Comm. ACM 6*, Oct. 1963, 616]
M. C. Pike
Statistical Research Unit of the Medical Research Council,
University College Hospital Medical School, London,
England

This procedure was tested on an Elliott 803 computer using the

standard Elliott Algol compiler. The expression

$$2 \times Gauss\ (x) - 1$$

was evaluated for $x = 0(.01)6$ and the answers checked with those
given in *Tables of Probability Functions, vol. II*, U.S. National
Bureau of Standards, Washington, D.C., 1942, where they are
given to 15 decimal places. There was a maximum error of 1 in
the 8th decimal place.

REMARKS ON:
ALGORITHM 123 [S15]
REAL ERROR FUNCTION, ERF(x)
   [Martin Crawford and Robert Techo *Comm. ACM 5*
   (Sept. 1962), 483]

ALGORITHM 180 [S15]
ERROR FUNCTION—LARGE X
   [Henry C. Thacher Jr. *Comm. ACM 6* (June 1963), 314]

ALGORITHM 181 [S15]
COMPLEMENTARY ERROR FUNCTION—
LARGE X
   [Henry C. Thacher Jr. *Comm. ACM 6* (June 1963), 315]

ALGORITHM 209 [S15]
GAUSS
   [D. Ibbetson. *Comm. ACM 6* (Oct. 1963), 616]

ALGORITHM 226 [S15]
NORMAL DISTRIBUTION FUNCTION
   [S. J. Cyvin. *Comm. ACM 7* (May 1964), 295]

ALGORITHM 272 [S15]
PROCEDURE FOR THE NORMAL DISTRIBUTION
FUNCTIONS
   [M. D. MacLaren. *Comm. ACM 8* (Dec. 1965), 789]

ALGORITHM 304 [S15]
NORMAL CURVE INTEGRAL
   [I. D. Hill and S. A. Joyce. *Comm. ACM 10* (June
   1967), 374]

I. D. Hill·and S. A. Joyce (Recd. 21 Nov. 1966)
Medical Research Council,
Statistical Research Unit, 115 Gower Street, London
   W.C.1., England

These algorithms were tested on the ICT Atlas computer using
the Atlas Algol compiler. The following amendments were made
and results found:

ALGORITHM 123
   (i) **value** x; was inserted.
   (ii) $abs(T) < {}_{10}-10$ was changed to $Y - T = Y$
      both these amendments being as suggested in [1].

(iii) The labels 1 and 2 were changed to $L1$ and $L2$, the **go to** statements being similarly amended.

(iv) The constant was lengthened to 1.12837916710.

(v) The extra statement $x := 0.707106781187 \times x$ was made the first statement of the algorithm, so as to derive the normal integral instead of the error function.

The results were accurate to 10 decimal places at all points tested except $x = 1.0$ where only 2 decimal accuracy was found, as noted in [2]. There seems to be no simple way of overcoming the difficulty [3], and any search for a method of doing so would hardly be worthwhile, as the algorithm is slower than Algorithm 304 without being any more accurate.

ALGORITHM 180

(i) $T := -0.56418958/x/exp(v)$ was changed to $T := -0.564189583548 \times exp(-v)/x$. This is faster and also has the advantage, when v is very large, of merely giving 0 as the answer instead of causing overflow.

(ii) The extra statement $x := 0.707106781187 \times x$ was made as in (v) of Algorithm 123.

(iii) **for** $m := m + 1$ was changed to **for** $m := m + 2$. $m+1$ is a misprint, and gives incorrect answers.

The greatest error observed was 2 in the 11th decimal place.

ALGORITHM 181

(i) Similar to (i) of Algorithm 180 (except for the minus sign).

(ii) Similar to (ii) of Algorithm 180.

(iii) m was declared as **real** instead of **integer**, as an alternative to the amendment suggested in |4|.

The results were accurate to 9 significant figures for $x \leqslant 8$, but to only 8 significant figures for $x = 10$ and $x = 20$.

ALGORITHM 209

No modification was made. The results were accurate to 7 decimal places.

ALGORITHM 226

(i) $10 \uparrow m/(480 \times sqrt(2 \times 3.14159265))$ was changed to $10 \uparrow m \times 0.000831129750836$.

(ii) **for** $i := 1$ **step** 1 **until** $2 \times n$ **do** was changed to $m := 2 \times n$; **for** $i := 1$ **step** 1 **until** m **do**.

(iii) $-(i \times b/n) \uparrow 2/8$ was changed to $-(i \times b/n) \uparrow 2 \times 0.125$.

(iv) **if** $i = 2 \times n - 1$ was changed to **if** $i = m - 1$

(v) $b/(6 \times n \times sqrt(2 \times 3.14159265))$ was changed to $b/(15.0397696478 \times n)$.

Tests were made with $m = 7$ and $m = 11$ with the following results:

x	Number of significant figures correct		Number of decimal places correct	
	$m = 7$	$m = 11$	$m = 7$	$m = 11$
-0.5	7	11	7	11
-1.0	7	10	7	10
-1.5	7	10	8	10
-2.0	7	9	8	10
-2.5	6	9	8	11
-3.0	6	7	8	9
-4.0	5	7	10	11
-6.0	2	1	12	10
-8.0	0	0	11	9

Perhaps the comment with this algorithm should have referred to decimal places and not significant figures. To ask for 11 significant figures is stretching the machine's ability to the limit, and where 10 significant figures are correct, this may be regarded as acceptable.

ALGORITHM 272

The constant .99999999 was lengthened to .9999999999.

The accuracy was 8 decimal places at most of the points tested, but was only 5 decimal places at $x = 0.8$.

ALGORITHM 304

No modification was made. The errors in the 11th significant figure were:

$abs(x)$	$x > 0 \equiv upper$	$x > 0 \not\equiv upper$
0.5	1	1
1.0	1	2
1.5	21ᵃ(5)	2
2.0	25ᵃ(0)	4
3.0	0	0
4.0	2	3
6.0	6	0
8.0	14	0
10.0	23	0
20.0	35	0

[a] Due to the subtraction error mentioned in the comment section of the algorithm. Changing the constant 2.32 to 1.28 resulted in the figures shown in brackets.

To test the claim that the algorithm works virtually to the accuracy of the machine, it was translated into double-length instructions of Mercury Autocode and run on the Atlas using the EXCHLF compiler (the constant being lengthened to 0.39894228040143267793946). The results were compared with hand calculations using Table II of [5]. The errors in the 22nd significant figure were:

$abs(x)$	$x > 0 \equiv upper$	$x > 0 \not\equiv upper$
1.0	2	3
2.0	7	1
4.0	2	0
8.0	8	0

Timings. Timings of these algorithms were made in terms of the Atlas "Instruction Count," while evaluating the function 100 times. The figures are not directly applicable to any other computer, but the relative times are likely to be much the same on other machines.

INSTRUCTION COUNT FOR 100 EVALUATIONS

$abs(x)$	Algorithm number							
	123	180	181	209	226 $m = 7$	272	304[a]	304[b]
0.5	58			8	97	24	25	24
1.0	65[c]			8	176	24	29	29
1.5	164	128	127	9	273	25	35	35
2.0	194	78	90	8	387	24	39	39
2.5	252	54	68	10	515	24	131	44
3.0		42	51	9	628	25	97	50
4.0		27	39	9	900[d]	25	67	44
6.0		15	30	6	1400[d]	16	49	23
8.0		9	28	7	2100[d]	18	44	11
10.0		10	25	5	2700[d]	16	38	11
20.0		9	22	5	6500[d]	16	32	11
30.0		9	9	5	10900[d]	16	11	11

[a] Readings refer to $x > 0 \equiv upper$.

[b] Readings refer to $x > 0 \not\equiv upper$.

[c] Time to produce incorrect answer. A count of 120 would fit a smooth curve with surrounding values.

[d] 100 times Instruction Count for 1 evaluation.

Opinion. There are advantages in having two algorithms available for normal curve tail areas. One should be very fast and reasonably accurate, the other very accurate and reasonably fast. We conclude that Algorithm 209 is the best for the first requirement, and Algorithm 304 for the second.

Algorithms 180 and 181 are faster than Algorithm 304 and may be preferred for this reason, but the method used shows itself in Algorithm 181 to be not quite as accurate, and the introduction of this method solely for the circumstances in which Algorithm 180 is applicable hardly seems worth while.

Acknowledgment. Thanks are due to Miss I. Allen for her help with the double-length hand calculations.

REFERENCES:

1. THACHER, HENRY C. JR. Certification of Algorithm 123. *Comm. ACM 6* (June 1963), 316.

2. IBBETSON, D. Remark on Algorithm 123. *Comm. ACM 6* (Oct. 1963), 618.

3. BARTON, STEPHEN P., AND WAGNER, JOHN F. Remark on Algorithm 123. *Comm. ACM 7* (Mar. 1964), 145.

4. CLAUSEN, I., AND HANSSON, L. Certification of Algorithm 181. *Comm. ACM 7* (Dec. 1964), 702.

5. SHEPPARD, W. F. *The Probability Integral*. British Association Mathematical Tables VII, Cambridge U. Press, Cambridge, England, 1939.

ALGORITHM 210
LAGRANGIAN INTERPOLATION
GEORGE R. SCHUBERT*
University of Dayton, Dayton, Ohio
   * Undergraduate research project, Computer Science Program, Univ. of Dayton.

```
procedure LAGRANGE (N, u, X, Y, ANS);  real array X, Y;
  integer N;  real u, ANS;
comment This procedure evaluates an Nth degree Lagrange
  polynomial, given N + 1 data coordinates, and u the value
  where interpolation is desired. X is the abscissa array and Y
  the ordinate array. ANS is the resultant value of the function
  at u. The notation is that used in R. W. Hamming, Numerical
  Methods for Scientists and Engineers, pp. 94-95 (McGraw-Hill
  Book Company, Inc., 1962);
begin integer i, j;  real L;
  ANS := 0.0;
  for j := step 1 until N+1 do
begin L := 1.0;
  for i := step 1 until N+1 do
begin if i ≠ j then L := L × (u−X[i])/(X[j]−X[i])
end;
  ANS := ANS + L × Y[j]
end end
```

ALGORITHM 211
HERMITE INTERPOLATION
GEORGE R. SCHUBERT*
University of Dayton, Dayton, Ohio
* Undergraduate research project, Computer Science Program, Univ. of Dayton.

procedure *HERMITE* $(n, u, X, Y, Y1, ANS)$; **real array** $X, Y, Y1$;
  integer n; **real** u, ANS;
comment This procedure evaluates $a(2n+1)$th degree Hermite polynomial, given the value of the function and its first derivative at each of $n + 1$ points. X is the abscissa array, Y the ordinate array, and $Y1$ the derivative array. ANS is the interpolated value of the function at u. REFERENCE: R. W. Hamming, *Numerical Methods for Scientists and Engineers*, pp. 96–97 (McGraw-Hill Book Company, Inc., 1962);
begin integer i, j; **real** h, a;
  $ANS := 0.0$;
  for $j := 1$ **step** 1 **until** $n + 1$ **do**
begin $h := 1.0$; $a := 0.0$;
  for $i := 1$ **step** 1 **until** $n + 1$ **do**
begin if $i = j$ **then go to** *out*;
  $h := h \times (u-X[i])\uparrow 2/(X[j]-X[i])\uparrow 2$;
  $a := a + 1.0/(X[j]-X[i])$;
out: **end**;
  $ANS := ANS + h \times ((X[j]-u)\times(2\times a\times Y[j]-Y1[j])+Y[j])$
end end

CERTIFICATION OF ALGORITHM 211
HERMITE INTERPOLATION [George R. Schubert, *Comm. ACM*, Oct. 1963]
THOMAS A. DWYER
Argonne National Laboratory, Argonne, Ill.

The body of *HERMITE* was transcribed for the Dartmouth SCALP processor for the LGP-30 computer and ran successfully without corrections. It was tested using the error function and its derivatives. Roundoff error in the LGP-30 began to appear for values of n greater than 3. For n equal to 2 (third degree polynomial) the interpolated value agreed with the function within machine limitations (six significant figures) for steps in the argument data of 0.005.

ALGORITHM 212
FREQUENCY DISTRIBUTION
MALCOLM D. GRAY
The Boeing Co., Seattle, Wash.

procedure *FREQUENCY* (N, A, B, IUL, K, X, KA);
integer N, IUL; **integer array** KA; **real** A, B, K;
real array X;
comment Given a set X of variables in some interval $I = [a, b]$
  such that $a \leqq \min x, \max x \leqq b$, *FREQUENCY* determines the
  frequency distribution of X over k equal, half open subintervals
  of I. The interval I is transformed to the interval $J = [0, k]$
  with unit subintervals by $x' = (x_i-a)/[(b-a)/k]$, $i = 1, 2, \cdots,$
  n, and considering $x' = L \times M$, L and M integers. The value
  L then immediately determines the subinterval and M is used
  for boundary points. If $IUL = 0$, the subintervals are open
  on the upper end, except the kth. On entry, the array KA is
  assumed identically zero; on return, $KA[i]$ contains the frequency of X in the ith subinterval;
begin integer i, L; **real** BAK, XP;
  $BAK := (B-A)/K$;
  for $i := 1$ **step** 1 **until** N **do begin**
    $XP := (X[i]-A)/BAK$;
    $L := entier (XP)$;
    if $XP = L$ **then go to** $p2$ **else** $L := L + 1$; **go to** $p5$;
$p2$: **if** $IUL = 0$ **then go to** $p3$ **else if** $L = 0$ **then** $L := L + 1$;
      go to $p5$;
$p3$: **if** $XP \neq K$ **then** $L := L + 1$;
$p5$: $KA[L] := KA[L] + 1$;
  end;
end *FREQUENCY*

ALGORITHM 213
FRESNEL INTEGRALS
MALCOLM D. GRAY
The Boeing Co., Seattle, Wash.

real procedure FRESNEL (w, S, C); **value** w; **real** S, C;
comment FRESNEL computes the Fresnel sine and cosine integrals $S(w) = \int_0^w \sin [(\pi/2)t^2] \, dt$ and $C(w) = \int_0^w \cos [(\pi/2)t^2] \, dt$ using the series expansions

$$S(w) = w \sum_{i=1}^{\infty} \frac{(-1)^{i+1} x^{2i-1}}{(4i-1)(2i-1)!} \quad \text{and}$$

$$C(w) = w \sum_{i=1}^{\infty} \frac{(-1)^{i+1} x^{2i-2}}{(4i-3)(2i-2)!}$$

for $|w| < \sqrt{22/\pi}$ and $x \equiv \pi w^2/2$, and using the asymptotic series

$$S(w) = \alpha - \frac{1}{\pi w} [P(x) \sin (x) + Q(x) \cos (x)],$$

$$C(w) = \alpha - \frac{1}{\pi w} [P(x) \cos (x) - Q(x) \sin (x)]$$

where $|w| \geq \sqrt{22/\pi}$, $x \equiv \pi w^2/2$,

$$Q(x) = 1 - \sum_{i=2}^{\infty} \frac{(-1)^i (4i-5)!!}{(2x)^{2i-2}}, \quad P(x) = \sum_{i=1}^{\infty} \frac{(-i)^{i+1}(4i-3)!!}{(2x)^{2i-1}},$$

and $n!! \equiv n(n-2)(n-4)\cdots 1$. If $w \geq 0$, then $\alpha = \frac{1}{2}$, or if $w < 0$, then $\alpha = -\frac{1}{2}$.

This algorithm is a translation of a FAP coded subroutine currently in use on the IBM 7094 at the Boeing Company. The FAP program yields the following errors when tested at 0.05 increments of x:

x		ΔS	ΔC
0.00,	1.00	$<1 \times 10^{-7}$	$<1 \times 10^{-7}$
1.05,	8.65	$<1 \times 10^{-6}$	$<1 \times 10^{-6}$
8.70,	10.30	3×10^{-6}	2×10^{-6}
10.35,	11.00	5×10^{-6}	4×10^{-6}
11.05,	12.15	$<1 \times 10^{-6}$	3×10^{-6}
12.20,	15.00	$<1 \times 10^{-6}$	$<1 \times 10^{-6}$

where ΔS and ΔC are the approximate average absolute deviations (over the range) from the reference. The user must supply $S(w) = C(w) = \pm\frac{1}{2}$ if $w \to \pm \infty$. REFERENCES: ALGORITHMS 88–90, J. L. Cundiff, *Comm. ACM*, May 1962, Born, M. and Wolf, E., *Principles of Optics*, Pergamon Press (1958), pp. 369–431;

```
begin real x, x2, eps, term; integer n; eps := 0.000001;
  x := w × w/0.6366198;
  x2 := −x × x; if x ≥ 11.0 then go to asympt;
  begin real frs, frsi;
      frs := x/3;  n := 5;  term := x × x2/6;
      frsi := frs + term/7;
loops: if abs(frs−frsi) ≦ eps then go to send; frs := frsi;
      term := term × x2/n/(n−1);  frsi := frs + term/(n+n+1);
      n := n + 2;  go to loops;
send:  S := frsi × w;  end;
```

```
begin real frc, frci;
      frc := 1;  n := 4;  term := x2/2; frci := 1 + term/5;
loopc: if abs(frc−frci) ≦ eps then go to cend; frc := frci;
      term := term × x2/n/(n−1);  frci := frc + term/(n+n+1);
      n := n + 2;  go to loopc;
cend:  C := frci × w;  end;  go to aend;
asympt: begin real S1, S2, half, temp;  integer i;
      x2 := 4 × x2;  term := 3/x2;  S1 := 1 + term;  n := 8;
      for i := 1 step 1 until 5 do begin n := n + 4;
          term := term × (n−7) × (n−5)/x2;  S1 := S1 + term;
          if abs(term) ≦ eps then go to next;  end i;
next:  for i := 1 step 1 until 5 do begin n := n + 4;
          term := term × (n−5) × (n−3)/x2;  S2 := S2 + term;
          if abs(term) ≦ eps then go to final;  end i;
final: if w < 0 then half := −0.5 else half := 0.5;  term := cos(x);
      temp := sin(x);  x2 := 3.1415927 × w;
      C := half + (temp×S1−term×S2)/x2;
      S := half − (term×S1+temp×S2)/x2;
      end;
aend:  end FRESNEL
```

CERTIFICATION OF ALGORITHM 213 [S20]
FRESNEL INTEGRALS [M.D. Gray, *Comm. ACM 6* (Oct. 1963), 617]
Malcolm Gray (Recd. 29 May 1964 and, revised, 11 June 1964)
Computer Science Div., Stanford U., Stanford, Calif.
(now at The Boeing Company, Seattle, Wash.)

Necessary changes to the algorithm are:
(1) in the first line, replace
      real S, C; with **real** w, S, C;
(2) in the formula for $P(x)$, replace $(-i)^{i+1}$ with $(-1)^{i+1}$
(3) the statement beginning
    loopc: *if abs(frc−frci)*
should read
    loopc: **if** *abs(frc−frci)*
(4) in the body, replace the line
    next: **for** $i := 1$ **step** 1 **until** 5 **do begin** $n := n + 4$;
with the lines
    next: *term* := $S2 := 0.5/x$; $n := 4$;
        for $i := 1$ **step** 1 **until** 5 **do begin** $n := n + 4$;
The procedure (with the above changes) was executed on the Burroughs B5000 at Stanford University and gave results as indicated in the algorithm.

Communications from Helmut Lotsch of the W. W. Hansen Laboratories, Stanford University, and from Harold Butler of the Los Alamos Scientific Laboratory, Los Alamos, New Mexico, state that they found these same errors, and after the corrections were made, similar results were obtained. Mr. Lotsch's work was done on the B5000 and Dr. Butler's work was done on the IBM 7090.

ALGORMITH 214
q-BESSEL FUNCTIONS $I_n(t)$
J. M. S. Simões Pereira
Gulbenkian Scientific Computing Center, Lisbon, Portugal

procedure qBessel (t, q, n, j, s); **integer** n, j; **real** t, q, s;
  array s;
comment This procedure computes values of any q-Bessel function $I_n(t)$ for n integer (positive, negative or zero) by the use
  of the well-known expansion

$$I_n(t) = \sum_{k=0}^{\infty} \frac{q^{\frac{1}{2}k(k-1)+\frac{1}{2}(n+k)(n+k-1)}t^{n+2k}}{(q)_k(q)_{n+k}}$$

where $|q| < 1$, $(q)_n = (1-q)(1-q^2) \cdots (1-q^n)$, $(q)_0 = 1$ and
$1/(q)_{-n} = 0$ $(n=1, 2, 3, \cdots)$. (See L. Carlitz, The product of
q-Bessel functions, *Port. Math. 21* (1962), 5–9.) Moreover, j
denotes the number of terms (at least 2) retained in the summation, and $s[i]$ stands for the sum of the first $i+1$ terms of the
expansion. This procedure has been translated into Fortran
for the IBM 1620 and run successfully;
begin integer k, m, p; **real** c, u; $m := abs(n)$; $c := 1$; **if**
  $n = 0$ **then go to** A;
for $p := 1$ **step** 1 **until** m **do** $c := c \times (1-q{\uparrow}p)$; **if** $n < 0$ **then**
  go to B;
A: $u := q \uparrow (n\times(n-1)/2) \times (t{\uparrow}n)/c$; $s[0] := u$;
for $k = 1$ **step** 1 **until** j **do**
begin $u := u \times q \uparrow (n+2\times k-2) \times (t{\uparrow}2)/((1-q{\uparrow}k)\times(1-q{\uparrow}(n+k)))$;
  $s[k] := s[k-1] + u$ **end**;
B: $u := q{\uparrow}((m-1)\times m/2) \times t \uparrow (n+2\times m)/c$; $s[m] := u$;
for $k := m + 1$ **step** 1 **until** j **do**
begin $u := u \times q \uparrow (n+2\times k-2) \times (t{\uparrow}2)/((1-q{\uparrow}k)(1-q{\uparrow}(n+k)))$;
  $s[k] := s[k-1] + u$ **end**
end

REMARK ON ALGORITHM 214
q-BESSEL FUNCTIONS $I_n(t)$ [J. M. S. Simões Pereira,
  Comm. ACM 6 (Nov. 1963), 662]
J. M. S. Simões Pereira (Recd 6 Jan 1964)
Gulbenkian Scientific Computing Center, Lisbon, Portu-
  gal

  Corrections:
1. Insert a dummy statement labeled C just before the final **end**.
2. Add a statement **go to** C just before the label B.
3. Add a colon in the clause **for** $k := 1$ **step** 1 **until** j **do** ...

ALGORITHM 215
SHANKS
HENRY C. THACHER, JR.*
Argonne National Laboratory, Argonne, Ill.
* Work supported by the U. S. Atomic Energy Commission

```
procedure  Shanks (nmin, nmax, kmax, S);
value  nmin, nmax, kmax;
integer  nmin, nmax, kmax;
array  S;
comment
```

comment This procedure replaces the elements $S[nmin]$ through $S[nmax-2 \times kmax]$ of the array S by the $e[kmax]$ transform of the sequence S. The elements $S[nmax-2 \times kmax+1]$ through $S[nmax-1]$ are destroyed. The $e[k]$ transforms were discovered by D. Shanks (*J. Math. Phys. 34* (1955), 1–42). $e[1]$ is equivalent to the (*delta*) $\uparrow 2$ transformation. The $e[k]$ transforms are particularly valuable in estimating B in sequences which may be written in the form $S[n] = B + \sum a[i] \times q[i] \uparrow n$ $(i=1, 2, \cdots, k)$.

The transformation is carried out by the epsilon algorithm (Wynn, P., *M.T.A.C 10* (1956), 91–96). ALGOL procedures for applying the algorithm to series of complex terms are given by Wynn (*BIT 2* (1962), 232–255).

The body of this procedure has been tested using the Dartmouth Self-Contained ALGOL Processor for the LGP-30 computer. It gave the following results on the sequence for the smaller zero of the Laguerre polynomial, $L[2](x)$:

n	$S[n]$	$e[1](S[n])$	$e[2](S[n])$	$e[1]^2(S[n])$
0	0.0000000	0.5714285	0.5857432	0.5857616
1	0.5000000	0.5851059	0.5857854	0.5857859
2	0.5625000	0.5857318	0.5857861	0.5857861
3	0.5791016	0.5857816		
4	0.5838396	0.5857859		
5	0.5852172			
6	0.5856198	*True Value* 0.5857864375		

These results are in satisfactory agreement with those given by by Wynn (1956);

```
begin integer j, k, limj, limk, two kmax;
  real T0, T1;
  two kmax := kmax + kmax;
  limj := nmax;
for j := nmin step 1 until limj do
  begin T0 := 0;
  lim := j − nmin;
  if limk > two kmax then limk := two kmax)  limk := limk − 1;
  for k := 0 step 1 until limk do
    begin T1 := S [j−k] − S [j−k−1];
      if T1 ≠ 0 then T1 := T0 + 1/T1 else
      if S [j−k] = 1099 then T1 := T0 else
        T1 := 1099;
      comment  1099 may be replaced by the largest number
        representable in the computer;
      T0 := S [j−k−1];
      S [j−k−1] := T1
    end for k
  end for j
end Shanks
```

CERTIFICATION OF ALGORITHM 215
SHANKS [H. C. Thacher, Jr., *Comm. ACM 6* (Nov. 1963), 662]
LARRY SCHUMAKER (Recd. 16 Dec. 63)
Computation Ctr., Stanford U., Stanford, Calif.

Algorithm 215 was coded in Extended ALGOL for the Burroughs B-5000 and was tested on a large number of sequences. One apparent typographical error was noted. The statement $lim := j − nmin$ should have read $limk := j − nmin$. The following tables were reproduced exactly: (a) tables on p. 5 and p. 33 of [1]; (b) Table I on p. 95 of [2]; (c) Tables III and IV on p. 28 of [3].

REFERENCES:
1. SHANKS, D. Non-linear transformations of divergent and slowly convergent sequences. *J. Math. Phys. 34* (1955), 1–42.
2. WYNN, P. On a device for computing the $e_m(S_n)$ transformation. *MTAC 10* (1956), 91–96.
3. WYNN, P. On repeated application of the ϵ-algorithm. *Chiffres 4* (1961), 19–22.

ALGORITHM 216
SMOOTH
RICHARD GEORGE*
Argonne National Laboratory, Argonne, Ill.
* Work supported by the U. S. Atomic Engergy Commission.

```
procedure  SMOOTH (Data) which is a list of length:  (n);
  integer n;  real array Data;
  begin
      comment This procedure accomplishes fourth-order smooth-
        ing of a list using the method given by Lanczos, Applied
        Analysis (Prentice-Hall, 1956). This algorithm requires only
        one additional list for temporary storage;
      real Factor, Top; integer Max I, I, J; array Delta [1 : n];
      Factor := 3.0/35.0;
      Max I := n − 1;
      for I := 1 step 1 until Max I do
        Delta [I] := Data [I+1] − Data [I];
      for J := 1 step 1 until 3 do
        begin
          Top := Delta [1];
          Max I := Max I − 1;
          for I := 1 step 1 until Max I do
            Delta [I] := Delta [I+1] − Delta [I]
        end;
      Max I := n − 2;
      for I := 3 step 1 until Max I do
        Data [I] := Data [I] − Delta [I−2] × Factor;
      Data [1] := Data [1] + Top/5.0 + Delta [1] × Factor;
      Data [2] := Data [2] − Top × 0.4 − Delta [1]/7.0;
      Data [n] := Data [n] − Delta [n−3]/5.0 + Delta [n−4] × Factor;
      Data [n−1] := Data [n−1] + Delta [n−3] × 0.4 − Delta
        [n−4]/7.0
  end;
```

ALGORITHM 217
MINIMUM EXCESS COST CURVE
WILLIAM A. BRIGGS
Marathon Oil Co., Findlay, Ohio

procedure *MINIMUM EXCESS COST CURVE* (*nodes, links, source, sink, I, J, crash, normal, slope, node, lij, ERROR*);
value *nodes, links, source, sink;*
integer *nodes, links, source, sink;*
integer array *I, J, crash, normal, slope, node, lij;*
comment This procedure utilizes a network-type description of a project to compute the minimum cost involved in expedition of the project completion date. Project tasks are identified and completion order specified by the vector pair I, J, which contain node numbers of the events starting and ending each task. The tasks are parameterized within the vectors *crash, normal*, and *slope*—which contain the crash or minimum task completion times, the normal task completion times, and the increased cost per unit decrease in task duration (the slope of the time-cost curve), which must be a nonzero integer. The procedure initially determines the normal-duration critical path, then successively reduces the durations of the tasks with the flattest cost slope, adjusting the critical path, until minimum durations are reached. The FORD-FULKERSON labeling technique is utilized. Each task must proceed from a lower-numbered node to a higher-numbered one—if not, exit to the nonlocal label *ERROR* is made. Nodes should be numbered sequentially, starting at the initial event (*source*) and continuing to the final event (*sink*). The maximum node number is equivalent to the value *nodes*, while the value *links* denotes the total number of tasks. The arrays are of dimensions I, J, *crash, normal, slope, lij* [1:*links*] and *node* [1:*nodes*];
begin integer $m, n, tb, nji, nij, lex, kf, nj, ni, ntv, ord, infinity, temp;$
integer array $labl[1:nodes,1:3], f[1:links,1:2];$
comment *infinity* is herein used to represent the largest available integer;
for $m := 1$ **step** 1 **until** *links*-1 **do**
  if $I[m] \geq J[m] \lor I[m] > I[m+1] \lor J[m] > J[m+1]$ **then go to** *ERROR*;
if $I[links] \geq J[links]$ **then go to** *ERROR*;
for $n := 1$ **step** 1 **until** *nodes* **do** $labl[n, 1] := labl[n, 2] := labl[n, 3] := node[n] := 0;$
for $m := 1$ **step** 1 **until** *links* **do**
  begin $f[m, 1] := f[m, 2] := 0;$
    $temp := node[I[m]] + normal[m];$
    if $node [J[m]] < temp$ **then** $node [J[m]] := temp$
  end;
  $ntv := ord := 0; \quad tb := node[sink];$
$A:$ $labl[source, 1] := source; \quad labl[source, 3] := infinity;$
  for $m := 1$ **step** 1 **until** *links* **do**
    begin if $labl[I[m], 1] = 0$ **then go to** B;
      if $labl[J[m], 1] \neq 0$ **then go to** C;
      $nji := node[J[m]] - node[I[m]];$
      if $nji \neq normal[m]$ **then go to** $A1$;
      $lex := slope[m] - f[m, 1];$
      if $lex \leq 0$ **then go to** $A1$;
      $kf := 1; \quad$ **go to** $A2$;
$A1:$ **if** $nji \neq crash[m]$ **then go to** C;
    $lex := infinity; \quad kf := 2;$
$A2:$ $labl[J[m], 1] := I[m]; \quad labl[J[m], 2] := kf;$

$\quad$ **if** $labl[I[m], 3] > lex$ **then go to** $A3$;
$\quad$ $labl[J[m], 3] := labl[I[m], 3]; \quad$ **go to** $A4$;
$A3:$ $labl[J[m], 2] := lex;$
$A4:$ **if** $J[m] = sink$ **then go to** D **else go to** C;
$B:$ **if** $labl[J[m], 1] = 0$ **then go to** C;
$\quad$ $nij := node[I[m]] - node[J[m]];$
$\quad$ **if** $nij \neq normal[m] \lor f[m, 1] = 0$ **then go to** $B1$;
$\quad$ $lex := f[m, 1]; \quad kf := -1; \quad$ **go to** $B2$;
$B1:$ **if** $nij \neq normal[m] \lor [m, 2] = 0$ **then go to** C;
$\quad$ $lex := f[m, 2]; \quad kf := -2;$
$B2:$ $labl[I[m], 1] := J[m]; \quad labl[I[m], 2] := kf;$
$\quad$ **if** $[labl[J[m], 3] > lex$ **then go to** $B3$;
$\quad$ $labl[I[m], 3] := labl[J[m], 3]; \quad$ **go to** $B4$;
$B3:$ $labl[I[m], 3] := lex;$
$B4:$ **if** $I[m] = sink$ **then go to** D;
$C:$ **end**;
$\quad$ **for** $n := 1$ **step** 1 **until** *nodes* **do if** $labl[n, 1] = 0$ **then** $node[n] := node[n] - 1;$
$F:$ **for** $n := 1$ **step** 1 **until** *nodes* **do** $labl[n, 1] := labl[n, 2] := labl[n, 3] := 0;$
$\quad$ **go to** A;
$D:$ **if** $labl[sink, 3] = infinity$ **then go to** OUT;
$\quad$ $ntv := ntv + labl[sink, 3]; \quad nj := sink;$
$G:$ $ni := labl[nj, 1];$
$\quad$ **if** $labl[nj, 2] > 0$ **then go to** $G1$;
$\quad$ **for** $m := 1$ **step** 1 **until** *links* **do if** $I[m] = nj \land J[m] = ni$ **then** $f[m, -labl[nj, 2]] := f[m, -labl[nj, 2]] + labl[sink, 3];$
$G1:$ **for** $m := 1$ **step** 1 **until** *links* **do if** $I[m] = ni \land J[m] = nj$ **then** $f[m, labl[nj, 2]] := f[m, labl[nj, 2]] + labl[sink, 3];$
$\quad$ **if** $ni = source$ **then go to** OUT;
$\quad$ $nj := ni; \quad$ **go to** G;
$OUT:$ **for** $m := 1$ **step** 1 **until** *links* **do**
$\quad$ **begin** $lij[m] := node[J[m]] - node[I[m]];$
$\quad$ **if** $lij[m] > normal[m]$ **then** $lij[m] := normal[m]$
$\quad$ **end**;
$\quad$ $ord := (tb-node[sink]) \times ntv; \quad tb := node[sink];$
$\quad$ **if** $labl[sink, 3] = infinity$ **then** $ntv := infinity;$
$ANS:$ **comment** as control passes through here—
$\quad$ *ord* is the ordinate of the minimum project excess cost curve at a total project duration of *node[sink]*,
$\quad$ successive values of *ord* plotted versus *node[sink]* generate the minimum project excess cost curve.
$\quad$ *node[1:nodes]* contains the event times at each node
$\quad$ *lij[1:links]* contains the durations of each task
$\quad$ *ntv* is the slope of the cost curve back in time from total duration *node[sink]*.
$\quad$ these values should be printed in some readable form;
$\quad$ **if** $ntv < infinity$ **then go to** F;
$\quad$ **end** *MINIMUM EXCESS COST CURVE*;

REMARK ON ALGORITHM 217 [H]
MINIMUM EXCESS COST CURVE [William A. Briggs,
  Comm. ACM 6 (Dec. 1963), 737]
JOHN F. MUTH.(Recd. 26 Dec. 1967)
Michigan State University, East Lansing, MI 48823

KEY WORDS AND PHRASES: critical path scheduling, PERT,
cost/time tradeoffs, network flows
CR CATEGORIES: 3.59, 5.41.

Algorithm 217 was transliterated into FORTRAN and successfully
run on the CDC 3600 system at Indiana University after the following changes were made:

(1) In the first Boolean expression of the program the term:
$$J[m] \geq J[m+1]$$
was replaced by the term:
$$(I[m] = I[m+1] \wedge J[m] \geq J[m+1])$$

(2) The line:
$$A3:\ labl[J[m], 2] := lex;$$
was replaced by:
$$A3:\ labl[J[m], 3] := lex;$$

(3) In the statement labeled $B1$, the symbols:
$$[m,2] = 0$$
were replaced by:
$$f[m, 2] = 0$$

(4) Two statements before the statement labeled A was replaced
by
$$ntv1 := ntv := ord := 0$$
where $ntv1$ was an additional integer variable. The third
statement before ANS was replaced by:
$$ord := (tb\text{-}node[sink]) \times ntv1 + ord;\quad ntv1 := ntv;$$

ALGORITHM 218
KUTTA MERSON
PHYLLIS M. LUKEHART*
Argonne National Laboratory, Argonne, Ill.

procedure *KuttaMerson* $(n, t, y, eps, h, fct, first)$;
value n, eps;
integer n;
real t, eps, h;
real array y;
Boolean *first*;

* Work supported by the U. S. Atomic Energy Commission.

procedure *fct*;

comment This procedure integrates the system of ordinary first-order differential equations $y[i] = f[i](t, y[1], y[2], \cdots y[n])$ from $t = t$ to $t = t + h$ by the Kutta-Merson method (L. Fox, *Numerical Solution of Ordinary and Partial Differential Equations*, p. 24, Pergamon Press, 1962). The working interval of calculation is adjusted by the procedure so that the maximum absolute error of the dependent variables is less than *eps*. For optimum error control, the equations should be scaled so that all dependent variables have approximately the same magnitude. Input variables for the procedure are n, the number of equations, t, initial value of the independent variable, y, array of initial values of dependent variables, *eps*, allowable error, h, the total interval, *fct*, a procedure evaluating the derivatives, and *first*, a Boolean variable which indicates whether the working interval has been adjusted to secure the desired accuracy. On the initial call of the procedure for a given system, *first* should be **true**. It will be set **false** by the procedure, and the proper working interval determined. The procedure *fct* has as formal parameters the simple real variable t, and the real arrays y and f. For $i = 1, 2, 3, \cdots, n$ it must assign to $f[i]$ the value of the first derivative of $y[i]$ appropriate to the values of t and y. The body of this procedure has been tested using the Dartmouth SCALP compiler for the LGP-30 computer. For the equation $dy/dt = -2ty^2$ and input data $t = 1, y = .5, h = 1$, $eps = .0001$, the average error was .000003 and the time was 30 min. For the linear boundary value problem $d^2y/dt^2 = -1 - (t^2+1)y$, $y(\pm1) = 0$, the maximum error was .0000024 (L. Collatz, *The Numerical Treatment of Differential Equations*, pp. 145, 225, Springer-Verlag, Berlin, 1960) and the time, 90 min. More accuracy may be achieved by using a smaller value of *eps*;

begin integer i, loc;
  real *error*;
  array $y0, y1, y2, f0, f1, f2[1:n]$;
  own integer *ploc*;
  own real *hc*;
  Boolean *increase*;
  if *first* **then begin** $hc := h$; $ploc := 1$;   *first* := **false end**;
  $loc := 0$;
next: $fct(t, y0, f0)$;
    for $i := 1$ **step** 1 **until** n **do**
  $y1[i] := y0[i] + hc/3 \times f0[i]$;
  $fct(t + hc/3, y1, f1)$;
    for $i := 1$ **step** 1 **until** n **do**
  $y1[i] := y0[i] + hc/6 \times f0[i] + hc/6 \times f1[i]$;
  $fct(t + hc/3, y1, f1)$;
    for $i := 1$ **step** 1 **until** n **do**

$y1[i] := y0[i] + hc/8 \times f0[i] + 3 \times hc/8 \times f1[i]$;
$fct(t + hc/2, y1, f2)$;
for $i := 1$ **step** 1 **until** n **do**
$y1[i] := y0[i] + hc/2 \times f0[i] - 3 \times hc/2 \times f1[i] + 2 \times hc \times f2[i]$;
$fct(t + hc, y1, f1)$;
for $i := 1$ **step** 1 **until** n **do**
$y2[i] := y0[i] + hc/6 \times f0[i] + 2 \times hc/3 \times f2[i] + hc/6 \times f1[i]$;
increase := **true**;
for $i := 1$ **step** 1 **until** n **do**
begin *error* $:= abs(.2 \times (y1[i] - y2[i]))$;
  comment To test on relative error change this expression to $abs(.2 - .2 \times y2[i]/y1[i])$;
  if *error* $>$ *eps* **then**
  begin $hc := hc/2$;
    $ploc := 2 \times ploc$;
    $loc := 2 \times loc$;
    go to *next*
  end;
  if *error* $\times 64 >$ *eps* **then** *increase* := **false**
end i;

$t := t + hc$;
for $i := 1$ **step** 1 **until** n **do**
$y0[i] := y2[i]$;
$loc := loc + 1$; $'$
if $loc <$ *ploc* $\wedge$ *increase* $\wedge$ $loc = loc \div 2 \times 2 \wedge ploc > 1$ **then**
begin $hc := 2 \times hc$;
  $loc := loc \div 2$;
  $ploc := ploc \div 2$
end;
go to *next*
end *KuttaMerson*

CERTIFICATION OF ALGORITHM 218 [D2]
KUTTA MERSON [Phyllis M. Lukehart, *Comm. ACM* 6 (Dec. 1963), 737]

KAREN BORMAN PRIEBE (Recd. 10 Feb. 1964)
Woodward Governor Company, Rockford, Illinois

Algorithm 218 was translated into FAST for the NCR 315 and gave satisfactory results with the following corrections, if the equations were scaled as recommended in the comment of the original algorithm. Ignoring this scaling can lead to results that do not satisfy the intended error criterion.

1. **procedure** *KuttaMerson* $(n, t, y, eps, h, fct, first, x)$;
instead of
    procedure *KuttaMerson* $(n, t, y, eps, h, fct, first)$;

2. **real array** y, x;
instead of
    real array y;

3. **if** *first* **then begin for** $i := 1$ **step** 1 **until** n **do** $y0[i] := y[i]$;
    $hc := h$;
instead of
    if *first* **then begin** $hc := h$; $\cdots$

4.   **if** $loc < ploc$ **then**
　　　begin
　　　　if $increase \wedge loc = (loc \div 2) \times 2 \wedge ploc > 1$ **then**
　　　　begin
　　　　　$hc := 2 \times hc;$
　　　　　$loc := loc : 2;$
　　　　　$ploc := ploc \div 2$
　　　　end;
　　　　go to $next$
　　　end;
　　　for $i := 1$ **step** 1 **until** n **do** $x[i] := y0[i];$
　　end $KuttaMerson$
instead of
　　　if $loc < ploc \wedge increase$ $\cdots$
　　end $KuttaMerson$

5.   The following sentences should be added to the initial comment of the procedure:

　The values of the dependent variables at $t + h$ are placed in the array x. Note that the values of t and $first$ are changed as side-effects of the procedure. {As originally written, $KuttaMerson$ seemed unable to obtain the values of the solution at t or to transmit the values of the solution at $t + h$ to the outside program!—Ed.}

6.   Change $array$ to **array** in the body of the procedure.

7.   Insert after **own integer** $ploc$;

$$\textbf{own array } y0[1{:}n];$$

Delete $y0$ from the existing array declaration.

REMARK ON ALGORITHM 218 [D2]
KUTTA-MERSON [Phyllis M. Lukehart, *Comm. ACM 6*
　(Dec. 1963), 737]·
G. Bayer (Recd. 25 Oct. 1965)
Technische Hochschule, Braunschweig, Germany

　Successive calls of $Kutta\ Merson$ with $first \equiv$ **false** do not reach the upper bound $t+h$ if the interval h is unequal to the interval h of the first call with $first \equiv$ **true**.
　Proposed correction:
1) declaration **real** hc, instead of **own real** hc;
2) **if** $first$ **then begin for** $i := 1$ **step** 1 **until** n **do** $y0[i] := y[i];$
　　　　　$hc := h;$ $ploc := 1;$ $first :=$ **false**
　　　　end else $hc := h/ploc;$
　instead of **if** $first$ **then begin** $\cdots$ **end**;

ALGORITHM 219
TOPOLOGICAL ORDERING FOR PERT NET-
WORKS
Robert H. Kase
Atlantic Refining Co., Philadelphia, Penn.

```
procedure Topological Ordering (i, j, tri, n, ne);
integer n, ne;  integer array i, j, tri;
comment  Nodal points i and j represent activities in a PERT
  network. n is the number of activities. tri is a tape record index
  vector locating where additional data for each activity is stored.;
begin integer a, b;  integer array ni, nj, event [1:n];
  comment  An event vector is set up containing ne events.
    New nodal numbers ni and nj are assigned for all activities.;
  ne := ni[1] := 1;  event[1] := i[1];
  begin for a := 2 step 1 until n do
    begin for b := 1 step 1 until ne do
    if i[a] = event[b] then begin ni[a] := b;
    go to repeat 1 end;
    end;
  ni[a] := ne := ne + 1;  event[ne] := i[a];
repeat 1:  end;
  begin for a := 1 step 1 until n do
    begin for b := 1 step 1 until ne do
    if j[a] = event[b] then begin nj[a] := b;
    go to repeat 2 end;
    end;
  nj := ne := ne + 1;  event[ne] := j[a];
repeat 2:  end;
  begin integer t, bigtal;  integer array rank, con[1:ne];
  comment Event ranking (topological ordering);
  for a := 1 step 1 until ne do
    begin rank[a] := 1;  con[a] := 0 end;
  bigtal := 1;
pass:  t := 0;
  for a := 1 step 1 until n do
    begin if rank[nj[a]] ≤ rank[ni[a]] then
    rank[nj[a]] := rank[ni[a]] + 1 else go to fill;
    if rank[nj[a]] > bigtal then
    bigtal := rank[nj[a]];  t := 1
fill:  con[rank[ni[a]]] := con[rank[nj[a]]] := 1
    end;
  if t ≤ 0 then go to new;
  for a := 1 step 1 until bigtal do
    begin if con[a] = 0 then go to Loop end;
    comment  Loop should be a label of a procedure statement
      which calls a subroutine to detect those events which may
      be in a loop in the PERT network or the label of a print out
      indicating that loop(s) exist in the network. In any case
      a loop exists and further problem processing is impossible.;
  for a := 1 step 1 until bigtal do con[a] := 0;
  go to pass;
  comment  Reassignment of a new nodal number, ni, to all
    activities;
new:  t := 1:
  for a := 1 step 1 until bigtal do
    begin for b := 1 step 1 until ne do
    if rank[b] = a then begin event[b] := t;  t := t +1 end
    end;
  for a := 1 step 1 until n do ni[a] := event[ni[a]]
  end;
  comment  Using the new nodal number, ni, activities (i and j)
    and their corresponding tri may now be arranged in topological
    sequence with conventional sort routines. Sorting should be
    done on ni.;
end
```

REMARK ON ALGORITHM 219

Topological Ordering for PERT Networks

[R. H. Kase, *Comm. ACM 6*, 12 (Dec. 1963), 738–739]
Dennis Tenney [Rec 31 Jan. 1977 and 14 March 1976]
Knutson and Associates, 1700 North 55th St., Boulder, CO 80301.

ACM Algorithm 219 has been implemented successfully with two necessary modifications:

(1) change

```
  end;
  ni[a] := ne := ne +1; event[ne] := i[a];
  repeat 1: end;
```

to

```
  ni[a] := ne := ne+1; event[ne] := i[a];
  repeat 1: end;
  end;
```

(2) change

```
  end;
  nj := ne := ne+1; event[ne] := j[a];
  repeat 2: end;
```

to

```
  nj[a] := ne := ne+1; event[ne] := j[a];
  repeat 2: end;
  end;
```

ALGORITHM 220
GAUSS-SEIDEL
PETER W. SHANTZ
University of Waterloo, Waterloo, Ontario, Canada

```
procedure GAUSS-SEIDEL (n, A, B, tol);
  value n, tol;  array A, B;  real tol;  integer n;
```

comment *GAUSS-SEIDEL* solves a system, $Ax = B$, of n simultaneous linear equations in n unknowns. A is the matrix of coefficients, B an inhomogeneous vector. The standard Gauss-Seidel iterative technique is employed until $|x_K^{(i)} - x_K^{(i-1)}| < tol$ for all K, where $x_K^{(i)}$ denotes the ith iterant of the unknown x_K. (Cf. Ralph G. Stanton, *Numerical Methods for Science and Engineering*, Ch. 8);

```
begin array X, Y[1:n];  integer i, j, K;
  for i := 1 step 1 until n do X[i] := Y[i] := 0;
START:  for i := 1 step 1 until n do
            begin Y[i] := B[i];
              for j := 1 step 1 until n do
              Y[i] := if i = j then y[i] else
                Y[i] − A[i, j] × Y[j];
              Y[i] := Y[i]/A[i, i]
            end i;
comment   Now test for convergence;
  for K := 1 step 1 until n do
  if abs(Y[K] − X[K]) ≥ tol then
  begin for i := 1 step 1 until n do
    X[i] := Y[i];  go to START
  end convergence test;
end GAUSS-SEIDEL
```

2. Permit the user to initialize the array X to an appropriate value at the start of the iteration.

3. Modify *tol* to be a relative error, rather than an absolute error.

4. Incorporate a guard against nonconvergence.

CERTIFICATION OF AND REMARK ON
  ALGORITHM 220
GAUSS-SEIDEL [P. W. Shantz, *Comm. ACM 6* (Dec. 1963), 739]
A. P. BATSON (Recd 6 Jan. 1964)
University of Virginia, Charlottesville, Va.
NIKLAUS WIRTH (Recd 6 Jan. 1964)
Computer Science Div., Stanford U., Stanford, Calif.

[EDITOR'S NOTE. Two substantially equivalent contributions were received on the same day, and so the editor has merged them.—G.E.F.]

The following errors were detected.

1. The procedure cannot communicate the solution to the outside block unless X (or Y) is made a parameter of the procedure.

2. The identifier *GAUSS-SEIDÉL* may not contain a hyphen.

3. In the fourth line after the label *START* change $y[i]$ to $Y[i]$.

With the above errors corrected, *GAUSS SEIDEL* was successfully run on the Stanford 7090 computer in Wirth's Extended ALGOL, and on the Virginia ALGOL compiler for the Burroughs 205.

The following improvements would be desirable.

1. Avoid repeated reference to the subscripted variable $Y[i]$ inside the j loop.